Imperfect Competition and International Commodity Trade

Imperfect Competition and International Commodity Trade

Theory, Dynamics, and Policy Modelling

MONTAGUE J. LORD

CLARENDON PRESS · OXFORD
1991

Oxford University Press, Walton Street, Oxford OX2 6DP
Oxford New York Toronto
Delhi Bombay Calcutta Madras Karachi
Petaling Jaya Singapore Hong Kong Tokyo
Nairobi Dar es Salaam Cape Town
Melbourne Auckland
and associated companies in
Berlin Ibadan

Oxford is a trademark of Oxford University Press

Published in the United States
by Oxford University Press, New York

© Montague J. Lord 1991

All rights reserved. No part of this publication may be reproduced, stored in a retrieval system, or transmitted, in any form or by any means, electronic, mechanical, photocopying, recording, or otherwise, without the prior permission of Oxford University Press

This book is sold subject to the condition that it shall not, by way of trade or otherwise, be lent, re-sold, hired out or otherwise circulated without the publisher's prior consent in any form of binding or cover other than that in which it is published and without a similar condition including this condition being imposed on the subsequent purchaser

British Library Cataloguing in Publication Data
Lord, Montague J.
Imperfect competition and international commodity trade:
theory, dynamics, and policy modelling.
1. Primary commodities. 2. International cooperation
I. Title
382.17
ISBN 0-19-828347-4

Library of Congress Cataloging in Publication Data
Lord, Montague J.
Imperfect competition and international commodity trade: theory, dynamics, and policy modelling / Montague J. Lord.
Includes index
1. Exports–Developing countries. 2. Commercial products–Developing countries.
3. Imperfect competition–Developing countries. 4. Commercial policy.
5. International trade.
I. Title
HF4055.L67   1991   382'.09172'4–dc20   90-41915
ISBN 0-19-828347-4

Set by Keytec Typesetting Ltd., Bridport, Dorset, UK
Printed in Great Britain by Biddles Ltd
Guildford and King's Lynn

To Greta

Preface

This book offers a unified treatment of the theory, econometric modelling, and policy evaluation of international commodity trade. This approach has been made possible by recent advances in those three areas of economics. Progress has been made in the specification of consumer preferences for differentiated products, and in the determination of equilibrium conditions in markets with imperfect competition; strategies for model selection have been devised, and ways of using time-series analysis to represent dynamic behaviour have been developed; finally, greater recognition has been given to the importance of the dynamics underlying adjustment processes in international trade.

One of the most important developments in trade theory in the past decade has been the analysis of trade in the presence of imperfectly competitive markets. The focus of recent trade theories has been the industrialized countries, where trade between countries with rival industries has been analysed primarily against the background of small-group competition. However, these theories have also included imperfectly competitive markets in which trade takes place under conditions of large-group competition. As a result, empirical research can now be based on econometric models that capture the dynamics underlying trade and price formation in those markets.

Another significant advance has been the establishment of procedures for the selection of models for use in empirical econometric analysis, specifically criteria for the formal evaluation of those models (theory consistency, statistical and economic significance, parameter constancy, and model encompassing). Their application to a set of commodities allows the legitimacy of the postulated theory to be examined, and national and international policies to be assessed. The importance of these policies is growing as the developing countries struggle to expand their export earnings in an effort to deal with their external debt and sustain their economic growth. At the same time, in the industrialized countries intra-industry trade in primary commodities has expanded, which has given rise to concerns about access to the markets of these countries. There is thus a greater need for detailed insights into the specification, estimation, and simulation of international commodity trade that can be used to evaluate trade issues of the developing and the industrialized countries.

This book does not purport to present a general theory of international trade in the presence of imperfect competition. On the contrary, the analysis of commodity trade with imperfect competition in this work underscores how closely related trade relationships are to the assumptions about the behaviour of economic agents. Thus, for the market structure of international commodity trade, the analyses in this work depend on one of several sets of assumptions. Nevertheless, it is hoped that a unified treatment of the theory, dynamics, and policy modelling of international commodity trade with imperfect competition will provide a useful framework for analysing trade in goods of this type, and, by extension, for analysing trade in goods of other types.

The book is designed to be used by a wide range of international trade economists. They include those involved in research on trade theories, on commodity trade relations, on development trade, on intra-industry trade in primary commodities by the industrialized countries, and on the construction, analysis, and use of econometric models of trade in primary commodities and of the world markets for them. It will also interest professional economists and policy-makers responsible for the formulation of trade policies. Finally, it can be used in courses in applied econometrics, international economics, and development trade economics. The mathematical and statistical knowledge required of the reader is modest: a familiarity with calculus and some acquaintance with statistical inference and regression analysis. Much of the more advanced material in the book is illustrated by numerical examples drawn from actual case studies.

Since each major part of this book deals with a separate, though interrelated, area of international commodity trade—theory, dynamic specification, model selection and estimation, and policy analysis—the equations are numbered in sequence within each part. The chapter number precedes the equation number. For example, equation (21) of Chapter 7 is denoted equation (7.21).

Part I of this study describes the approach used for analysing commodity trade and its implications for modelling trade. No one theory or particular form of dynamic specification can provide a complete description of commodity trade in the real world. What is essential is that key features of the process be represented in the system used to characterize international commodity trade. The resulting system can therefore be viewed as an interpretation of the process by which trade takes place.

Part II lays out the general theoretical framework used in this study for analysing commodity trade. The framework is built on recent theories of trade in the presence of imperfectly competitive markets, in particular those in which trade takes place under conditions of large-group competition. A major advantage of this type of competition is that empirical research can be based on econometric models that capture the dynamics underlying trade and price formation in commodity markets.

Part III specifies the dynamics underlying the adjustment processes in commodity trade. The specification strategy adopts the findings of recent studies on dynamic time-series models that explain observed disequilibria in the light of steady-state solutions of behavioural relationships. The error correction mechanism, which is based on the theory of cointegral processes, provides a particularly appropriate specification for the demand relationships in the system of equations. In the supply relationships, a stochastic difference equation is used to characterize the different lag structures of commodities. The steady-state solutions of the dynamic relationships are shown to encompass the system of equations derived in Part II.

Part IV presents the empirical results of the estimated system of equations for Latin America's major commodity exports and assesses the results in the light of well-established development trade theories. The estimates are unique in their level of disaggregation and are shown to yield much higher price and income elasticities than those of previous studies that have relied on aggregated data. This part ends with an empirical investigation of price formation in commodity markets that underscores the importance for trade of dynamic feedback effects in these markets.

Part V examines the impact on the exports of developing countries of changes in the international trade environment, and the use of econometric models for the assessment of trade policies. Two types of policies are considered: those that influence the long-term growth of exports, and those that have a one-time impact on exports. The first chapter in this part traces the effect of income changes in the industrialized countries on the exports of the developing countries. The incorporation into the system of equations of the feedback between economic activity and commodity market price formation has important implications for well-established theories of this relationship. The following two chapters examine optimal export policies of the developing countries and trade liberalization policies of the industrialized countries. The analyses provide important insights into the impact on trade policies of feedback effects in commodity markets and dynamic adjustment processes in commodity trade. The study ends with a summary of the findings and the main conclusions.

M.J.L.

Acknowledgements

Throughout the preparation of this book, I received helpful suggestions and support from many people. In this regard, I am particularly grateful to David Hendry for many useful discussions on dynamic specification in econometric modelling. His approach to applied econometrics has greatly influenced the treatment of the subject in this work. The econometric methodology used also owes much to Andrew Harvey and Grayham Mizon. The chapters on the theory of trade benefited from comments and suggestions made by participants in sessions at the 1985 and 1987 meetings of the American Economic Association, the Eighth World Congress of the International Economic Association, the 1988 North American Meeting of the Econometric Society, and the 1989 Conference on International Trade of the Applied Econometrics Association. The empirical analysis in this study benefited from the help of several individuals. Greta Boye provided invaluable support at all stages of the work. Foremost among her many contributions was the construction of the econometric models, specifically the estimation and validation of the models, the design of programs used to calculate the empirical results, and the simulations of the models. For the management and programming of data bases for estimating the trade models, I am indebted to Mark Walsh and Kevin Johnson, both of whom brought to the work their enthusiasm and expertise in systems design and applications.

Finally, I am grateful to all the colleagues and friends who commented on various parts or drafts of the manuscript, in particular John Chadburn, who commented upon the manuscript in its entirety, Glenn Westley, who provided valuable observations on parts of an earlier version of the study, and Walter Labys and an anonymous referee, who offered numerous helpful comments and suggestions. Needless to say, the views expressed in this work do not necessarily reflect those of these individuals or of any institution with which I have been associated during the preparation of this work.

M.J.L.

Contents

List of Figures	xv
List of Tables	xvii
List of Abbreviations	xix
List of Symbols	xx

PART I INTERNATIONAL COMMODITY TRADE

	Introduction	3
1	The Determinants of International Commodity Trade	6
2	The Empirical Analysis of Commodity Trade	15

PART II COMMODITY TRADE THEORY

	Introduction	27
3	The Demand for Traded Commodities	30
4	The Supply of Traded Commodities	48
5	Equilibrium Conditions in Commodity Trade	58
6	Summary	65

PART III DYNAMICS OF COMMODITY TRADE

	Introduction	71
7	Specification Strategy	73
8	The Demand Functions	86
9	The Supply Functions	103
10	Summary	111

PART IV EMPIRICAL ANALYSIS OF COMMODITY TRADE

	Introduction	115
11	Testing Development Trade Theories	117
12	Commodity Exports of Latin America	129
13	The Demand for Commodity Imports	139
14	The Demand for Commodity Exports	157
15	Characterization of Commodity Export Supply	172
16	Commodity Market Price Formation	189
17	Summary	214

Part V Modelling Trade Repercussions and Trade Policies

	Introduction	219
18	The International Transmission of Income Changes	220
19	Optimal Export Policies	242
20	Trade Liberalization	257
21	Summary	284

Annexes

A	Data and Estimation Procedure	291
B	Dynamic System of Equations	302
C	Regression Results of Model Equations	306
D	Derivation of Equations	372

References 386

Author Index 407

Subject Index 413

List of Figures

3.1	Illustration of export and import demand schedules for copper	44
4.1	Illustration of export supply schedule for coffee	54
6.1	Schematic diagram of commodity trade model	65
7.1	Dynamic adjustment process of coffee export demand function	83
8.1	Dynamic adjustment process of coffee export demand function with relative prices	96
8.2	Long-run equilibrium solution of coffee export demand function	97
8.3	Long-run equilibrium solution of coffee export demand function with different relative price elasticities	98
9.1	Illustration of lag coefficients for price variable in export supply functions	106
13.1	Average normalized lag distribution of import prices in major markets	150
13.2	Lag distribution of import prices in some major markets for cocoa	153
15.1	Average lag coefficients of export supply functions	173
15.2	Hypothetical response profiles of export supply functions	185
16.1	Schematic diagram of price formation in commodity markets with stocks	191
16.2	Response profiles of production to one-time price change	198
16.3	The cattle cycle	200
16.4	Estimated response of coffee prices to a major supply disturbance	211
18.1	Response path of market prices to one-time 1 per cent increase in world economic activity	226
18.2	Illustration of market price response to demand and supply changes	228
18.3	Response path of exports to one-time 1 per cent increase in world economic activity	236
19.1	Illustration of 5-year export subsidy needed to achieve 7 per cent target growth rate of exports	253

20.1	Sugar prices of the world, the USA, the EEC, and Japan	266
20.2	(a) Illustration of sustained multilateral decrease in *ad valorem* tariff; (b) Beef prices after multilateral trade liberalization	275
A1	Structure of trade data	291

List of Tables

11.1	Growth rates of real values of trade and economic activity, 1960–1987	118
12.1	Percentage contribution of major commodity exports of Latin America to total exports, 1970–1987 average	130
12.2	Distribution of Latin America's major commodity exports by principal geographic market	133
12.3	Growth rates of real values of imports and economic activity of the industrialized countries and of major commodity exports of Latin America, 1961–1987	136
13.1	Income distribution elasticities	142
13.2	Import demand functions of principal export markets	143
13.3	Aggregate import demand functions of selected commodities in principal markets	146
13.4	Average lag response of import prices to a one-time change in market price	152
14.1	Elasticities of substitution for commodity exports	158
14.2	Distribution of error-correction coefficients	163
14.3	Dynamic equilibrium solution of export demand function	165
15.1	Export supply response to price change	174
15.2	Initial lag structures of export supply functions	178
15.3	Illustration of summary statistics to describe lag distribution	184
16.1	Consumption functions	192
16.2	Short-term price and income elasticities of consumption	194
16.3	Production functions	196
16.4	Stock demand functions	204
18.1	Total import response to one-time 1 per cent increase in income	228
18.2	Total export response to one-time 1 per cent increase in foreign income	230
19.1	Export subsidy required to achieve 3 per cent higher export growth rate	252

List of Tables

19.2	Export tax needed to achieve 3 per cent higher growth rate of export earnings	254
20.1	Protection in selected commodities by the USA, the EEC, and Japan, 1984–1986	262
20.2	*Ad valorem* tariff-equivalent levels in selected commodities in the USA, the EEC, and Japan, 1979–1986	267
20.3	Effect of trade liberalization on US, EEC, and Japanese imports	269
20.4	Simulated world price and quantity effects of trade liberalization in the USA, the EEC, and Japan	278
A1	Latin America: principal geographic markets of regionally significant non-fuel exports	293
A2	Description of major commodity exports of Latin America	294
A3	Description of world market prices for commodities	297
C1	Regression results of import demand equation (8.12)	307
C2	Regression results of import price equation (9.50)	316
C3	Regression results of export demand equation (8.17)	322
C4	Regression results of export supply equation (9.41)	345
C5	Regression results of consumption equation (8.29)	366
C6	Regression results of production equation (9.47)	368
C7	Regression results of stock demand equation (8.38)	370

List of Abbreviations

ACP	African, Caribbean, and Pacific (countries)
ARIMA	autoregressive integrated moving average
ARMA	autoregressive moving average
CAP	Common Agricultural Policy
CBT	Chicago Board of Trade
CES	constant elasticity of substitution
CGE	computable general equilibrium (model)
c.i.f.	cost, insurance, freight
CMS	constant market shares (analysis)
CSE	consumer subsidy equivalent
ECM	error correction mechanism
EEC	European Economic Community
FAO	Food and Agriculture Organization (of the United Nations)
f.a.s.	free alongside ship
f.o.b.	free on board
GATT	General Agreement on Tariffs and Trade
GDP	gross domestic product
IBRD	International Bank for Reconstruction and Development (World Bank)
ICCO	International Cocoa Organization
ICO	International Coffee Organization
IMF	International Monetary Fund
IPC	Integrated Programme for Commodities
ISA	International Sugar Agreement
LDCs	less developed countries
MDCs	more developed countries
MFN	most-favoured nation (treatment)
MTNs	multilateral trade negotiations
NICs	newly industrialized countries
NTBs	non-tariff barriers
OECD	Organisation for Economic Co-operation and Development
OPEC	Organization of Petroleum Exporting Countries
PSE	producer subsidy equivalent
SITC	Standard International Trade Classification
UN	United Nations
UNCTAD	United Nations Conference on Trade and Development
USDA	United States Department of Agriculture

List of Symbols

A	labour
B	capital
C	consumption
D	deflator
E	cost of production
F	foreign exchange earnings
H	foreign exchange reserves
I	interest rate
K	stocks
M	imports
N_o	numeraire
P	price
Q	production, or exportable output
R	relative price (of exporter of interest to that of imports in the geographic market)
S	market share (also used to denote a subsidy in Part V)
T	secular trend (also used to denote a tax or a tariff in Part V)
V	factor input cost
W	major disturbance
X	exports
Y	economic activity
Z	export demand of country of interest relative to export demand of competitors

Subscripts

i	exporter of interest
j	geographic export market
k	others

Superscripts

d	demand
s	supply

Greek symbols
In Part II:

α	substitution parameter in overall indifference schedule
β	substitution parameter in import indifference schedule
γ	price elasticity of export supply, or of rest-of-world supply
δ	distribution parameter in production schedule

List of Symbols

ε_m^p	price elasticity of import demand for a commodity
ε_n^p	price elasticity of demand for the numeraire good
ε_x^p	price elasticity of export demand for a product
θ	substitution parameter in production schedule
ξ	elasticity of stock demand with respect to real rate of interest and expected price*
π	distribution parameter for indifference schedule
ρ	elasticity of import supply with respect to prices and foreign exchange holdings*
σ	efficiency and shift parameters in production schedule
τ	returns to scale
ϕ	price elasticity of demand in consuming country
ψ	elasticity of export price with respect to variables in reduced form relationship

In Part III:

ε_m^p	price elasticity of import demand
ε_m^y	income elasticity of import demand
ε_x^p	price elasticity of export demand
ε_x^r	elasticity of substitution
ε_s^r	elasticity of substitution for market shares
γ	price elasticity of export supply

(Other symbols are used as coefficients in the dynamic specification of postulated relationships.)

---

* Numeric subscripts distinguish elasticities of variables.

Part I

International Commodity Trade

Introduction

International trade in primary commodities dominates the export performance of the economy of the developing world. Despite efforts to diversify exports, most developing countries continue to rely on a relatively small number of primary commodity exports for a large portion of their foreign exchange earnings. According to United Nations Conference on Trade and Development trade data, the share of primary commodities in total earnings from non-fuel exports in 1986 was more than 70 per cent in Latin America, more than 50 per cent in Africa, and almost 25 per cent in developing Asia (UNCTAD 1989). Within these regions, 27 countries, or 42 per cent of the countries for which data are available, derive more than two-thirds of their export earnings from primary commodities. This dependence is underscored by the large number of developing countries that are virtually mono-export, or so-called mono-culture, economies. According to information from the International Monetary Fund (IMF 1989), one-third of the non-oil developing countries for which data are available each derive more than half their export earnings from just one commodity.

For this reason, the performance of commodity trade has been of major concern to producers and economic policy-makers in the developing countries. Although it is not the only way to achieve economic progress, export growth does provide foreign exchange earnings with which to import the capital and intermediate goods needed for domestic production. However, since the beginning of the 1980s the amount of export earnings channelled to such imports for the purpose of sustaining economic growth has been severely limited by debt-servicing obligations. In 1989, for instance, for every dollar earned from exports by the developing countries as a whole, $0.13 had to be used to service the interest on their external debt; in Latin America interest-servicing obligations have amounted to $0.23 of each dollar earned from exports (International Bank for Reconstruction and Development 1989). At the same time, the 1981–2 world recession reduced access to markets for the manufactured goods exported by the developing countries, since an important market for goods of this type had been other developing countries and these were forced to cut imports dramatically. As a result, whereas the volume of primary commodity exports by the developing countries remained virtually unchanged during the recession, the volume of their manufactured exports declined. Since then, the annual growth in commodity exports by developing countries has often exceeded that of their exports of manufactured goods. Between 1984 and 1986, for example, the volume of developing countries' exports of manufactured goods grew by less than the 8 per cent average annual growth of primary commodity exports by this group of countries (United Nations 1989). Faced with reduced access to foreign markets for manufactured exports and with increased debt-servicing obligations, many developing countries have turned their attention to the performance of their primary commodity exports.

Notwithstanding recent developments, primary commodity trade has historically lagged behind trade in manufactured goods. During the last quarter-century, the average annual growth rate of commodity trade has been about two-thirds of that of trade in manufactured goods (United Nations 1989). As a result commodities have been considered a poor motor of export growth, and consequently a disappointing handmaiden of economic growth. Explanations of the export performance of the developing countries have been based in some cases on the relationship between economic activity in the industrialized countries and the demand for primary commodity exports of the developing countries, and in other cases on the relationship between commodity prices and the supply of exports from the developing countries.

In an attempt to reduce the disparity in growth rates, developing countries have adopted national and international policies aimed at stimulating international commodity trade. In the past, they sought to reduce the disparity between the growth of commodity trade and that of trade in manufactured goods through international commodity arrangements, either between producing and consuming countries or between producing countries alone. While the developing countries have viewed such arrangements as a means of stabilizing and expanding their exports, the industrialized countries have been motivated to participate in them by their desire to ensure access to supplies, particularly since the oil embargo imposed by the Organization of Petroleum Exporting Countries (OPEC) in the first part of the 1970s. However, the 1981–2 world recession sharply curtailed trade and led to the adoption of inward-oriented policies by the industrialized countries. As a result, the developing countries now tend to focus on national policy initiatives, in the form of export policies and trade intervention measures in their principal foreign markets, the industrialized countries.

The major constraints placed on commodity imports by the industrialized countries have been non-tariff barriers (NTBs) to trade and domestic subsidies to producers. Producer subsidies are provided by the US government through both price and income support programmes, and by the European Economic Community (EEC) through domestic price support programmes, which have created significant surpluses in sugar, grains, and beef. A prime example of the effects of these policies is the transformation of the EEC from a net importer of sugar into the world's largest sugar exporter. The developing countries have argued that the reduction or elimination of NTBs and domestic subsidies in the industrialized countries would offer them an opportunity of expanding their exports and, as a result, of reducing the growth differential that has existed between commodities and manufactured goods in international trade.

Paralleling proposed foreign policy prescriptions for commodity importing countries are initiatives that exporting countries could take in order to influence their exports. Such policy initiatives apply both to individual countries and to developing countries as a whole, since the industrialized countries often compete in many of the same commodities, and the developing countries could regain the market shares they lost to the industrialized countries in the period following the Second World War. The export market share of the developing countries in total world trade of primary commodities has gradually declined from 34 per cent in the early 1960s to 28 per cent in the first half of the 1980s, while the market share of the industrialized countries in these goods has risen from 55 to 63 per cent during the same period. Thus, the developing countries would appear to have considerable scope for influencing their exports.

Ultimately, the challenge to each country lies in its capacity to manage its exports in the face of the inherent characteristics of its export markets, which may or may not reflect conditions in the world commodity market. An analysis of international commodity trade in relation to individual producing and consuming countries can provide greater insight into the interest of the developing countries in trade policy measures, particularly those that afford them concrete opportunities for directly influencing their exports.

International commodity trade issues can be analysed only in the light of the structure and dynamics underlying that trade. Despite the considerable progress that has recently been made in the international trade literature in identifying the major determinants of trade, such progress has been almost wholly limited to the analysis of exports by the industrialized countries. This study brings together recent theories of the structure of trade and advances in applied econometrics in order to provide a unified treatment of trade in the major types of goods exported by the developing countries. To this end, it develops a general theoretical framework for commodity trade, specifies the dynamics of the system of equations used, and tests its validity by applying it to trade at a highly disaggregated level. Once it has been ascertained that the system provides a good representation of the adjustment process underlying commodity trade, it is used to assess alternative policies in the developing and the industrialized countries.

1
The Determinants of International Commodity Trade

The analysis of international commodity trade has been concerned primarily with the effects of income and price changes on the level of such trade. This concern has given rise to considerable empirical research on trade policies that could be used to compensate the developing countries for the slower growth and greater instability of their primary commodity exports relative to exports of manufactured goods by the industrialized countries. In this study care has been taken to distinguish between the estimation of the behavioural relationships in commodity trade and the testing of hypotheses concerning commodity trade. Most of the study focuses on the behavioural relationships in trade and on the interaction of these relationships in the structure of trade. However, in so far as international commodity trade is considered to be imperfectly competitive, we shall also be concerned with testing trade theories. The empirical results of this study in both areas have a direct bearing on key issues in development trade. The magnitude of income and price elasticities in international commodity trade determines not only the size of the transmission of changes in economic activity and prices in foreign markets, but also the impact of trade policies on the level of trade.

This chapter examines some of the essential features that need to be considered in the characterization of international commodity trade, one being the conditions that give rise to imperfect competition; it also explores some of the possible reasons for product heterogeneity in international commodity trade. The following chapter reviews empirical methods in international economics and considers issues related to empirical tests of trade theories and policies that have a bearing on international trade in primary commodities.

1.1 Theoretical Foundations of the Determinants of Trade

The fundamental theory of international trade has focused on the determination of what goods will be traded by countries, and has paid little attention to the determination of how much trade will take place.

The modern theory of international trade explains trade by differences between countries in the relative abundance of factors (see Dixit and Norman 1980: ch. 4). However, this explanation derives from a model that comprises few countries, few factors, and few goods; consequently multilateral trade flows are excluded, and transportation costs are seldom considered. In a multi-dimensional framework, Deardorff (1984: 492–3) has noted the obstacles encountered by empirical research in quantifying trade levels because of the inherent difficulty of obtaining needed information, such as the factor endowments of different countries in the Heckscher–Ohlin model. As a result, it has been difficult to arrive at conclusions about the levels of trade that would take place in accordance with factor proportions.

Even if such data were available, multi-dimensional trade theory based on factor proportions yields rather unrealistic conclusions about the level and distribution of trade. It implies, on the one hand, specialization of production and, on the other, trade flows based on transportation costs. Yet in reality, the developing countries export primary commodities mainly to the industrialized countries, which themselves produce and export many of the same goods. Moreover, trade patterns are not predicated on location alone. In their critical survey of agricultural trade models, for example, Thompson and Abbott (1983) found that spatial equilibrium models, which rely on transportation costs, provided inadequate explanations of observed trade flows.

Both Thompson (1981) and Deardorff (1984) have noted a number of other factors that explain levels of trade: product heterogeneity by country of origin, importers' diversification of supply sources, historical and political ties between trading partners, and switching costs to importers. Moreover, the key assumptions of the factor proportions theory appear to be implausible. In a series of papers in Feenstra (1988), assumptions regarding identical and homothetic preferences, linearly homogeneous production functions, factor mobility within countries, and perfect competition were tested and rejected in a number of cases. With respect to preferences, Hunter and Markusen (1988) have provided evidence against the homotheticity assumption, which implies unitary income elasticities for traded goods. The seminal work by Paul Armington (1969), which hypothesizes that importers have different demands for the same good originating from different foreign suppliers, offers a way of deriving well-defined import demand functions, since an imported good is considered to be different from the same good produced domestically. Armington's assumption has offered a theoretical basis for computations of import demand functions, the results of which have demonstrated that preferences are neither identical nor homothetic across countries (e.g. the surveys by Goldstein and Khan 1985; and Stern *et al.* 1976).

More importantly, the Armington assumption provides an explanation of the observed trade flows between countries that would not have been predicted by spatial equilibrium models. As a consequence, that assumption has often been adopted in empirical studies on international trade; its applications range from the modelling of trade (e.g. Grennes *et al.* 1978 for models of agricultural trade) to the evaluation of government policies (e.g. Whalley 1985 for an analysis of the impact of trade liberalization). Use of this approach implies a departure from a perfectly competitive market structure of international trade. The fact that the elasticity of substitution between competing foreign suppliers to a market is less than infinity when importers differentiate supply sources means that each exporting country can exert some, albeit small, influence on the demand for its exports through relative-price variations. For instance, a sector-oriented policy that increases the supply of exports, and as a result shifts the supply curve to the right, would reduce the relative price of the exporting country and would increase the quantity of exports demanded. Product differentiation therefore gives rise to some degree of market power. When that market power is negligible, so that the actions taken by one exporting country do not give rise to reactions by its competitors, the market structure can be described as monopolistic competition.

In monopolistic competition the specific characteristics of international trade can be summarized as follows: (1) there are many countries exporting differentiated goods; (2) the actions of exporters in each country have a negligible impact on a market in the sense that exporters in other countries do not react to decisions taken by their competitors about the quantity to be exported (the Cournot assumption); (3) the ability of exporting countries to influence the price at which they sell their products gives rise to a downward-sloping demand curve, so that the equilibrium price is greater than the marginal cost; and (4) free entry drives any pure profit, of at least the marginal exporter, to zero.

The first of these characteristics, namely many exporting countries in the market, is distinguished in the recent literature on monopolistic competition by the 'large-group' case and the 'small-group' case. The large-group case refers to situations in which all the exporting countries are small relative to the aggregate market. The small-group case refers to markets with a relatively small number of exporters. As the number of exporting countries increases, the equilibrium price can approach the competitive equilibrium solution. However, the competitive equilibrium solution is not ensured in the large-group case. Small exporters can still influence their export price when importers discriminate between exporters of a commodity. Moreover, importers often have imperfect information about the conditions under which a commodity is traded, and this lack of information can reduce the effective substitution of small suppliers that have very similar export characteristics. The range of solutions in monopolistic competition is useful for the development of a general theoretical framework with which to characterize commodity trade. On the one hand, monopolistic competition encompasses most commodity markets that contain a large number of exporters, without precluding those markets in which an exporter is sufficiently large that its actions have a perceptible effect on market prices (e.g. Brazil in coffee). On the other, a general interpretation of monopolistic competition allows the empirical analysis in later chapters to test whether equilibrium in individual commodity markets is near the competitive equilibrium, near the monopoly equilibrium, or between the two.

Whether or not a producing country that engages in price competition induces other producing countries to respond to actions taken by it determines the way trade is characterized. Where changes in output by one producing country cause other producing countries to adjust their output, there will be a reaction function for each country. This condition is described by Cournot oligopoly. The opposite situation is where each producing country takes the price of its competitors as given. Here there are two possibilities. If entry to the market is restricted, for instance because of large fixed costs, producing countries may have economic profits. This situation is described by Bertran oligopoly. If, in contrast, there is free entry into the market, economic profits will be driven to zero. This is the Chamberlinian case and is the one that will be used here, because it is most widely applicable to the analysis of commodity trade.

In the 1980s research on product differentiation and imperfect competition in international trade increased sharply. Helpman and Krugman (1985) have provided a synthesis of this research, although they have not attempted to unify all the recent developments in international trade theory, since the results depend on the particular type of market structure being considered. The expansion of this research, according to Deardorff (1984), has been due less to the questionable assumptions of the factor proportions theory than to the need to explain observed trade patterns that are inconsistent with that theory. Dominant among these observed patterns has been the growth of intra-industry trade. Grubel and Lloyd (1975) calculated that this type of trade represented as much as one-half of the amount of all goods traded by the industrialized countries, and intra-industry trade in primary commodities accounted for 30 per cent of it.

Although intra-industry trade in primary commodities has also been observed, recent work on product differentiation and imperfect competition in international trade has focused almost entirely on trade in manufactured goods. Primary commodity trade theory and its empirical analysis have continued to be based on the traditional theory of trade, even though empirical tests have not supported its use. As Thompson (1981) has noted in his survey of agricultural trade models, 'The principal contribution to date [of quantitative analysis] has probably been in testing theory. The conventional assumption of product homogeneity has been tested and found wanting' (p. 45). Yet, he found that 'the theoretical foundations for several of the approaches are either non-existent or of doubtful validity. As a result, their usefulness for prediction and for policy prescription purposes is severely limited' (p. 45). In the next section we explore some of the possible reasons for this product differentiation in international commodity trade.

1.2 Sources of Product Heterogeneity

The topic of product differentiation has been treated extensively in the trade literature and a common framework has emerged for its analysis. The literature has followed Chamberlin (1933: ch. 4) in distinguishing two types of product differentiation. The first deals with differentiation of goods that are themselves perfectly homogeneous but are none the less differentiated because of variations in the attributes of the manner in which they are sold. In international trade, perceived differences in exports in this case are due to those attributes of the export that relate either to the agents undertaking the transaction or to the transaction itself. The second type deals with differentiation of commodities arising from quality distinctions that are inherent in the good itself, compared with the same type of good sold by competitors. In international trade, differences in exports in this case are due to the physical attributes of the exported commodity. Both types of differentiation arise in traded commodities when importers perceive that exports of a commodity from one country are not perfect substitutes for those of the same commodity exported from another country.

Commodities are said to be *horizontally differentiated* when importers differ in their choice of the geographic origin of the good even though its quality does not vary from country to country. Importer distinctions of homogeneous products from different exporting countries arise because of attributes related to the export of the product. Hotelling (1929) introduced this concept in his model of spatial competition, and Lancaster's (1979) work on market structures, based on the characteristics approach to consumer preferences, has focused on this type of differentiation. In international trade, differences between exporters of a commodity can relate either to the conditions surrounding the sale or to established relationships between trading partners.

There are at least six reasons why primary commodities can become horizontally differentiated in international trade. These reasons are not commodity-specific; rather, they are inherent in trade processes. First, adjustment costs are involved in switching from one supplier to another. These costs include loss of 'loyalty' preferences given by exporters to established buyers and loss of reliability of supply sources. There are also learning costs involved in purchases made from new foreign suppliers.

Second, the reliability of supply sources often plays a dominant role for buyers of primary commodities. In coffee, for example, Marshall (1983: 20–1) observed that buyers prefer to purchase large volumes of standard-quality beans from reliable sources rather than undertake the necessary tasting and evaluating of lots obtained from different exporters. This trend has grown as manufacturers of coffee have become larger and small-scale enterprises have formed cooperative syndicates. The larger the scale of the transaction, the greater the need for supply reliability. In cocoa, Curtis *et al.* (1987: 20) has noted the importance given by US chocolate manufacturers to Brazil's ability to supply large quantities of cocoa on a regular basis. B. D. Stonehouse has said that the greater reliability of copper suppliers in Chile compared with those in Zambia or Zaire has been important in the decision made by buyers concerning the country of origin of this good.[1]

Third, exporters offer their products on different marketing and customs regulation conditions, and these conditions influence the choices made by importers about which sources of a commodity they prefer. In the sale of a commodity, exporters can establish consumer preferences for their product that lead to brand loyalties. Marketing initiatives can also take the form of contracts that provide favourable terms for delivery dates, credits, or other services.

Fourth, diversification of supply sources arises either because of the preference of importers for a variety of attributes offered by different exporters (the Dixit–Stiglitz (1977) assumption that underlies the Helpman–Krugman (1985) analysis of international trade), or because importers want to reduce the risk of supply disturbances associated with dependence on a few foreign supply sources. Hazilla and Kopp (1984) have empirically assessed the concerns of the USA about its vulnerability to disruptions of raw material supply from foreign sources and have found that disruption can be severe when substitution possibilities become exhausted. Diversification of supply sources designed to reduce the risk of foreign supply disruptions has become particularly important since the early 1970s when several industrialized countries, most notably the USA, were cut off from traditional foreign petroleum suppliers in the Middle East.

Fifth, historical and political ties with countries can also cause a certain reluctance by importers to switch suppliers of a commodity. In such cases the exporting country could alter its relative export price somewhat without driving away buyers. As a result, these ties could give rise to export demand functions that are less than perfectly elastic with respect to price. Such functions would not emerge when preferential trade arrangements existed, however, since prices are then established by the importers; an exporting country entering into a preferential trade arrangement would increase its market share as a result of an upward shift in its export demand schedule, rather than because of a lower relative export price. For example, the EEC Lomé Convention established preferential trade arrangements with African, Caribbean, and Pacific (ACP) countries; and the sugar import quotas of the USA provide a policy-determined price received by the foreign suppliers that are part of the arrangement.

Finally, Thursby *et al*. (1986) and Deardorff (1984) have suggested that differences in the harvest periods of the countries can affect trade patterns. The growth in exports of fruits and vegetables from the Southern Hemisphere countries to the industrialized countries demonstrates the importance of this determinant of trade for agricultural goods.

Although a large proportion of trade takes place in terminal markets, these trading centres are used mainly as hedging facilities. As a result, few contracts run to the delivery of the physical commodity. For example, the Chicago Board of Trade (CBT), which handles over half of the world grain trade, has registered delivery by traders of only between 1 and 3 per cent of the volume traded on that exchange (CBT 1982). Instead, the bulk of physical trade occurs between the manufacturer or processor and the supplier, which enables them to maintain control over the source of the good purchased. The major exception to the type of transaction normally undertaken in terminal markets is that which occurs in the London Metal Exchange, where copper and several other metals are traded. This is a physical market where metals are bought and sold at guaranteed quality levels. Consequently, buyers are normally well informed about the source of the commodities they purchase. Nevertheless, as in the grain trade, the bulk of the copper trade takes place directly between producers and consumers (for details see Roberts 1985).

The second type of differentiation, known as *vertical differentiation*, arises from variations in the quality of a commodity. The work of Avner Shaked and John Sutton (see Shaked and Sutton 1982, 1983; Sutton 1986) has concentrated on market structures with this type of differentiation.

Vertical differentiation is an important determinant of the pattern of international trade in primary commodities. Exports of coffee, for instance, are divided by coffee type into arabicas or robustas. Each of these two types of coffee beans is graded according to its source, nature, and quality. Specification of the coffee type, according to Marshall (1983: 174–5), is essential to the exporter and the buyer. (For estimates of coffee import demand by variety, see Abaelu and Manderscheid 1968.) The quality standards are highly complex and depend on whether the exporter is contracting with a dealer, who is concerned primarily with the green appearance of the bean, or with the roaster buyer, who is concerned mainly with quality characteristics. Coffee is specifically differentiated by country of origin in terminal markets. The New York exchange establishes tolerances for differences between suppliers at origin of Mild Arabica, based on the grade of beans and the flavour of the coffee (Roberts 1985: 286–9).

Cocoa beans are specifically differentiated by country of origin. The origin of the bean is a critical determinant of its quality, both because of climatic and soil conditions and because of the inherent characteristics of the bean in the producing country. Standardization of flavour is essential to chocolate manufacturers, and as a consequence the flavour of the product is closely linked to the country of origin of the cocoa bean (Curtis *et al.* 1987: 16–20). The extent to which traditional supply sources are an important determinant of trade in this commodity is demonstrated by the fact that a producer of chocolate would be more likely to remove a product from the market than to switch sources of supply of cocoa for the manufacture of a specific product. The cocoa exchange in New York has a grading system similar to that described for coffee; it is based on tolerances for defects, bean quality, and other characteristics. Though less important than flavour, these too play an important role in the choice of cocoa beans purchased by buyers.

In the wheat trade, Grennes *et al.* (1978: 30–1) have noted the lack of perfect substitutability of the wheat varieties of different countries in producing bread, pastry products, and other processed goods. Grains are divided into classes and sub-classes according to shape, texture, colour of the kernel, and source. (For a description of the grading system for grains, see CBT 1982: 17–21.) Quality variations have been shown by Bale and Ryan (1977) to be significant in determining the demand for different varieties of wheat.

Cotton fibre quality depends on fibre strength, fineness, maturity, and length uniformity (Atkin 1983: 79–87). Beef is differentiated by degree of softness and fat content; in the industrialized countries, domestically produced beef is too soft and high in fat content for processing into manufactured goods, so range-reared beef for this use is imported from developing countries (IBRD 1981). Iron ore grades differ widely by mine and geographic region (Franz *et al.* 1986). Even sugar, when traded internationally in its raw form, is differentiated by refiners according to its quality (Fry 1985).

The important point is that physical differences need to be taken into account by the importer. Ladd (1983), in his review of studies relating the demand for agricultural goods to product characteristics, mentions the preference of Italians for Argentine maize, which when used as feed gives what they consider to be a 'healthier' looking colour to poultry. All these distinctions result in commodity de-standardization in international trade.

In this study, explicit recognition is given to possible product de-standardization in the determination of international commodity trade. The implications of this assumption, as well as of the desire of importing countries to diversify supply sources, will be described in Chapter 3, where the exports of a country are defined by a set of objectively measurable characteristics in the preference functions of importers. The characteristics approach to preference ordering has a long history in the economics of consumer behaviour and has only recently been extended to the theory of international trade. In the characteristics approach to consumer behaviour, goods are valued by importers for their utility-bearing attributes. The importer attributes a specific quantum of characteristics to a country's exports. Each importer has a preference structure for characteristics, and variety, or diversity, is one of the characteristics desired. All these preferences are considered in their implicit form, so that demand for traded commodities is described in Part II in familiar Marshallian demand functions and is easily incorporated into the structure of the dynamic models of commodity trade specified in Part III.

Note

1 Personal interview with B. D. Stonehouse, formerly a trader on the London Metal Exchange, 30 April 1988.

2
The Empirical Analysis of Commodity Trade

Empirical models of trade have usually adopted rather standard model structures, although the various approaches to modelling trade have been divided along methodological lines. These differences in methodologies are dictated largely by the purposes for which the models have been constructed. Several studies have surveyed these various methodologies and their applications. The early work by Leamer and Stern (1970: chs. 6 and 7) provides one of the most comprehensive studies of trade modelling methodologies. Grennes *et al.* (1978: ch. 3) have also provided a critical survey of trade models. These include some techniques that are either empirically or theoretically unfounded and have therefore contributed little to our understanding of the nature and structure of markets and trade relationships between countries. Excellent critiques of the different analytical approaches to agricultural trade modelling have been provided by Sarris (1981), Thompson (1981), and Thompson and Abbott (1983). Finally, Labys and Pollak (1984) have made a comprehensive survey of methodologies for modelling commodity markets, primarily as they relate to market price formation, and Labys (1987) has compiled a bibliography of international trade aspects of commodity markets.

This chapter examines empirical issues in modelling trade. The first section surveys different methodologies employed for modelling trade and describes previous empirical econometric analyses of trade. The second section considers different approaches to testing trade theories and policies. The chapter concludes with an overview of the analytical framework used in this study of international commodity trade.

2.1 Modelling Trade

One of the earliest attempts to simulate trade flows was based on gravity models. This approach, developed by Tinbergen (1962), Poyhonen (1963), and Linnemann (1966), explained trade by the income of each of the trading partners and the distance between them. (See Demler and Tilton 1980, and references therein, for applications to mineral trade.) Models of this type have been able to provide reasonable measures of trade, but they have lacked a theoretical foundation. As a result, their usefulness for policy inferences has been unexploitable. However, these early models established separate demand functions for the trade of a country with each of its trading partners and thus implicitly presupposed product heterogeneity in international trade.

At the other extreme are the probability models developed by Savage and Deutsch (1960). They have been applied by Uribe et al. (1966) to describe trade as randomly generated flows, and by Dent (1967) to model commodity trade. These models estimate the demand for imports of a country, and allocate foreign supply sources among the potential suppliers on a random basis. Since patterns of trade are usually rather stable, dummy variables have been used to explain observed preferences for products originating from particular countries.

In between these two approaches there has been a variety of other methodologies. Until recently, the constant market shares (CMS) method had been widely used to analyse exports. (Balassa 1978, 1979 has been largely responsible for the spread of this technique.) The CMS method partitions the performance of a country's exports of a good into the structural components, which are associated with the growth in foreign demand and the geographic distribution of exports, and the share adjustment effect, which is associated mainly with the country's export competitiveness. Richardson (1971a, 1971b) has shown that several problems are inherent in the empirical applications of this technique, which lacks a theoretical foundation. It has therefore been used primarily to provide an accounting framework for assessing past export performances. Even then, it has not offered a means of determining why the export market shares of a country have varied.

Spatial equilibrium models have already been examined as part of multi-dimensional trade theory based on factor endowments. Supply and demand are estimated for trading regions, and prices that include transportation costs link trade between countries. The equilibrium solution that minimizes the transportation cost for the good is then calculated by a quadratic programming procedure. The recent work by Greenhut et al. (1987) in this area has brought together much of the theoretical analysis of product differentiation and the spatial approach to modelling trade. However, there are several problems associated with models of this type. The major criticisms of this approach, according to Sarris (1981), are (1) that trade flows are highly sensitive to changes in transportation costs and to the price elasticities of demand and supply; (2) that price changes are insensitive to large shifts in trade because price formation depends on the aggregates of production, consumption, and, where appropriate, stocks; and (3) that, as mentioned earlier, the explanation of trade flows is inadequate.

Another approach to modelling trade, the application of which has grown rapidly, is based on computable general equilibrium (CGE) models. Shoven and Whalley (1984) have described the advances made in the use of models of this type and have reviewed their applications. In these models trade is always assumed to be in equilibrium. Any change in the system generates a sequence of equilibria through time. The model is 'calibrated' so that the data set, including parameter values, can produce a 'benchmark' equilibrium from which different solutions are generated as a consequence of changes in policies or economic conditions. The parameter values include exogenously specified elasticities drawn from the literature. Although rich in detail and theory, these models do not lend themselves to validation since the equations are deterministic. Moreover, as pointed out by Shoven and Whalley (1984: 1045–6), the results of these models are largely predetermined, in so far as the choice of parameter values and model structures are left to the researchers.

Despite the limitations of CGE models, the development of imperfect competition in international trade theory has necessitated the general equilibrium approach provided by Dixit and Norman (1980), and empirical studies in this area have required the adoption of a similar approach. CGE models have usually explained trade flows by the Armington assumption of product heterogeneity by country. Not only does the assumption allow for intra-industry trade, it also eliminates the tendency of early models of this type to generate excessive specialization in the production of goods that would occur if trade were based on factor endowments. This type of model has lent itself to the evaluation of international trade policies (e.g. Srinivasan and Whalley 1986). In Chapter 20 we shall examine some of the critical differences between CGE models and econometric models for evaluating trade policies, in particular those that predict the effects of trade liberalization policies.

Econometric models of trade were originally concerned with the transmission of changes in economic activity between countries. (See the surveys of Gana *et al.* 1979, and Helliwell and Padmore 1985.) These models have typically linked individual macroeconomic models of countries or regions through a trade share matrix. The elements in the matrix specify trade between countries or regions. The vectors for imports indicate the share of imports obtained by each country from each of its foreign suppliers so that, by construction, the elements of the vectors sum to unity. The international transmission of changes in the levels of economic activity of the different countries through international transactions is then measured by the trade share matrix. In particular, when the level of imports of a country is altered by a change in that country's economic activity, the amount of the change transmitted to its trading partners is calculated in accordance with their respective market shares.

Most of the econometric models dealing with the international transmission of economic activity include a variable that measures the export competitiveness of a country. Helliwell and Padmore (1985: App.) provide a comparison of the different methods used in these models for accounting for changes in export market shares through vsriations in relative export prices. The models usually relate the export shares of a country in a particular import market to relative export prices. However, there is a tendency to adopt practical, rather than theoretically sound, approaches to the share determination process. In particular, there is a noticeable absence in all the models of export supply and demand equations for determining export prices. For instance, in the INTERLINK model of the Organisation for Economic Co-operation and Development (OECD), the prices of food and raw materials are functions of demand in the industrialized countries, as measured by changes in their real gross domestic product (GDP) and GDP deflator (Larsen et al. 1983: 65–91).

Commodity market models have been incorporated into the LINK project (see Adams 1979) and the IBRD's SIMLINK system (Tims and Waelbroeck 1982: 80–115). In addition, a number of country-specific models have been constructed in order to analyse national commodity policies. These analyses cover coffee in Brazil, Central America, and the Ivory Coast (Priovolos 1981; Siri 1980; and Pobukadee 1980) and copper in Chile and Zambia (Lasaga 1981; and Obidegwu and Nziramasanga 1981). (For a synopsis of these studies, see Adams and Behrman 1982.) In all these cases, the commodity models serve to determine the world market prices of the commodities, which are then translated into export prices by using the historical relationship between the world market price and the unit value of exports. However, they do not determine commodity trade at the national level, and, as pointed out by Adams (1978), they do not attempt to reconcile the differences between aggregate trade in commodities and trade disaggregated by producers and consumers of the commodities predicted in the international trade models.

The present study attempts to fill this gap by developing a theory-based econometric modelling framework for international commodity trade. Its usefulness to ongoing work in modelling trade lies in the representation of dynamic adjustments of behavioural relationships to their long-run equilibrium relationships suggested by economic trade theory. The models are specified and estimated in their structural forms, rather than in their reduced forms, for individual exporting and importing countries. The motivation for this approach lies not only in my own interest in measuring income and price elasticities in international commodity trade, but also in the capacity of models in their original structural form to assess the effects on developing countries of changes in economic activity in the industrialized countries and to measure the impact of trade policies.

2.2 Analysing Trade Theories and Policies

One of the major issues in the development trade literature has been the link between the export performance of the developing countries and the economic growth of the industrialized countries. Analytical interest in income and price elasticities in international commodity trade stems from concerns about the performance of the developing countries' exports both in the existing international trade environment and under conditions brought about by various policy initiatives. Given the commodity export performance of the developing countries, three separate issues need to be considered. The first is the long-term relationship between the growth rate of commodity exports from the developing countries and the rate of economic growth in the industrialized countries. The other two issues relate to the capacity of countries to influence the level of trade. The first of these depends on evidence of product heterogeneity, which would suggest that countries can alter the demand for their exports through relative-price changes. The second concerns the magnitude of the effect that national policy initiatives, in either the major import markets or the exporting countries, could have on commodity trade.

2.2.1 *Evaluating Development Trade Theories*

The tie of commodity export growth to the economic growth of the industrialized countries has been the subject of heated debate in development trade literature since the early 1950s. The evolution of this debate during the last three decades has been eloquently described by Little (1982), and his account leaves little room for doubt that it has been a highly charged emotional issue among policy-makers and academicians alike.

Quantitative international economics has contributed to the debate by providing empirical evidence, on the one hand, of the relationship between economic activity in the industrialized countries and the demand for the primary commodity exports of the developing countries, and, on the other, of the relationship between commodity prices and the supply of exports from the developing countries. As might be expected, studies on these topics have moved from rather broad generalizations about these relationships to specifics about differences in the comparative export performance of commodities and countries. The development of empirical analysis in this field is described in Chapter 11, and it provides a useful framework in which to consider the empirical results of the individual relationships estimated in this study. The analysis of a highly disaggregated level of trade lends itself to generalizations about development trade theories.

The empirical results of individual relationships can be misleading, however, since they do not consider feedback effects within the system of equations. Changes in economic activity in the industrialized countries have a direct impact on the demand for exports and an indirect effect on exports arising from the accompanying changes in world market prices. If the interaction of the relationships underlying international trade and the world markets for primary commodities is examined as part of a complete market system, the examination permits consideration of dynamic feedback effects in trade and in the market prices for these goods. Part V begins with an investigation at a highly disaggregated level of trade of the transmission of changes in economic activity and of the markets for several commodities. When feedback effects between economic activity and market prices are considered, the results provide valuable insights into the relationships between economic growth in the industrialized countries and export growth in the developing countries.

2.2.2 Testing the Product Homogeneity Assumption

Models of trade often assume, either implicitly or explicitly, that product heterogeneity exists. They do so in order to explain observed patterns of trade that do not conform to the factor proportions theory of trade. Empirical analyses of this assumption have been based on tests of the assumption of commodity arbitrage underlying perfect competition in international trade. Commodity arbitrage refers to the tendency of prices of internationally traded goods to be the same for all countries except for differences in transportation costs, and it is what gives rise to the 'Law of One Price'. Isard (1977) has argued that the Law of One Price cannot be presumed to hold for all goods traded by a country. He found that in manufactured goods substantial exchange rate variations have brought about changes in the relative dollar-equivalent price of narrowly defined goods traded by West Germany and Japan—but not by Canada—relative to the price of those same goods traded by the USA.

None the less, conflicting evidence has been presented on the extent to which commodity arbitrage prevails in primary commodities and in manufactured and semi-manufactured goods. Richardson (1978) found that there had been imperfect commodity arbitrage between the USA and Canada, even for relatively homogeneous commodity groups.[1] The existence of commodity arbitrage was rejected at the 5 per cent level of significance for such products as fruits and vegetables and pig iron, as well as for non-alcoholic beverages. In contrast, commodity arbitrage was also found to be present in tractors and animal feeds (e.g. maize). Consequently, his findings did not support generalizations about perfect substitutability of primary commodities with respect to country of origin through commodity arbitrage any more than it did about perfect substitutability of processed goods.

In a critique of the work of Isard (1977) and Richardson (1978), Crouhy-Veyrac et al. (1982) argued that previous tests of the Law of One Price had been inadequate since transportation costs were assumed to remain constant over time. As part of this argument, they first conducted a test for cocoa and coffee in order to find evidence that supported the Law of One Price. They found the ratios of the dollar-equivalent prices of coffee and cocoa in the USA and France to approximate unity over the sample period, and the coefficient estimates of the Richardson-type equation to indicate proportionality between French and US prices for the two goods (for quarterly data, but not for monthly data). After deducing that their approach was valid, they proceeded to apply the same test to 38 primary commodities and 8 semi-processed goods. Only one food product was found to be perfectly substitutable with respect to country of origin, as evidenced by the fact that, on average, relative prices were the same across countries. Their evidence also supported the Law of One Price for one semi-processed good. Among the minerals, perfect substitutability by country of origin was found in 9 of the 25 products; however, the evidence mainly supported metallic products and alloys rather than primary metals, in contrast to what would otherwise have been expected. Accordingly, the results of their test were inconclusive.

Further evidence on the Law of One Price has been provided by Protopapadakis and Stoll (1983, 1986). Exchange rate changes were found substantially to alter the relative dollar-equivalent prices of commodities by country of origin in the short run (e.g. in copper and coffee), but in the long run prices usually adjusted to one another. The results are not surprising, since the data were based on prices in the commodity exchanges located in the UK and the USA, so that differences in market prices were purely a function of exchange rate differences between the pound sterling and the dollar. Thus, the test did not reveal whether or not relative-price variations arise from product heterogeneity in the narrowly defined commodity exports of different foreign suppliers. In fact, the data were explicitly selected so as to avoid the possibility of product differentiation by country of origin, and instead were explicitly related to the international market prices of the commodities.

Finally, Thursby et al. (1986) have shown that price data for as narrowly defined a commodity as wheat violate the Law of One Price.

They tested the weaker version of the law whereby changes in the relative common-currency price would be expected to approximate zero. The data were for exporters of wheat from the USA, Australia, and Canada to the Netherlands and Japan. There were few instances in which relative-price variations were less than 1 per cent; overall, the mean average relative-price change was over 5 per cent. From this evidence they concluded that product differentiation exists in the international wheat trade.

Product differentiation in international trade has also been measured by price differences in exports of a narrowly defined good originating from different supplying countries. Hufbauer (1970) developed an index that calculates the standard deviation of export unit values of individual items in a product category for a particular year from the unweighted mean of their unit values in the same year. Besides containing the measurement errors noted by Gray and Martin (1980), which arise from the use of an unweighted mean and from the use of export unit value data as proxies for export prices, the Hufbauer index is not, in fact, a measure of commodity differentiation, since it calculates price differences only at a given period of time. These price differences, according to Greenaway (1984), can arise for a variety of reasons that are unrelated to product differentiation. For example, they may simply reflect price supports under preferential trade arrangements, such as those that exist in the US sugar market and in the EEC with ACP member-countries under the Lomé Convention. Instead, it is the effects of *changes in relative prices* on the demand for exports that reflect commodity differentiation. Evidence for or against differentiation in commodity trade must therefore come from empirical analyses that test whether export demand is influenced by relative export price changes. The results of such an analysis in Part IV, at a highly disaggregated level of trade, lend themselves to generalizations about this issue.

2.2.3 Assessing Trade Policies

Evidence of the long-term gradual loss in market shares of the developing countries in overall primary commodity trade suggests that the below-average growth of their export earnings in goods of these types has been, at least in part, a consequence of trade policies and not a consequence of conditions beyond the control of the developing countries. Against the background of trade policy intervention measures of either importing countries or exporting countries, two separate aspects of trade determinants are considered in this study: the first relates to the magnitude of price elasticities in international commodity trade and the degree of intervention in the markets; the second concerns the dynamics of trade policy measures.

Commodity trade, in particular, is characterized by long delays in the adjustment of economic agents to changes in market conditions. For this reason, the impact of trade policies on the market can take a considerable amount of time to occur. Moreover, the price path is determined by the dynamics underlying the adjustment processes in commodity markets, and it seldom, if ever, follows a smooth response; instead, it follows a cyclical response. Therefore, trade policy analyses undertaken within a comparative-static framework, such as that used by CGE models, yield misleading results in so far as they do not consider the long periods of adjustment in commodity markets that account for oscillations in the adjustment processes. The contrast between the results of CGE models and those of the econometric models in this study will be examined in Part V in relation to trade liberalization policies, on which a considerable amount of work with comparative-static models has already been done. Similar considerations regarding the dynamics of optimal export policies of the developing countries are also examined in Part V.

2.3 Methodological Framework

The approach to the analysis of international commodity trade adopted in this study is one that builds from theory and dynamic specification to estimation and validation, and finally to policy analyses. As such, it develops a theory-based econometric model with which to analyse commodity trade policies. The theory formulated embodies important recent advances in consumer preferences that give rise to product heterogeneity in international trade; it describes equilibrium conditions in such a market; and it makes explicit the constraints that need to be imposed if complete systems of trade are to be formulated and estimated (Part II). The dynamic specification of the system applies the recent work on dynamic time-series models that explain observed disequilibria in the light of long-run, or steady-state, solutions that are theory-consistent (Part III). The application of the model to a set of developing countries dependent on selected commodities demonstrates the approach to statistical inference and tests the legitimacy of the theoretical and empirical interpretation of commodity trade (Part IV). Finally, the implications of the results for trade policy analysis are examined (Part V).

Note

1 The test consisted in the estimation of an equation that relates the local currency price of a good originating from one country to (*a*) the local

currency price of the same good originating from another country, (b) the exchange rate between the two countries, and (c) transportation costs. The null hypothesis was that the sum of the estimated coefficients of the log-linear form of the relationship was approximately equal to unity.

Part II

Commodity Trade Theory

Introduction

In any commodity market analysis that seeks to model international price behaviour, imperfect competition between trading partners is dealt with by implicit acceptance as long as it does not affect the global results. In commodity trade analysis, however, such a disposal of imperfect competition is not an option open to the analyst. Imperfect competition arising from product differentiation underlies the modelling framework of this study. Such differentiation prevails because commodities differ greatly in the way they are exported, and importers either have different preferences for attributes or perceive differences in the quality of a particular commodity. When an importer differentiates between a commodity exported by one country and the same commodity exported by another, the exporting countries can exert some control over the price at which they sell their products, and the market structure is one of imperfect competition. When there are many exporters of a commodity that is differentiated by importers, monopolistic competition prevails.

In this study no attempt is made to develop a general theory of commodity trade with imperfect competition. Imperfect competition can have different causes, each of which may have different effects. Several attempts have recently been made to provide a general framework for the analysis of imperfect competition in international trade (e.g. Helpman and Krugman 1985; Jones 1984). None the less, as Dixit and Norman (1980: 265) have pointed out, 'To arrive at a general theory of trade with imperfect competition is ... impossible.' What must be sought instead are the conditions that give rise to a market structure that most closely characterizes trade in the particular types of goods being modelled.

The first step in this process is to explain why commodity differentiation may arise and lead to a negatively sloped export demand schedule for producing countries. Accordingly, the first chapter in this part specifies the importer's preference structure for a differentiated commodity. Imperfect substitutability of the same commodity originating from different countries is interpreted as the importer's desire for characteristics that distinguish a product exported by one country from those of others. This interpretation is based on the characteristics approach to commodity differentiation in international trade. However, it extends the approach in such a way that importers can select various supply sources of a commodity. Imperfect substitution of exports of the same commodity by different suppliers can therefore reflect the importer's preference for diversification of supply sources. Imperfect information also reduces the degree of substitutability of exports of a particular commodity that have very similar characteristics.

The next chapter (Chapter 4) derives the supply schedules of exporting and importing countries. In the recently elaborated theory of monopolistic competition, economies of scale provide the upper limit to the amount of commodity differentiation, defined as the number of firms or countries in the industry. When varieties of a product are defined along a finite space (for example, a circle) and consumers differentiate on the basis of the country of origin of the good, the number of countries in the world determines the upper limit to the amount of commodity differentiation. Returns to scale may therefore be increasing, constant, or decreasing in this situation.

Specification of a preference function that is consonant with differentiation of traded commodities provides the basis for the formulation of the monopolistically competitive market structure. Unlike the given demand schedules of either a pure monopolist or a pure competitor, the demand schedule of the exporter of a differentiated commodity depends on the price of close substitutes. The market for differentiated commodities is accordingly characterized by a non-cooperative price equilibrium. The conditions of such an equilibrium are examined in Chapter 5.

The theory of commodity trade formulated in this part of the study is framed in terms of static states; its immediate use is limited to comparative statics. In Part III, representations of lags that exist in adjustments of trade to long-run dynamic equilibrium solutions, or steady-state growth paths, are introduced. The resulting dynamic system provides a characterization of the underlying process whereby data are generated in international commodity trade. Part IV contains an empirical investigation of the validity of the inferences about the behavioural relationships and their system interaction in characterizing international commodity trade.

The model developed seeks to represent the essential features that underlie the process of international commodity trade—as distinct from an attempt to describe the complete trade system—in such a way as to permit the quantitative characterization of interrelationships between importers and exporters. Accordingly, the set of causal relationships is parsimonious and is based on the following assumptions: (1) all exporters have the same production function; (2) all importers have the same utility function; and (3) commodity characteristic differences arise outside the production process, so production is independent of product specification. The first two assumptions provide symmetry in the commodity market; consequently, equilibrium in the market is ensured.

Terminology

To make the concept of differentiation explicit, a distinction is made between 'commodity' and 'product'. A *product* refers to a commodity that is distinguished by nature yet is intertemporally equivalent to itself. The nature of the commodity is defined by its location since each location embodies a set of characteristics. Hence the same commodity available from two different locations is considered to be two distinct products, each defined by its composite of characteristics. A *commodity* refers to a set of products having the same characteristics, although the proportions of shared characteristics may vary and give rise to commodity differentiation. Finally, the term 'commodity' is used interchangeably with 'good'.

3
The Demand for Traded Commodities

The fundamental interpretation of commodity differentiation in international trade is that an importer perceives commodity exports from alternative countries as embodying different proportions of characteristics. Hence commodity differentiation can arise in international trade because exports are perceived by importers to contain different proportions of characteristics. Commodity export characteristics include quality (size, condition, grade, uniformity, colour, variety), marketing conditions (delivery time, credit terms, reliability of supplies), and cultural, historical, or political ties between trading partners. A number of these characteristics have already been described in Chapter 1 for particular commodities that are widely traded at the international level.

This interpretation of commodity differentiation is based on the characteristics approach in the economics of consumer behaviour. It is their characteristics that give rise to differentiation in internationally traded commodities. As mentioned earlier, commodities become horizontally differentiable when importers differ in their choice of product types even though their quality may be the same; in contrast, commodities are said to be vertically differentiable when their product types differ only in quality and all importers have the same preference ordering. Naturally, commodities differ greatly in the way they are exported by countries. Some differences in characteristics have to do with horizontal differentiation, such as the range of marketing services offered by alternative exporters; others have to do with vertical differentiation, such as the quality grade of coffee exports.

Once traded commodities become differentiated, the importer may prefer to diversify the suppliers of a commodity rather than purchase that commodity from only one country. This situation is described by the variety approach to consumer preferences. Preferences are defined in this chapter in such a way that they encompass the characteristics and diversity approaches that have been alternatively adopted in recent theories of commodity differentiation in international trade. These preferences underlie much of the work on the 'new theory of international trade' (see Helpman and Krugman 1985) and motivate the analysis of importers' demand for particular commodities in terms of

Armington's (1969) approach, whereby products are distinguished by their geographic origin.

The first part of this chapter describes these two approaches. It argues that the diversity approach presupposes commodity differentiation, rather than explaining how it can arise in traded goods that are physically homogeneous. It is therefore more appropriate to goods that are vertically differentiated. However, when used in conjunction with the characteristics approach, the diversity approach provides a more adequate description of reality than does the characteristics approach alone. It enables importers to reduce risk by affording them an opportunity of diversifying supply sources in international trade, whereas in the characteristics approach each importer purchases a commodity from only one supplier. In the second part of this chapter conventional demand schedules are derived from the formulated preferences. The derivation is straightforward. Conventional demand schedules make it easier than in the past to use the characteristics approach in the analysis of commodity differentiation in international trade.

3.1 Structure of Importer Preferences

Importer preferences are assumed to be based on the characteristics of a commodity. Each importer has a subjective preference for a given combination of characteristics, which leads him to value the exports of one country over those of another. One of the characteristics of a traded commodity is that it can be obtained by an importing country from several sources. This characteristic is important since the importer values diversification of supply sources. Even if the exports of two countries have the same proportion of characteristics, they can still be imperfect substitutes because the importing country prefers to diversify its supply sources in order to increase the safety of its supplies.

Heterogeneity in physically homogeneous commodities is difficult to conceive. This difficulty arises because in the traditional theory of consumer behaviour the concept of 'commodity' is fundamental. For example, a typical formulation of a consumer's preference ordering would be based on the utility derived from coffee and from the numeraire non-coffee good. In this context, the commodity has no characteristics other than those that physically define it; thus, arabica coffee sold in a tin at a supermarket is not distinguishable from arabica coffee sold in pre-ground form at a coffee shop. Yet to the consumer the coffee may be different, and this difference can be measured by the price he or she is willing to pay for it.

The commodity characteristics approach to consumer behaviour takes the characteristics of a commodity as the fundamental concept. In this context a commodity becomes a collection of characteristics and is thus a derived concept. In the example for coffee, a consumer's subjective preference ordering would, in the economics of commodity characteristics, be defined by objective characteristics: the type and grade of coffee, whether the beans were milled or left to be ground by the consumer, the type of packaging, and the characteristics of the retail outlet.

Extension of the economics of product characteristics to international commodity trade is straightforward. In the coffee example, the importer's subjective preference ordering now takes the place of that of the general consumer; the characteristics of the exporter take the place of those of the retail outlet. It is the characteristics of the exporter that provide the focus. Even when a commodity is perfectly homogeneous in its physical characteristics, exporters may have objectively different characteristics that will cause an importer to differentiate a commodity according to the foreign supplier.

3.1.1 The Economics of Commodity Characteristics in Trade

The commodity characteristics approach has only recently had a major impact on international trade theory, though it emerged early in the economic literature. It was formally introduced into the analysis of imperfect market structures by Hotelling (1929) in his seminal work on spatial economic theory. His concept of a commodity was represented by a commodity space in which competition among products in the space depends on the similarity of their characteristics. One characteristic underscored by Hotelling was the geographic location of the firm. He attributed commodity differentiation in otherwise physically homogeneous commodities to transport costs in their spatial distribution, as well as to the 'mode of doing business', existing socioeconomic ties, or 'differences in service or quality'. The resulting commodity differentiation has been shown by Gannon (1977) to help explain why, in the context of spatial economic theory, firms will tend to concentrate at the centre of a spatial market. In the household production theory developed by Gorman in 1956 (since published in Gorman 1980) and by Becker (1965), utility is derived from the consumption of characteristics produced by the household from purchased goods. (For a detailed exposition, see Deaton and Muellbauer 1980: ch. 10.) This concept has been adopted in the characteristics approach, which has been most extensively developed by Kelvin Lancaster.

In Lancaster's (1966, 1971) formulation, the characteristics contained in a commodity are objectively defined, whereas the consumers' preferences for characteristics are subjective. Each consumer therefore derives a different level of utility from the consumption of those characteristics. A consumer's behaviour may then be explored without knowledge of his or her utility function. However, the validity of Lancaster's initial formulation was shown by Hendler (1975) and Lucas (1975) to depend on two critical assumptions: the non-negative marginal utility of characteristics, and their weak separability so that utility depends only on the total amount of characteristics in a commodity and not on their proportions in alternative supply sources. Neither of these assumptions was required of the characteristics models developed by Ladd and Suvannunt (1976) and Ladd and Zober (1977).

More recently, Lancaster (1979) has extended his approach to monopolistically competitive market structures, and has framed it in terms that are commonly used in industrial organization. The generalization lends itself to many applications in the analysis of consumer demand and non-competitive markets. In the field of international economics, for example, the welfare implications of markets whose goods are differentiated have been examined by Katrak (1975), Lancaster (1980), and Helpman (1981).

The demand for internationally traded commodities that are differentiated can be expressed in either explicit or implicit schedules. An explicit schedule attempts to measure directly the demand for characteristics of an exported good. But since prices for particular characteristics are not directly observable, their estimation has been based on *structural hedonic price schedules*. In this approach the price of a commodity is used as a measure of its characteristics, and that price is related in an empirical manner to the measurable characteristics of the commodity. Structural hedonic price schedules have a long and voluminous history (see Deaton and Muellbauer 1980: 254–67). Ladd (1983) has surveyed attempts to estimate hedonic price schedules for agricultural products, and Rosen (1974) has provided a theoretical basis to explain the hedonic results of empirical estimates.

An implicit export demand schedule, on the other hand, does not separate the characteristics of a commodity. The measurable characteristics are considered together—in their reduced form—by the total quantity exported of the good, and the observed price of an export is again used as a measure of all its characteristics. This implicit demand schedule is the familiar Marshallian one that relates the quantity of the export demanded to its price. Although the approach is traditional, the notion underlying it is different: export demand represents demand for the characteristics of the export rather than for the product itself. Since our interest is in using the characteristics approach rather than in actually estimating the demand for the individual characteristics of exported goods, the implicit construct of demand schedules for exports will be used in the development of a model of international commodity trade.

The characteristics approach offers an intuitively appealing explanation of why importers differentiate between supplying countries even though the commodity itself is physically homogeneous. Unfortunately, it suffers from the inherent limitation of yielding a solution in which each buyer purchases from only one supplier. Diversification of supply sources arises only from aggregation of consumers (Lancaster 1975: 571–2). Thus, a country's diversification of different exporters of a particular commodity is simply interpreted as the summation within the country of purchases by individual agents from a single supplier. Clearly, this approach is unrealistic. We shall therefore examine an alternative approach to diversification of supply sources by importers.

3.1.2 The Economics of Variety in Trade

The analysis of consumer preferences that include the desire for variety has been developed along a completely separate vein from that of the characteristics approach. Nevertheless, this approach has been used to analyse commodity differentiation. It was used by Stern (1972), whose work has been illustrated by Meade (1974), to examine the issue of optimal product variety. Spence (1976) and Dixit and Stiglitz (1977) used the diversity approach to examine the same issue and to frame the analysis in more conventional terms. They showed that products that are near substitutes for one another will give rise to a monopolistically competitive market because of consumer preferences for product diversity. Since then, Sattinger (1984) and Hart (1985a, 1985b) have used the diversity approach to examine product variety in monopolistic competition.

According to this approach, a consumer's desire for diversity of supply sources arises from the convexity of the indifference curve for imperfectly substitutable products that constitute a commodity. This result is derived from one of the axioms of choice that reduces the consumer's decision to the constrained maximization of utility (Deaton and Muellbauer 1980: 26–30). It holds that a consumer's indifference curve for a given level of satisfaction in his or her ordinal utility function is strictly convex: any straight line drawn between two points of an indifference curve will always lie above any point along the curve.[1] As a result, a combination of products is always preferred to specialization in just one product.

In international trade, strict convexity of an importer's indifference curve will cause the importer to prefer a diversity of exporting countries to a single supplier-country. With free entry into the market, equilibrium will be one of monopolistic competition. Krugman (1979, 1981), Dixit and Norman (1980), Das (1982), Lawrence and Spiller (1983), Horn (1984), Hart (1985a, 1985b), Helpman and Krugman (1985), and Venables (1987) have used the preference for diversity implied by strictly convex indifference curves to derive a monopolistic competitive model of international trade for examining the welfare implications of markets with differentiated goods.

However, convexity of indifference curves is a weak axiom. In certain situations there will be an upper limit to the amount an importer is willing to give up from one supplying country in order to obtain more from another. For instance, once an importer buys a sufficiently large amount from one supplying country, there could be such favourable marketing arrangements with it that the importer would be unwilling to diversify his supply sources. When there are two suppliers, this situation would be depicted graphically by an indifference curve that intersects the axis of the primary supplier, at which point no amount is purchased from the other supplier. Once it intersects the axis, the indifference curve is no longer convex, and utility maximization for an importer can occur at a point where only one product is purchased; hence specialization in one supply source, rather than diversification among several sources.

A strictly convex indifference curve is even less likely to hold for *all* traded goods. If an importer believes that a product originating from one country is a perfect substitute for a product originating from another country, the importer's indifference curve for alternative suppliers will be a straight line. For instance, an importer of a commodity that is physically homogeneous, such as sugar, may view exports from alternative suppliers as perfect substitutes for one another. This limitation arises because, in the approach used to explain the diversity, the traditional concept of a commodity, rather than the characteristics of a commodity, is adopted as the fundamental concept.

In short, preference for diversity does not explain why an importer differentiates between a commodity originating from one supplier and the same commodity originating from another; what it does provide is an explanation of the observed diversification, by importers, of foreign suppliers of products that are near substitutes for one another.

3.1.3 Diversity as a Characteristic

A fundamental difference between the characteristics and the diversity approaches lies in the way in which preferences are structured. Lancaster assumes that each consumer has a most preferred product and that these most preferred products are uniformly distributed among all consumers. Diversification of supply sources in the Lancasterian view arises solely from aggregation of demand curves of individuals having different most preferred products. In the application of this approach to international trade, each buyer in a particular geographic market is dependent on a single supplying country, so that diversification of exporters by the geographic market is a reflection of buyer aggregation. In contrast, in the diversity approach the desire for supplier diversification exists at the level of the individual consumer; it originates from the consumer's preference for a mixture of products that constitute a commodity over that for any single product whose quantity alone would otherwise provide the same satisfaction. In the application of this approach to trade, imports directly reflect a nation's desire for diversification of supply sources.

Redefining Lancaster's most preferred product, which is an 'ideal' product, to include the characteristic of diversity eliminates the difference. In doing so, we introduce a new concept, namely that of *commodity type*. A commodity type is formed by the consumer's choice of characteristics from a variety of products. It is thus composed of a mixture of products in a commodity group.

In the present trade context, differentiation of a commodity exists at the level of the exporter. It arises from differences in the proportion of characteristics shared by exporters, not from physical differences; it is therefore associated with the source of supply characteristics. Importer preferences are defined by such characteristics, which include that of diversification of supply sources. An import is therefore a commodity type that is composed of product exports that form part of a commodity class.

3.2 Specification of Importer's Preference Structure

Let a commodity $Q = (Q_1, \ldots, Q_n)$ be a vector of products from sources $1, \ldots, n$. Copper, coffee, and sugar are examples of commodities. Exports of a commodity from different sources, such as Chile and Peru, are considered to be distinct products, say Q_i and Q_k. Similarly, the domestic output of an industry in a particular geographic market is considered to be a differentiated product. The products share the same characteristics, but the proportions of them differ.[2] Commodity differentiation, then, reflects differences in the proportion of characteristics that make up a commodity supplied by a country.

3.2.1 Preference Ordering

The most appropriate means of describing the importer's preference ordering for commodities differentiated by country of origin is the utility tree.[3] On the first level, a decision is made about how much to consume of commodity Q and all other goods whose composite forms the numeraire N_o; the decision is based on total expenditures and prices of the goods. At the next level, a choice is made about how much to consume of the differentiated products $Q_1, \ldots, Q_n$; the choice is based on the expenditures allocated to the commodity Q and on the relative prices of the products from different country sources. The importer's preference ordering at each level must be independent of that at other levels.

It is assumed that the satisfaction derived from consumption of commodity Q is weakly separable from the satisfaction derived from consumption of the numeraire good N_o. This weak separability ensures independent decision-making at each level and means that substitution can take place between goods Q and N_o, but not between a differentiated product Q_i within the commodity Q and the numeraire good N_o. As a result, changes in income and relative prices can bring about *intersectoral* substitution between Q and N_o, while changes in the relative price of Q_i can bring about only *intrasectoral* substitution between that differentiated product and other products of the same commodity, so that preferences for the numeraire good N_o remain unaffected.

The separability assumption is adopted by Lancaster (1979: 25) in his characteristics approach to consumer demand. In the context of international trade, once the total import quantity of a commodity is determined, the importer's decision about how much to purchase from alternative supply sources is made on the basis of the characteristics contained in the product exported by countries, instead of on the basis of the product itself. Moreover, it is subjective preferences that make an importer select a product exported by one country rather than one exported by another. For example, the UK may choose to buy coffee from Jamaica rather than from another supplier because of the domestic consumer appeal of the brand name of that coffee-producing country, credit terms, cultural ties with its former colony, and other characteristics of the exporter. (See Roemer 1977 for empirical evidence of historical and cultural linkages between countries as a determinant of trade in manufactures.) Although each product contains the same proportion of characteristics for all importers, each importer derives a different level of satisfaction from them. Thus, the importer makes a selection from among alternative exporters on the basis of the proportion of characteristics available in the particular products of the exporting countries.

We can incorporate the diversification of supply sources into the Lancasterian characteristics framework by defining it as a characteristic that is desired by importers. Each importer j ($j = 1, \ldots, m$) is associated with a commodity type Q_j. Each importing country is assumed to have a concept of an ideal commodity type, denoted Q_j^*, that embodies its most preferred characteristics mixture. The most preferred coffee imports by the Federal Republic of Germany may be a specific mixture of Jamaican and Guatemalan coffee; France will not necessarily have the same most preferred mix. When diversification of supply sources is a characteristic valued by importers, the ideal commodity type will include it. Thus Q_j^* always represents a composite of products in the commodity vector.

There is a trade-off between how much of a commodity can be obtained by the importer and from how many sources. The rate of substitution between volume q and variety v of commodity types is given by $-(\partial Q_j^*/\partial q)/(\partial Q_j^*/\partial v)$. The reason for the trade-off is as follows. If all supplies originated from one source, the importer would receive a greater amount of a product, and consequently greater satisfaction because of scale economies, but at the cost of the greater satisfaction he would otherwise derive were he to obtain the products from a variety of sources and enjoy greater security of supplies.

The importer may not obtain his most preferred commodity type, either because such a characteristics combination is impossible or otherwise unavailable, or because the information available to the importer is incomplete. If the information is incomplete and there are positive information acquisition costs, the importer will search for the most preferred commodity type only as long as the marginal benefit from the search exceeds the marginal cost of the search. Stigler's (1961) seminal work on information has demonstrated that, although the cost can remain constant, the benefits from the search diminish as the search becomes more complete. Hence an equilibrium will be reached at the point when the search is discontinued. This search for information about differentiated products has been shown by Salop (1976), Donnenfeld (1984), and Wolinsky (1986) to give rise to monopolistic competition. The cost associated with the acquisition of information limits the ability of the importer to search for the range of characteristics he seeks in his purchase of a particular commodity from a country. As a result, there is a natural tendency for products to remain imperfect substitutes for one another even though the characteristics of a commodity originating from one supply source are very similar to those of another supplier. Even if the equilibrium reached is one at which complete information is obtained, the importer might be unable to purchase his ideal commodity type because his combination of desired characteristics cannot be obtained from existing producers. In such a case, the importer will select the most similar commodity type available.

Whether or not the commodity type contains the particular proportion of characteristics most preferred by an importing country is likely to be a matter of indifference, provided the country receives compensation. In accordance with Lancaster (1979: 37–52), it is assumed that, for the importer, there is nothing to choose between the compensated difference d_j for a commodity type Q_j—which may not embody his most preferred characteristics mixture—and his ideal commodity type Q_j^*, such that

$$Q_j = Q_j^* d_j, \qquad (3.1)$$

where $d_j \geq 1$. The compensation d_j gives the amount of a commodity type needed to provide the importer with the same amount of satisfaction obtained from one unit of the ideal commodity type.

We can thus express the importer's choice in unit-equivalents of the most preferred commodity type. Let us assume that importer j selects a commodity type Q_j rather than his most preferred commodity type Q_j^*, so that the selected commodity type is worth Q_j/d_j units of the importer's ideal commodity type. The *quantity of imports*, denoted M, can then be expressed in unit-equivalents of the most preferred commodity type:

$$M_j = Q_j/d_j. \qquad (3.2)$$

Because of the transitivity property of equalities, the *quantity of exports*, denoted X, to geographic market j can be expressed in unit-equivalents of the most preferred commodity type:

$$\begin{aligned} X_{1j} + \ldots + X_{nj} &= (Q_{1j} + \ldots + Q_{nj})/d_j \\ &= \frac{Q_{1j}}{d_j} + \ldots + \frac{Q_{nj}}{d_j}. \end{aligned} \qquad (3.3)$$

Equations (3.2) and (3.3) express the amount of commodity imports M and the amount of product exports X to market j in conventional terms, although the characteristics approach to preference ordering is retained.

We can now proceed to specify the *indifference schedules* of the importer in conventional expressions for the imported commodity and alternative export products to that market. The assumption of separability in the preference ordering means that the amount to spend on the imported commodity M and all other goods, whose composite forms the numeraire N_o, is independent of how the amount spent on M is allocated among the different export products $X_1, \ldots, X_n$. For application to the estimation of a system of demand equations, it will be assumed that both intersectoral substitution of M and N_o and intrasectoral substitution of alternative export products $X_1, \ldots, X_n$ take place in the constant elasticity form. The importer's overall utility schedule is thus given by

$$U(M_j, N_{o,j}) = [\pi_j M_j^\alpha + (1 - \pi_j) N_{o,j}^\alpha]^{1/\alpha}, \qquad (3.4)$$

where $\alpha < 1$ and $0 < \pi_j < 1$. Let subscript i refer to a particular supplying country of interest and let k refer to each of the $n - 1$ other foreign supplying countries. Then the importer's sub-utility schedule for intrasectoral substitution in the imported commodity M is given by

$$U_m(X_1, \ldots, X_n) = \left(\pi_{ij} X_{ij}^\beta + \sum_{k=1}^{n-1} \pi_{kj} X_{kj}^\beta\right)^{1/\beta}, \qquad (3.5)$$

where $\beta < 1$ and $0 < \pi < 0.5$ such that $\pi_{ij} + \Sigma_k \pi_{kj} = 1$. Although the value of the distribution parameter π usually lies between zero and one, it is restricted here because, since the market is one of monopolistic competition, the relative market share of each exporter is small. This restriction will be shown to have practical advantages, namely that the export market share of any country always lies between zero and one.

Intersectoral and intrasectoral substitutions in (3.4) and (3.5) take place in terms of generalized constant elasticity of substitution (CES) preference functions. The CES function was introduced by Brown and Heien (1972) to overcome two restrictions of the linear expenditure system, which was first used by Klein and Rubin (1948). The restrictions in the linear expenditure system are, first, that the own-price elasticities of demand cannot exceed (minus) unity and, second, that cross-price elasticities are zero. In (3.4) and (3.5), both complementary and substitution effects are represented. The exponents α and β are interpreted to mean that, when the goods or product exports are perfect substitutes, their values approach unity; when the goods or product exports are non-substitutable, their values approach $-\infty$. Since product exports must be more closely substitutable for one another than for the numeraire good, the restriction $\alpha < \beta$ must be imposed. In the sub-section that follows, it will be shown that the own-price elasticity can lie between 0 and $-\infty$.

Specification of the indifference curve for intrasectoral substitution within the imported commodity in the CES form incorporates a preference for diversity. It can be shown that the indifference curve in (3.5) is strictly convex for positive values of X_{ij} and for $\beta < 1$. (For a proof, see Chiang 1984: 427.) The rate at which the importer is willing to substitute exports from a supplier of interest i for those from any other supplier k, where $k = 1, \ldots, n - 1$, in order to maintain a given amount of total commodity imports M, is equal to the marginal rate of substitution:

$$\frac{\partial X_{kj}}{\partial X_{ij}} = \left(\frac{\pi_{ij}}{1 - \pi_{ij}}\right)\left(\frac{X_{ij}}{X_{kj}}\right)^{\beta-1}. \qquad (3.6)$$

Consequently, the importer will prefer to obtain some supplies from both sources i and k rather than all supplies from just one source.

The effect of supplier diversification on the importer's preference function is to ensure that exporters consistently behave as monopolistic competitors. Salop (1979) has shown that, when there is no preference for diversity of supply sources, the demand schedule is kinked. The kink arises because the commodity type contains only one element, say exports X_{ij} (Lancaster 1979: 34–5). In such a situation, the market power of an exporting country would equal that of a pure monopolist as long as its effective price P_{ij} to the importer were less than the effective price P_{kj} of competitors. This condition defines the monopoly *market width* of the exporter.[4] Only if the effective price were at least equal to that of competitors would the exporter begin to trade as a monopolistic competitor. In contrast, when there is diversity in the preference structure, all exporters $X_{1j}, \ldots, X_{nj}$ in the vector M_j compete for market shares, and the exporter X_{ij} does not have exclusive control anywhere along his demand schedule. In such a situation, the elasticity of substitution is independent of the compensated difference, d, between product exports. In addition to a more comprehensive representation of importer preferences, the introduction of diversification of supply sources into the preference structure allows a more conventional approach to the derivation and analysis of the demand schedules.

3.2.2 Import Demand

Given the importer's preference ordering, it is now possible to derive the importer's demand schedule as well as the export demand schedules of its foreign suppliers. Separability of preferences in the utility tree approach allows the decision at each level to be considered an independent utility maximization problem. The first level of decision maximizes the overall utility function subject to the budget constraint; the next level maximizes the utility function for alternative supply sources of the commodity subject to the allocation of expenditures for imports of that commodity determined at the first decision-making level.

The utility maximization problem for the first level of decision by geographic market j, given a commodity import price P and a level of nominal dollar income Y^n, is

$$\max [\pi_j M_j^\alpha + (1 - \pi_j) N_{o,j}^\alpha]^{1/\alpha}$$
$$\text{subject to } P_j M_j + N_{o,j} = Y_j^n, \qquad (3.7)$$

where $\alpha < 1$ and $0 < \pi < 1$. The solution to the foregoing problem yields the overall demand schedules for commodity imports M and the numeraire N_o of importer j:

$$M_j^d = k_1 Y_j \left(\frac{P_j}{D_j}\right)^{\epsilon_m^p} \qquad (3.8a)$$

and

$$N_{o,j} = (1 - k_1)Y_j^n \left(\frac{P_j}{D_j}\right)^{\epsilon_h^p}, \qquad (3.8b)$$

where $\epsilon_m^p = 1/(\alpha - 1)$ and $\epsilon_n^p = \alpha/(\alpha - 1)$; $k_1 = [(1 - \pi_j)/\pi_j]^{1/(1-\alpha)}$, with expected sign $k_1 > 0$; $D = (1 + k_1 P^{\alpha/(\alpha-1)})^{(\alpha-1)/\alpha}$ is the deflator; and $Y = Y^n/D$ is constant dollar income.[5]

The demand schedules have two important properties.

1. The *income elasticities* are equal to unity, a hypothesis that will later be tested.
2. The *price elasticity of demand for commodity imports* (ϵ_m^p) can take on any value between 0 and $-\infty$. (Recall that $\epsilon_m^p = 1/(\alpha - 1)$ and $\alpha < 1$, so that $-\infty < \epsilon_m^p < 0$.)

3.2.3 Export Demand

Once the level of expenditures Y_m^n for the imported commodity M has been determined, the utility maximization problem of how much of the commodity to purchase from alternative suppliers—let us say an exporter of interest i and its competitors k to market j whose corresponding export prices are P_i and P_k—may be expressed as

$$\max [\pi_{ij} X_{ij}^\beta + (1 - \pi_{ij}) X_{kj}^\beta]^{1/\beta}$$
$$\text{subject to } P_{ij} X_{ij} + P_{kj} X_{kj} = Y_{m,j}^n, \qquad (3.9)$$

where $\beta < 1$ and $0 < \pi_{ij} < 0.5$. Then the *export demand schedules* for the country of interest i and its competitors k are (see Annex D for the derivation)

$$X_{ij}^d = k_2 M_j \left(\frac{P_{ij}}{P_j}\right)^{\epsilon_x^p}$$

and

$$X_{kj}^d = (1 - k_2) M_j \left(\frac{P_{kj}}{P_j}\right)^{\epsilon_x^p} \qquad (3.10)$$

where

$$\epsilon_x^p = 1/(\beta - 1),$$
$$k_2 = [(1 - \pi_{ij})/\pi_{ij}]^{1/(1-\beta)},$$
$$P_j = (P_{ij}^{\beta/(\beta-1)} + P_{kj}^{\beta/(\beta-1)})^{(\beta-1)/\beta}$$

is the import price of the commodity, and $M_j = Y_{m,j}^n/P_j$.

Demand for Traded Commodities

The export demand schedules have the following desired properties.

1. Export demand has a unitary elasticity with respect to the level of import demand in the geographic market, which is theoretically consistent: a change in the level of import demand in the foreign market will, *ceteris paribus*, cause a proportionate change in the demand for the exports of all supplying countries to that market.
2. The price elasticity of export demand (ϵ_x^p) has a value that lies between 0 and $-\infty$.
3. The constant k_2, which has the value $0 < k_2 < 1$, measures the exporter's market share.

The foregoing system of intersectoral and intrasectoral demand schedules in (3.8) and (3.10) lends itself to empirical application since the exponential form of the equations can be converted into double-logarithmic equations whose estimated coefficients are directly interpreted to be elasticities. Moreover, the use of CES preference functions for both intersectoral and intrasectoral substitution does not impose undue restrictions on the own-price and cross-price elasticities. Their values are consistent with those that would be expected for normal goods and product exports.

3.3 Illustration

Let us consider the export demand schedule of Chile's copper exports to the US market and the import demand schedule of that market. Since all importers are assumed to have the same utility function, it will be convenient for notational simplicity to make the subscript reference to the importer implicit in this section. The overall indifference curve of the USA and its indifference curve for copper from alternative supply sources are respectively represented by the following two equations:

$$U = [0.99 M^{0.33} + (1 - 0.99) N_o^{0.33}]^{1/0.33} \quad (3.11a)$$

and

$$U_m = [0.6 X_i^{0.71} + (1 - 0.6) X_k^{0.71}]^{1/0.71}, \quad (3.11b)$$

where X_i are Chile's exports and X_k are competing suppliers to the US market. Note that the requirement that intrasectoral substitution be greater than intersectoral substitution is met because $0.71 > 0.33$, which means that copper supplies from different country sources are closer substitutes than are US copper and the numeraire good, N_o.

The numerical values of the parameters in the demand schedules can be derived from the parameter values given for the indifference curves in (3.11). Recall that the price elasticity of import demand, ϵ_m^p, is equal to $1/(\alpha - 1)$. Since $\alpha = 0.33$ in this instance, $\epsilon_m^p = -1.5$ is the price elasticity of import demand. The price elasticity of export demand, ϵ_x^p, is equal to $1/(\beta - 1)$. In this case $\beta = 0.71$, so $1/(\beta - 1) = -3.5$. The constant terms take on values of $k_1 = 22{,}830$ for the import demand curve of the USA, and $k_2 = 0.25$ for the export demand curve of Chile. (In particular, $k_2 = [(1 - 0.6)/0.6]^{1/(1-0.71)}$.)

Then the US import demand curve and the export demand curve of Chile are, respectively,

$$M = 22{,}830 Y \left(\frac{P}{D}\right)^{-1.5} \qquad (3.12a)$$

and

$$X_i^d = 0.25 M \left(\frac{P_i}{P}\right)^{-3.5}. \qquad (3.12b)$$

These two demand curves are shown in Fig. 3.1. The slope of the US market demand curve is steeper than that of the demand curve for Chilean copper. The steeper slope results from the requirement that intrasectoral substitution be greater than intersectoral substitution in the US market. The two curves intersect at a price of about $1,150 a metric ton. At that price, the US export demand for Chilean copper is equal to

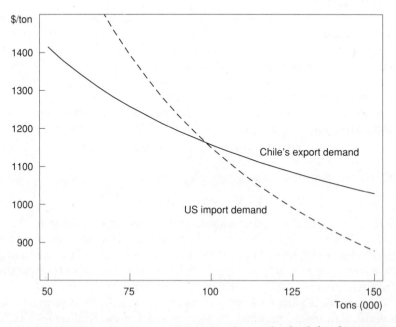

FIG. 3.1 Illustration of export and import demand schedules for copper

100,000 metric tons, and the total US copper import demand equals 400,000 metric tons. Accordingly, Chile's export market share equals 25 per cent, which is the value taken on by the constant k_2. Therefore, the interpretation of the constant k_2 is that its value represents the market share of the exporting country in long-run static equilibrium. In this illustration no economic interpretation is ascribed to the constant term k_1 in the import demand schedule because it relates different units of account, namely metric tons of copper to constant US dollar income.

3.4 World Demand

The total world consumption of a commodity is the sum of the quantities demanded by economic agents in the market. Thus far we have derived the import demand schedules of the foreign markets $j = 1, \ldots, m$ of the n exporting countries $i = 1, \ldots, n$ that are of interest to us. Therefore a consumption schedule for all other economic agents in the market is required if the total world consumption is to be obtained.

The consumption by at least two other groups of economic agents in the market needs to be considered: the economic agents for the domestically produced good in foreign markets, and the economic agents in the exporting countries. Additionally, when only some of the world's exporters of the commodity are being considered, the derived schedules for import demand will represent only part of the total world import demand, so that the import demand of other markets will have to be included in the consumption schedule of the rest of the world.

Rest-of-world consumption, denoted C_k, therefore comprises the consumption of the domestically produced commodity in foreign markets, domestic consumption in the exporting countries, and the import demand of foreign markets other than those of the exporting countries of interest. It has been assumed that substitution of goods by all consumers takes place in the constant elasticity form. It therefore follows that the rest-of-world consumption schedule is analogous to the import demand schedule in (3.8a). Consumption, C_k, depends on the constant dollar price of the commodity and on the constant dollar income of those economic agents:

$$C_k = k_1 Y_k \left(\frac{P}{D}\right)^{\epsilon^p}, \qquad (3.13)$$

where P is the commodity price, D is the deflator, and $Y = Y^n/D$ is constant dollar income. The expected sign of ϵ^p, the price elasticity of demand, is negative.

Then the total world consumption of the commodity is the sum of the quantities demanded by the importers and all other economic agents in the market:

$$C = \sum_{j=1}^{m} M_j + C_k, \qquad (3.14)$$

where import demand is obtained from (3.8a) and the rest-of-world consumption is obtained from (3.13).

Since consumption of importers and all other economic agents has been assumed to be a monotonically decreasing function of price, total consumption is also a monotonically decreasing function of price. Similarly, since it is the sum of the individual demand curves derived under the assumption that substitution of goods by economic agents takes place in the constant elasticity form, aggregate consumption also has a unitary elasticity with respect to income. This assumption—that the elasticity of substitution of goods is constant—is highly restrictive and therefore subject to empirical verification. Part IV of this study presents estimates of demand functions to test whether the income elasticities of import demand and of rest-of-world demand for individual commodities differ significantly from unity.

Notes

1 Formally, an indifference curve $f(X)$ is strictly convex over the interval (a, b) if $f[\theta a + (1 - \theta)b] < \theta f(a) + (1 - \theta)f(b)$ for all $0 < \theta < 1$. The left-hand side of the foregoing equation gives the values of the indifference curve $f(X)$ evaluated along the various points on line segment (a, b); the right-hand side of the equation gives the values of the straight line that joins point a to point b.
2 The characteristics of foreign, as well as domestic, suppliers can be represented in a characteristics space for a commodity. The characteristics space is taken to be a circle of unit circumference, where it is assumed that suppliers are uniformly distributed along the circumference of the circle according to the proportion of characteristics of the commodity contained in their products. If there is a one-to-one correspondence between suppliers and points on the circumference, then the points define the products $Q_1, \ldots, Q_n$ and the distance between each product is $1/n$, the inverse of the number of products.
3 The utility tree approach to preference ordering was introduced by Sono (1945) and Leontief (1947). Its utility function properties were subsequently analysed by Gorman (1959) and Pearce (1961), while Katzner (1968) demonstrated its preference ordering relationship to the utility function.
4 In the commodity characteristics space described by a circle of unit circumference, the quantity each exporter can sell of its product, when there is a strict one-to-one correspondence between an exporter and an importer, depends

on the segment diameter of the circumference between its product and the corresponding segments of the adjacent products on either side. Changes in the market width occur as a result of both variations in the price of the product relative to the adjacent product and variations in the proportion of characteristics in the product.

5 Other studies on differentiation in internationally traded goods have usually adopted a utility function for the first-decision level of the Cobb–Douglas type, which yields a linear expenditure system, e.g. Dixit and Stiglitz (1977), Dixit and Norman (1980: 281–94), Lawrence and Spiller (1983), Dixit (1984), Horn (1984), Sattinger (1984), and Flam and Helpman (1987). However, the Cobb–Douglas function is a special case of the CES function. (For a proof, see Chiang 1984: 428–30.) In particular, the CES utility function in (3.4) is a Cobb–Douglas utility function when $\alpha = 0$, which gives the following demand schedules:

$$M^d = Y^n/P \qquad (3.8a')$$

and

$$N_o = Y^n. \qquad (3.8b')$$

As already noted, these demand schedules impose excessive constraints on the price elasticities. The first constraint is the *independence* assumption that the cross-price elasticities are all zero; the second is that the own-price elasticity cannot exceed unity. Krugman (1982), however, has adopted the more general CES utility function to analyse differentiation in internationally traded goods.

4

The Supply of Traded Commodities

This chapter derives the supply schedules for exporters and importers of a commodity. The rest-of-world production schedule is also specified, which allows the world market for a commodity to be analysed in later chapters. Import supply schedules have seldom been derived, the assumption being that, since the price elasticity of the supply of imports is infinite, the inverse supply schedule can be used to define the import price. In this chapter it will be shown how the inverse supply schedule for imports is obtained. The export supply schedule is of greater interest to commodity trade than the import supply schedule because the degree to which export supply responds to price changes has an important bearing on the volume of trade that takes place. Consequently, the export supply schedule will be the central focus of this chapter.

In the production function considered in this book, returns to scale can be decreasing, constant, or increasing, and the supply schedule that corresponds to each of these cases is upward-sloping, horizontal, or downward-sloping, respectively. In contrast, returns to scale are always increasing in the new theory of international trade dealing with monopolistic competition because they establish the upper limit to the number of producers in the market. (For a discussion on economies of scale in international trade and a survey of the literature, see Helpman 1984; and Helpman and Krugman 1985.) The upper limit to the number of producers is determined here by the number of countries in the world capable of producing the commodity. The reason for this limit, as described in the previous chapter, is that the producing country constitutes the basis for the differentiation of a commodity.[1] As a result, our characterization of international commodity trade contains a greater degree of flexibility than that of recent theories of monopolistic competition in trade based on differentiated commodities and economies of scale. It allows returns to scale to be empirically tested, rather than presupposed in the model. Such a test is performed in Chapter 15.

4.1 Export Supply

The export supply schedule is derived from the exporter's objective of maximizing net foreign exchange earnings by means of a cost-minimizing combination of the factor inputs used to produce the commodity. Net foreign exchange earnings are the difference between total revenue and total cost. Total revenue depends on the quantity exported, denoted X, and the price of the commodity export, P, which in turn is related to the quantity exported in monopolistic competition. In this case, total revenue equals $XP(X)$. Total cost also depends on the quantity exported. The cost schedule of the exporter is derived from the least-cost combination of the inputs required to produce a given level of exports. Hence the exporter seeks to minimize the outlay for the inputs required to produce a given level of exports, subject to the production schedule. In monopolistic competition the export supply schedule is different from that in perfect competition, where the export supply curve is the same as the marginal cost curve. It will be shown that the difference between the two export supply schedules is associated with the price elasticity of

To facilitate exposition of the material in this section, it is assumed that both the exporter of interest and its geographic market are representative of all the countries that export and import the commodity. This assumption implies that all the exporters and importers have identical production schedules and that the market supply schedule is the sum of the individual supply schedules of the countries. In accordance with the earlier notation, for export supply $X^s = X^s_{ij}$, and for import supply $M^s = M^s_j$. Subscript references are therefore implicit for both the exporter and its geographic market.

The production schedule of the exporter needs to be expressed in a specific functional form if a supply schedule that lends itself to empirical estimation is to be obtained. The particular form adopted is the CES function. The production schedule relating the amount of labour A and capital B needed to produce a given level of commodity export X is

$$X = k_3(A^\theta + B^\theta)^{\tau/\theta} \tag{4.15}$$

where $\theta < 1$, $\tau > 0$, and $k_3 = \exp(\sigma_0 + \sigma_1 T + \sigma_2 W)$.[2]

The production schedule in (4.15) has the following important characteristics:

1 The constant term incorporates an efficiency parameter, $e^{\sigma_1 T}$, which measures the state of technology in the production of the commodity export, and a shift parameter, $e^{\sigma_2 W}$, which measures major disturbances—such as natural disasters and labour disruptions—in the production of the commodity export.

2 The value of τ determines returns to scale: when $\tau = 1$ there are constant returns to scale and the production function is of the Cobb–Douglas type; when $\tau > 1$ there are increasing returns to scale; when $0 < \tau < 1$ there are decreasing returns to scale.

The cost of production, denoted E, for a constant unit cost of capital, V_1, and of labour, V_2, is

$$E = V_1 A + V_2 B. \qquad (4.16)$$

A country seeking the cheapest way to produce a given level of commodity exports faces the problem of minimizing its production cost in (4.16) subject to the production schedule in (4.15). The solution to this problem yields the following *cost schedule* (see Annex D for the derivation):[3]

$$E = k_3^{1/\tau} X^{1/\tau} (V_1^{\theta/(\theta-1)} + V_2^{\theta/(\theta-1)})^{(\theta-1)/\theta} \qquad (4.17)$$

where $\theta < 1$ and $\tau > 0$. It is an explicit function of (1) the commodity output level, X, and (2) the input prices of capital and of labour, V_1 and V_2, respectively.

The supply schedule is derived from the exporter's objective of maximizing foreign exchange earnings $P(X)X - E(X)$. The first-order condition yields the following *export supply schedule* (see Annex D):

$$X^s = k_4 \left(\frac{P}{D}\right)^\gamma \exp(\phi_1 T + \phi_2 W). \qquad (4.18)$$

Equation (4.18) states that export supply is related to (1) the constant dollar price of the product (P/D), where

$$D = (V_1^{\theta/(\theta-1)} + V_2^{\theta/(\theta-1)})^{(\theta-1)/\theta}$$

is the deflator; (2) a trend variable T; and (3) a disturbance variable W. The parameters in the supply equation have the following definitions: $\gamma = \tau/(1-\tau)$; $\phi_1 = \sigma_1/(\tau - 1)$; $\phi_2 = \sigma_2/(\tau - 1)$; and

$$k_4 = \exp[\sigma_0/(\tau - 1)] \tau^{\tau/(1-\tau)} \left(1 + \frac{1}{\epsilon_x^p}\right)^{\tau/(1-\tau)}.$$

The constant k_4 contains the term $(1 + 1/\epsilon_x^p)^{\tau/(1-\tau)}$, which accounts for the difference between the export supply curve in monopolistic competition and that in perfect competition. Since $\tau > 0$ and $\epsilon_x^p < 0$, the values of the constant terms are usually positive. (The constant term will have an indeterminate solution if export demand is inelastic with respect to price, i.e. if $-1 < \epsilon_x^p < 0$, which is why, in theory, the supplier does not operate along the inelastic portion of the demand schedule.) If the exporter were to operate under conditions of perfect competition, the term would have no effect on the constant, since when the price elasticity of demand ϵ_x^p approaches (negative) infinity the term is nearly equal to one and does not affect the constant. In such a situation the export supply curve is the same as the marginal cost curve. If, however, the exporting country operates under conditions of monopolistic competition, the export supply curve will lie above the marginal cost curve by a constant proportion of marginal cost. Hence we obtain the well-known solution that the monopolistically competitive exporter will offer less of the commodity and at a higher price than would a perfectly competitive exporter.

The variable W in (4.18) represents a shift variable that measures major random disturbances in export supply. These major disturbances are related primarily to natural disasters and labour disruptions. For example, in a study on major swings in commodity market prices, Chu and Morrison (1984) found that supply disruptions had an important influence in each phase of the 1975 and 1981 recessions. These disruptions intensified the demand-induced price volatility of commodity market prices. The relative effect of supply disruptions in individual commodity exporting countries can be as important for their export earnings volatility as major supply disruptions in world commodity markets are for all producers and consumers. Hence the expected sign of the coefficient for the shift variable, ϕ_2, is usually negative.

The variable T in (4.18) measures technological changes in the production and export processes. On the one hand, it includes innovations and techniques that are introduced to improve the level of production and export, and, on the other, it incorporates expansion of infrastructures that support production and export processes. The effects of these technological changes are to bring about a long-term, or secular, shift in the export supply schedule.

The price elasticity of export supply is given by the exponent γ, which defines the percentage change in export supply brought about by a 1 per cent change in the constant dollar price of the commodity. Since $\gamma = \tau/(1 - \tau)$ and $\tau > 0$, it can take on a positive, negative, or infinite value. Because the value of the price elasticity of supply is directly dependent on the value of τ, which measures returns to scale, the shape of the export supply schedule is determined by the returns to scale of the export industry in the country. When there are decreasing returns to scale, so that $0 < \tau < 1$ and hence $\gamma > 0$, the export supply schedule is strictly increasing; when there are constant returns to scale, so that $\tau = 1$ and hence $\gamma = \infty$, the supply schedule is constant; and when there are increasing returns to scale, so that $\tau > 1$ and hence $\gamma < 0$, the supply schedule is strictly decreasing.

Most countries are likely to exhibit diminishing returns to scale in the production of their traditional commodity exports, and thus to have an upward-sloping supply schedule. Once the commodity has become a traditional export, the country will probably have exhausted its opportunity to obtain cost reductions from increases in the size of the industry. Variable inputs of capital and labour in the production of the commodity will ultimately lead to diseconomies of scale when combined with fixed inputs such as arable land or mineral reserves. Diseconomies can also arise from large-scale management, particularly by producer associations and government agencies, and from rising transportation costs per unit exported to more distant geographic markets. An increase in the amount of variable inputs in the industry then generates a less-than-proportional increase in exportable output. Average costs rise with output as unit costs begin to rise. At the minimum point of the average cost curve the marginal cost curve rises. Then the export supply curve, which is related to the rising part of the marginal cost curve above the average cost curve, is strictly increasing. However, economies of scale can arise from lower factor input prices, greater access to financial capital, and increased acceptance of commodity types. As a result, some commodity exporting countries might have downward-sloping supply curves.

The slope of the export supply curve of commodity exporting countries is an empirical issue that will be examined in Part IV. What this section has demonstrated is that the slope of the curve depends on returns to scale in the industry of the exporting country when the production schedule takes a generalized CES form. In general, returns to scale are expected to be decreasing in commodity exports of developing countries because, once the industry has become established and the commodity has become a traditional export, a country will have exhausted its opportunity to obtain cost reductions from increases in the size of its industry. None the less, economies of scale, or increasing returns to scale, have recently been used to explain how imperfect competition occurs in international trade. (See Helpman 1984 and Das 1982 for surveys.) Fortunately, our use of the characteristics approach to consumer behaviour offers an explanation of how monopolistic competition arises in international commodity trade without the need to invoke economies of scale; for economies of scale have been shown to lead to a strictly downward-sloping export supply schedule when derived for use in empirical analysis from a production schedule of a generalized CES form—a condition that places undue restriction on empirical analysis.

4.2 Illustration

Colombia's coffee exports to the USA are assumed to be well represented by the production and cost schedules derived above. Since the production of coffee in Colombia is labour-intensive, it will facilitate

Supply of Traded Commodities

computation to assume that, for every ton of coffee produced and exported, the cost of labour, V_1, is $100 and the cost of capital, V_2, is only about $1. If the substitution parameter, θ, between labour and capital is equal to 0.5, then

$$(V_1^{\theta/(\theta-1)} + V_2^{\theta/(\theta-1)})^{(\theta-1)/\theta} = \frac{V_1 V_2}{V_1 + V_2} \simeq 1,$$

and the total cost schedule for the production of coffee in Colombia is

$$E = k_3^{1/\tau} X^{1/\tau}. \tag{4.17'}$$

Let us assume that output of coffee in Colombia has a less-than-proportional response to an increase in inputs, in particular that $\tau = 0.38$, and that the value of k_3 is 0.0044. Then the cost schedule is

$$E = (0.00000058) X^{2.65}. \tag{4.17''}$$

The average cost, AC, and marginal cost, MC, schedules are therefore

$$AC = E/X = (0.00000058) X^{1.65} \tag{4.19a}$$

and

$$MC = \partial E/\partial X = (0.0000015) X^{1.65}. \tag{4.19b}$$

The slope of the marginal cost schedule is always greater than that of the average cost curve because there are decreasing returns to scale. As a result, the condition that requires marginal cost to be greater than average cost is met.

Profit maximization behaviour by Colombian exporters will lead them to supply an amount of coffee to US buyers that equates their marginal revenue with their marginal cost. The marginal revenue can be derived from the export demand schedule. Prescience of the empirical results presented in Part IV would permit us to assume that the price elasticity of export demand is equal to -1.1, and that Colombia has a 16 per cent share of the US market. From (3.10), the US export demand schedule for Colombian coffee is

$$X^d = 0.16 M (P/P_j)^{-1.5}$$

where P_j is the price of coffee imports into the USA. In accordance with the first-order condition used to derive the supply schedule (see Annex D for the derivation),

$$MR = \left(1 + \frac{1}{\epsilon_x^p}\right) P = 0.09 P. \tag{4.20}$$

Then the supply of Colombian exports, X^s, of coffee to the USA that will maximize foreign exchange earnings is found from the solution for

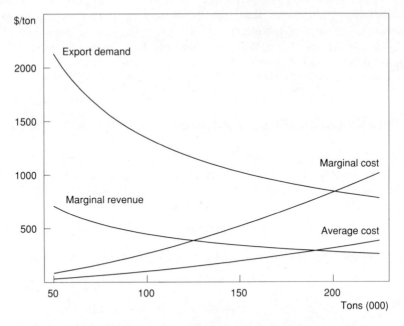

FIG. 4.1 Illustration of export supply schedule for coffee

X when marginal revenue in (4.20) is set equal to marginal cost in equation (4.19b):

$$X^s = 1732 P^{0.6}. \tag{4.18'}$$

The export supply curve arising from the profit maximization behaviour of Colombian exporters to the USA is shown in Fig. 4.1. (It contains the same parameters as those estimated in Part IV.) The marginal cost and revenue curves intersect each other at 125,000 metric tons, at which level of exports the price the USA is willing to pay for Colombian coffee is $1,160 a metric ton. In comparison, the amount that Colombia would have exported under conditions of perfect competition would have been 200,000 metric tons at less than $840 a ton. Thus, Colombia supplies less for more in monopolistic competition than it would in perfect competition.

4.3 Import Supply

When the importer has little or no influence on the market price of a commodity, it can be demonstrated that its import supply schedule has a

Supply of Traded Commodities

price elasticity that approaches infinity. Conversely, the import price can be obtained from the inverse of the import supply schedule.

The supply of imports depends on circumstances in both the foreign supplying countries and the import market itself. Foreign supplying countries encompass all possible producers except those in the importing country, who can reasonably be considered to represent a very small proportion of the world output. The amount that foreign producers are willing to supply is influenced primarily by the market price of the commodity P relative to the general price level D. The amount of the commodity that producers will supply to the importing country is influenced by the importer's foreign exchange availability, denoted H_j, and the ratio between the import price P_j and the world market price P, since any change in relative price will induce a change in import supply. The schedule for import supply, denoted M^s, to a geographical market j can accordingly be expressed as

$$M_j^s = k_5 (P/D)^{\rho_1} (P_j/P)^{\rho_2} H_j^{\rho_3}. \tag{4.21}$$

Equation (4.21) states that the import supply depends on the market price of the commodity, which influences supply availability in the world, and on the relative import price of the commodity and foreign exchange holdings, each of which influences foreign supply availability to the importer.

The import supply equation has two important features. The first is that individual commodity imports are unlikely to be affected by the overall foreign exchange position of an economy because changes in international reserves would have a negligible influence on the supply of a single imported good. The elasticity of import supply with respect to foreign exchange reserves is therefore considered to approach zero, i.e. $\rho_3 \simeq 0$, so that $H_j^{\rho_3} \simeq 1$.

The second and more important feature is that the relative price elasticity of import supply approaches infinity, i.e. $\rho_2 \simeq \infty$, since the importing country takes price as given. Any rise in the import price relative to the world market price would induce foreign producers to sell all supplies to the importer, whereas a fall in the relative price would cause them to withdraw all supplies from that particular geographic market.[4] The inverse supply schedule therefore shows that import price has a proportional response to movements in the market price of the commodity (see Annex D):

$$P_j = k_6 P, \tag{4.22}$$

where $k_6 = k_5^{(-1/\rho_2)}$. In the import price equation (4.22), the elasticity with respect to the world market price of the commodity is unity. The constant k_6 accounts for the difference between the two prices arising from transportation costs, varying methods of valuation, tariffs, and purchase agreements under long-term contracts. (See Chu and Krishnamurty 1978, and Parniczky 1974, for an examination of differences between market prices and trade prices.)

4.4 World Supply

The world supply schedule of the commodity is simply the sum of the exports of all the countries of interest plus production for the rest of the world. Rest-of-world production, denoted Q_k, comprises all production other than that for export by countries of interest. It is made up of production for domestic consumption in countries of interest plus production in all other countries. Since it has been assumed that the production function always takes a CES form, the schedule for rest-of-world production is analogous to that derived for export supply. Accordingly, rest-of-world production Q_k depends on the commodity's constant dollar price P/D, as well as on a secular trend T and major random disturbances W:

$$Q_k = k_4\left(\frac{P}{D}\right)^\gamma \exp(\phi_1 T + \phi_2 W), \qquad (4.23)$$

where $\gamma > 0$ and $\phi_1, \phi_2 \geq 0$.

Summation of the supply of exporters of interest X_i and rest-of-world production Q_k yields the aggregate supply schedule

$$Q = \Sigma_i X_i^s + Q_k, \qquad (4.24)$$

where export supply X_i^s is obtained from (4.18) and rest-of-world output Q_k is obtained from (4.23).

Notes

1 This interpretation is quite similar to the new theory of trade in which economies of scale are considered to be external to the firm but internal to the industry, and in which the country constitutes the basis for the industry (hence the term 'intra-industry trade' to describe trade in the same commodity between two countries). It suggests that, under these conditions, economies of scale are unnecessary for the analysis of a theory of international trade in which there is imperfect competition.
2 In order to simplify the presentation of the material in this section, I have adopted a restricted form of the generalized CES function:

$$X = k_3[\delta A^\theta + (1-\delta)B^\theta]^{\tau/\theta}, \qquad (4.15')$$

where $0 < \delta < 1$. The distribution parameter, δ, describes the share of factor inputs in the production of the commodity. For the interested reader, the results of the derivation of the supply schedule based on the generalized CES function will be presented in separate notes.

3   The cost schedule, based on the generalized CES production function in (4.15'), is

$$E = k_3^{1/\tau} X^{1/\tau} \left(\frac{V_1^{\theta/(\theta-1)}}{\delta^{1/(\theta-1)}} + \frac{V_2^{\theta/(\theta-1)}}{(1-\delta)^{1/(\theta-1)}} \right)^{(\theta-1)/\theta} \qquad (4.17')$$

In this case, the input prices of the capital and labour are weighted by the relative share of each input in the total production of the commodity.

4   The supposition of an infinite price elasticity of import supply is conventionally used in empirical studies of demand for imports (e.g. Turnovsky 1968; Houthakker and Magee 1969; Khan and Ross 1975; Murray and Ginman 1976; and Roussland and Parker 1981). The exception would occur if the importer were a monopsonist capable of influencing the market price of the commodity.

5
Equilibrium Conditions in Commodity Trade

The previous two chapters have described supply and demand in monopolistic competition in terms of a representative exporter and a representative importer of a commodity. In accordance with the theory of trade with monopolistic competition, the conventional practice has been adopted of assuming that all consumers have symmetric utility functions and all producers have identical production functions, as well as equal fixed and variable costs. When symmetry of utility and production functions exists, equilibrium in one exporting country implies that there is equilibrium in the commodity market. Symmetric equilibrium for a market in which there is monopolistic competition is also considered in Sattinger (1984) and Hart (1985a). This chapter applies their findings on the equilibrium implications of monopolistic competition to international commodity trade.

In the first part of the chapter, we derive the level of output that maximizes the foreign exchange earnings of countries that export a commodity. Market equilibrium for perishable and storable commodities is then described. The chapter ends with an examination of the properties of that equilibrium.

5.1 The Exporter

The key behavioural assumption of monopolistic competition in the determination of the equilibrium price and quantity of exporters is that the small market share of each exporting country causes producers to ignore the actions of their competitors and not to be concerned about the reaction of competing exporters to their own actions. As a result, an exporter that seeks to maximize net foreign exchange earnings under these conditions will do so by assuming that the prices of competing suppliers are given.

To simplify notations, but not to lose generality, let us assume that there is only one foreign market and that all producing countries are exporters to that market. Then for each exporter i, $i = 1, \ldots, n$, net foreign exchange earnings, denoted F_i, are the difference between total revenue, $P_i(X_i)X_i$, and total cost, $E_i(X_i)$:

$$F_i = P_i(X_i)X_i - E_i(X_i), \qquad (5.25)$$

so that the maximization of net foreign exchange earnings by each country yields the well-known first-order condition

$$\left(1 + \frac{1}{\epsilon_x^p}\right)P_i = E_i'(X_i) \qquad (5.26)$$

for all exporters $i = 1, \ldots, n$ to that market, where $\epsilon_x^p < 0$ is the price elasticity of export demand and $E_i'(X_i) = \partial E_i(X_i)/\partial X_i$ is the marginal cost.

From the solution obtained in (5.26), the export price of each country can be expressed as

$$P_i = \frac{E_i'(X_i)}{(1 + 1/\epsilon_x^p)}. \qquad (5.27)$$

Hence in monopolistic competition, $(1 + 1/\epsilon_x^p) < 1$ for $\epsilon_x^p < -1$. Since the exporting country will operate along the elastic part of its demand schedule, the equilibrium price is equal to a fractional mark-up over marginal cost, and the amount of the mark-up will depend on the price elasticity of export demand. Equation (5.27) can also be interpreted in such a way that the left-hand side of the expression represents the inverse of the export demand schedule, and the right-hand side the inverse of the export supply schedule.

When there is free entry into the market, the number of exporting countries will increase until there are no excess earnings. All net foreign exchange earnings will therefore be eventually driven to zero:

$$F_i = P_i X_i - E_i(X_i) = 0.$$

Then the export price of each exporter will be equal to the average cost:

$$P_i = \frac{E_i(X_i)}{X_i}. \qquad (5.28)$$

Like the exporting country that operates under conditions of pure competition, a country that operates in a market with monopolistic competition has an export price that equals its average cost. But, like that of the pure monopolist, its demand schedule will slope downward.

5.1.1 Export Price and Quantity

The equilibrium export price can also be derived from the equality between export demand in (3.10) and export supply in (4.18):

$$X_{ij}^d = X_{ij}^s. \tag{5.29}$$

The solution indicates that in monopolistic competition the export price of a country is related not only to market price P, but also to the general price level D, foreign market imports M, a secular trend t, and major random disturbances w:

$$P_i = k_6 P^{\psi_1} D^{\psi_2} M^{\psi_3} t^{\psi_4} w^{\psi_5} \tag{5.30}$$

where $0 < \psi_1 < 1$, $\psi_2 > 0$, $0 < \psi_3 < 1$, and $\psi_4, \psi_5 \lessgtr 0$.

The most interesting feature of the expression for the export price in (5.30) is its relationship to the market price. A change in the market price P induces a less-than-proportional response in the export price P_i, since $0 < \psi_1 < 1$. The reason for the less-than-proportional response is that the price elasticity of export demand is always greater than that of market demand (see Section 3.2 and Fig. 3.1 above). As a result, a change in the quantity demanded of a commodity will lead to a smaller change in the export price than in the market price. Hence the elasticity of export price with respect to market price is less than unity.

The price elasticity, $\psi_1 = \epsilon_x^p/(\epsilon_x^p - \gamma)$, that is, the elasticity of the export price P_i with respect to the market price P, equals the price elasticity of export demand ϵ_x^p divided by the difference between ϵ_x^p and the price elasticity of export supply γ. Usually the price elasticity of export supply for primary commodities is considered to be small, in which case the expected value of the elasticity of the export price with respect to the market price is near to, but less than, unity. Only if the price elasticity of export supply were to be completely inelastic with respect to price would a change in market price induce a proportional change in export price.

The export price has the anticipated response to its other determinants in (5.30). It moves in the same direction as a change in either the general price level D or foreign market imports M. The change in the export price will be greater than, the same as, or smaller than the change in the general price level, but there will always be a less-than-proportional response in the export price as a result of a change in foreign market imports. Major disturbances w, usually associated with natural disasters, industrial strikes, and the like, have the expected effect of bringing about a price rise.

Substitution of the equilibrium export price in (5.30) in either the export demand schedule in (3.10) or the export supply schedule in (4.18) yields the equilibrium quantity of exports. The resulting equilibrium solution for the quantity exported is not shown here because the notation would be excessively complicated, but the derivation is straightforward.

5.1.2 Second-Order Condition

For the equilibrium output of the exporting country to be a maximum, the sufficient condition is that $E_i''(X_i) < 0$. From (5.25), the second-order condition for a relative maximum is

$$\frac{\partial P_i}{\partial Q_i} < \frac{\partial E_i''(X_i)}{(1 + 1/\epsilon_x^p)}; \qquad (5.31)$$

that is, the slope of the inverse demand curve must be less than the slope of the inverse supply curve.

The slope of the export demand curve is always negative because, as may be recalled from Chapter 3, the marginal revenue curve is always downward-sloping under monopolistic competition. The slope of the export supply curve derived in Chapter 4 was shown to depend on returns to scale and can therefore be positive, constant, or negative. Specifically, the export supply curve is downward-sloping when there are economies of scale. As a result, both the export demand and the export supply curves can be downward-sloping. The second-order condition for a maximum will be satisfied as long as the slope of the supply curve is steeper than that of the demand curve.

5.2 The Commodity Market

Market equilibrium is reached in one of two ways, depending on whether or not the commodity can be stored. In the case of commodities that are perishable, the system is closed by equilibrium between consumption C and production Q:

$$Q = C. \qquad (5.32)$$

World consumption is specified in (3.14) and world production is given in (4.24). The solution for market price can be shown to have a proportional response to a change in the general price level. This proportionality implies that the terms of trade for the commodity remain constant in equilibrium. If they do not, either consumption or production is related to the nominal, rather than the constant, price of the commodity. In this case the market price will have a non-proportional response to a change in the general price level. The market price is also positively related to changes in world economic activity and major disturbances.

When a commodity can be stored for a long period of time, equilibrium in the market occurs when the supply, or actual availability, of stocks is equal to the demand for stocks by agents in the market.

Market prices are then usually determined by the 'stock-adjustment processes'. (For a description of the process, see Labys 1973: 91–103.) The stock-adjustment process determines an equilibrium price in the commodity market in which the supply of stocks equals the demand for stocks.

The supply of stocks is simply defined to be actual stocks on hand. The change in the supply of stocks, or inventories (denoted K^s), equals the difference between total world production Q and total world consumption C:

$$\Delta K^s = Q - C. \tag{5.33}$$

Demand for stocks K^d is primarily associated with transactional, precautionary, and speculative motives for holding stocks. Transactions and precautionary demand are related to the level of consumption or production and the cost of holding stocks (usually measured by the real rate of interest). The speculative demand for stocks is motivated by the desire of economic agents to profit from future price changes and is therefore related to the expected price P^e of the commodity:

$$K^d = k_7 C P^{e^{\epsilon_i}}, \tag{5.34}$$

where C denotes consumption of the commodity and the anticipated sign of the expected price elasticity is non-negative since the desire to hold stocks is positively related to expectations about the price level.

Equilibrium in the world market of a commodity with stocks is attained when the demand for stocks equals the supply of stocks:

$$K^d = K^s, \tag{5.35}$$

which yields a solution for market price.

5.3 Properties of Equilibrium

The equilibrium for commodity trade in monopolistic competition should satisfy three requirements: existence, uniqueness, and stability. Although stability of equilibrium is quite easy to ensure and is the most important for the purposes of empirical econometric application of the model, the literature on monopolistic competition has been unable to ensure existence or uniqueness without invoking stringent conditions. In particular, asymmetric utility and production schedules lead to equilibria with unequal prices. (For an analysis, see Salop and Stiglitz 1977; Miyao and Shapiro 1981; Perloff and Salop 1983; and Hart 1985b.) Here, as in Sattinger (1984) and Hart (1985a), we assume symmetric utility and production schedules, which yield equilibria in which prices are equal.

5.3.1 Existence and Uniqueness

Equilibrium for commodity trade in monopolistic competition exists when each exporting country satisfies the first-order condition for maximization of foreign exchange earnings described by (5.26). (For a formal proof see Jones 1984.) An equilibrium solution cannot be attained in empirical trade models when there is improper specification of the coefficient values of functions. The result is that convergence is not achieved when the models are simulated. In addition, Venables (1984) found that multiple equilibria can occur when the parameters in the model used to characterize trade in monopolistic competition are unstable. This finding is closely related to the stability of parameters issue in empirical econometric analysis.

When the first-order condition for maximization of foreign-exchange earnings described by (5.28) is satisfied, the equilibrium is unique. The export price is set at a level at which it equals average cost, so net foreign-exchange earnings (total revenue less total cost) are zero. A unique equilibrium is ensured by free entry into the market because positive net foreign-exchange earnings would induce new production to continue to come onstream until net earnings were reduced to zero. In the absence of free entry into the market, positive net foreign-exchange earnings could prevail and many different solutions would be possible.

The first-order condition is a necessary, but not a sufficient, condition for a unique solution. The second-order sufficient condition for a maximum described by (5.31) may not be satisfied if there are increasing returns to scale and the negatively sloped supply curve is steeper than the slope of the demand curve. The conditions that need to be imposed in order to ensure a unique equilibrium solution in monopolistic competition without the second-order condition being invoked are illustrated in Hart (1985a). In the present case, however, the second-order condition is invoked. Once that condition is satisfied, i.e. the slope of the supply curve is steeper than that of the demand curve, the sufficient condition for a maximum will be ensured and, in conjunction with the first-order condition for maximization of foreign exchange earnings, a unique solution will exist.

5.3.2 Stability

Whether an equilibrium is stable for an exporting country i can be evaluated from its excess demand schedule:

$$ED_i = X_i^d - X_i^s. \tag{5.36}$$

The necessary Walrasian condition for stability is that $\partial ED_i/\partial P_i < 0$. The condition will be satisfied if the export supply curve is positive. If it is negative, the slope must be less than that of the export demand curve. For the export demand and supply schedules in (3.10) and (4.18),

$$\frac{\partial ED_i}{\partial P_i} = (\epsilon_x^p P_i^{\epsilon_x^p}\{RXD\} - \gamma P_i^{\gamma}\{RXS\})/P_i \qquad (5.37)$$

where $\{RXD\}$ represents the 'rest of export demand' terms, and $\{RXS\}$ represents the 'rest of export supply' terms, whose respective values are both positive. The parameter ϵ_x^p is the export demand elasticity and the parameter γ is the export supply elasticity. Then Walrasian stable equilibrium occurs when

$$\epsilon_x^p P_i^{\epsilon_x^p} < \gamma P_i^{\gamma}. \qquad (5.38)$$

The condition is automatically satisfied if there are decreasing or constant returns to scale: the left-hand side of the inequality is negative since the slope of the export demand curve is always negative, and the right-hand side is positive (since $\gamma > 0$ when there are decreasing returns to scale) or (positive) infinite (since $\gamma = \infty$ when there are constant returns to scale). On the other hand, if there are increasing returns to scale, $\gamma < 0$. In this case $\epsilon_x^p < \gamma$ for stability to occur. Since the parameters in (5.38) are the export demand elasticity ϵ_x^p and the export supply elasticity γ, the stability condition is $\epsilon_x^p < \gamma$.

The symmetry assumption ensures that there will be stability of equilibrium in the market when there is stability of equilibrium for each agent. It will be recalled that all consumers have identical utility schedules and that all producers have identical production schedules. A stable equilibrium in one exporting country and one importing country implies stability for all countries. Hence when there is stability in the representative exporting and importing countries, there is stability of equilibrium in the commodity market.

6
Summary

Figure 6.1 provides a visual representation of the equations of the system used to describe the underlying features of international commodity trade. The schematic form is a modified version of a method suggested by Tustin (1953). (For an exposition of this method and

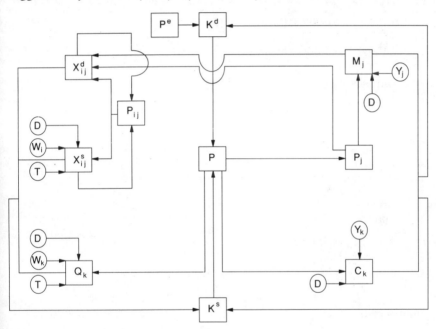

FIG. 6.1 Schematic diagram of commodity trade model

C_k	rest-of-world demand	P_{ij}	export price of country i to geographic market j
D	deflator		
K^d	desired stocks	Q_k	rest-of-world supply
K^s	actual supply of stocks	T	secular trend
M_j	import demand of geographic market j	W	major disturbance
		X^d_{ij}	export demand of country i by geographic market j
P	world market price		
P^e	expected market price	X^s_{ij}	export supply of country i to geographic market j
P_j	import price of geographic market j		
		Y	economic activity

others adopted from engineering problems, see Allen 1956: 281–303.) The modification of Tustin's method directs attention to the interdependence of variables and eliminates the constant multiple of each dependence. The variables are enclosed in a box if they are endogenous and in a circle if they are exogenous. The direction of dependence is shown by arrows. An arrow emanating from a box or circle indicates the influence of that variable on another; an arrow penetrating a box or circle shows what influences that variable.

The system of equations itself is summarized as follows.

General Notation

C	consumption
D	general price level
E	cost of production
F	foreign exchange earnings
K	stocks
M	imports
P	price
Q	production
T	secular trend
W	major disturbance
X	exports
Y	economic activity

Subscripts

i	exporter of interest
j	geographic export market
k	competing exporters (to country of interest i)

Superscripts

d	demand
s	supply

6.1 Foreign Market Imports

(a) Import demand

$$M_j^d = k_1 Y_j \left(\frac{P_j}{D_j}\right)^{\epsilon_m^p} \tag{3.8a}$$

where $\epsilon_m^p < 0$.

(b) Import price

$$P_j = k_6 P \tag{4.22}$$

6.2 Exports

(a) Export demand

$$X^d_{ij} = k_2 M_j \left(\frac{P_{ij}}{P_j}\right)^{\epsilon^p_x} \tag{3.10}$$

where $\epsilon^p_x < 0$, and $0 < k_2 < 1$.

(b) Export supply

$$X^s = k_4 \left(\frac{P_{ij}}{D_i}\right)^{\gamma} \exp(\phi_1 T + \phi_2 W) \tag{4.18}$$

where $\gamma > 0$, $\phi_1 \lessgtr 0$, $\phi_2 < 0$.

(c) Equilibrium condition

$$X^d_{ij} = X^s_{ij} \tag{5.29}$$

6.3 World Market

(a) Demand

  (i) Rest-of-world demand

$$C_k = k_1 Y_k \left(\frac{P}{D}\right)^{\epsilon^p} \tag{3.13}$$

  where $\epsilon^p < 0$.

  (ii) Total world consumption

$$C = \sum_{j=1}^{m} M_j + C_k \tag{3.14}$$

  (iii) Demand for stocks

$$K^d = k_7 C P^{e^{\epsilon_k}} \tag{5.34}$$

(b) Supply

   (i) Rest-of-world supply

$$Q_k = k_4 \left(\frac{P}{D}\right)^\gamma \exp(\phi_1 T + \phi_2 W) \qquad (4.23)$$

   where $\gamma > 0$ and $\phi_1, \phi_2 \leqq 0$.

   (ii) Total world production

$$Q = \sum_i X_i^s + Q_k \qquad (4.24)$$

   (iii) Supply of stocks

$$\Delta K^s = Q - C \qquad (5.33)$$

(c) Equilibrium condition

   (i) Perishable commodities

$$Q = C \qquad (5.32)$$

   (ii) Storable commodities

$$K^d = K^s \qquad (5.35)$$

Part III

Dynamics of Commodity Trade

Introduction

When changes in explanatory variables exert their full effect on dependent trade variables within the same period, the comparative analysis of equilibrium conditions before and after a change over a sequence of time periods reduces to a set of static problems. However, even in time-series data with annual periodicity, commodity markets adjust with lagged responses by economic agents. Suppliers of copper, coffee, and cocoa, for example, take at least six years to react fully to price changes. Only perishable product suppliers, like those of bananas, appear to conform to a static model by adjusting most of their production levels to price changes within the same year. On the demand side, apart from normal delays in consumer response, dynamic effects arise in the derived demand for raw materials and for basic, pre-processed foods from the filtering of price changes to the final demand for the commodity.

Non-instantaneous adjustment in either supply or demand gives rise to observations that can represent states of disequilibrium. If all changes in the explanatory variables were to cease, the dependent trade variable of a given relationship in the system would converge to its equilibrium state after a sufficient number of years had elapsed. Actual observations of these variables can therefore be considered to be related to their equilibrium states in a predictable way. Accordingly, the aim of dynamic specification is to describe the observed behaviour of the variables as an adjustment to long-run equilibrium states consistent with their postulated equilibrium relationships.

The purpose of this part of the study is to formulate the dynamic relationships that characterize the underlying processes of these adjustments in commodity trade. Put succinctly, it seeks to generalize the static, or timeless, relationships formulated in Part II and to apply them to their dynamic characterization of adjustment to long-run equilibrium. In this way it will be possible to apply the postulated theory of commodity trade to empirical observations of trader behaviour and, at the same time, to evaluate the validity of the theory.

The dynamic specification strategy common to all the equations in the system will first be presented. To facilitate understanding of the general strategy, it will be applied to a relationship containing only one explanatory variable; its extension to a more complex relationship is straightforward. After the general approach to dynamic specification has been described, the strategy is applied to the specific theoretical relationships in the system of equations derived in Part II so as to represent the essential features underlying international commodity trade.

7
Specification Strategy

In recent years several distinct methodologies for econometric analysis have been developed. It is therefore important at the outset to specify the methodology employed in formulating the dynamic relationships for the empirical econometric analysis of commodity trade. The approach follows the modelling strategy developed in a series of papers by Davidson *et al.* (1978), Hendry and Richard (1981), and Hendry (1986), which owe a great deal to Sargan's (1964) seminal paper; more generally, it reflects the traditional approach to econometric analysis as set forth in Harvey (1981). (For recent surveys of this and other methodologies, see Pagan 1987; Gilbert and Di Marchi 1989; and Pagan and Wickens 1989.)

The chapter begins with a general dynamic specification of a model that is used in Chapter 9 to describe the data-generating processes of supply relationships in commodity trade. It then moves to a more specific formulation of dynamic relationships, which in Chapter 8 is used to characterize the data-generating processes of demand relationships in international commodity trade, and is shown to be consistent with the theoretical relationships postulated in Part II. The final section presents an empirical illustration of the dynamic characterization of the adjustment processes in commodity trade described in the initial sections of this chapter.

7.1 General Dynamic Specification

The dynamic specification of any postulated theoretical relationship is based on the introduction of appropriate lags in the explanatory variables. However, when the adjustment of the dependent variable to a change in the value of one of the explanatory variables is gradual, the high correlation between successive lagged values of the explanatory variable can lead to multicollinearity, and imprecise estimates of the coefficients make the lag structure difficult to determine. To avoid this problem a stochastic difference equation is used in which the lag structure is approximated by directly introducing lagged values of the dependent variable. The rationale for this construct is that the dependent variable is regarded as a stochastic process in which observations evolve over time in accordance with some probabilistic law. Thus, the process that generates the data of the dependent trade variable in the current period, t, is one in which the dependent trade variable is related to its own past behaviour and to the present and past behaviour of the explanatory variables.

Consider the general form of a long-run equilibrium relationship for any of the behavioural functions postulated in Part II. A simple representation of such a relationship is one between two variables X and Y and is described by a nonlinear function of the form

$$X = kY^\epsilon, \qquad (7.1a)$$

where k is some constant. It can, for example, represent a country's exports X as a function of economic activity Y in its trading partner. When the foregoing relationship is expressed as a logarithmic function, the function becomes linear-in-logarithms and is consistent with the form of the equation used empirically to estimate the relationship

$$x = \alpha + \epsilon y, \qquad (7.1b)$$

where lower-case letters denote the logarithms of upper-case letters; i.e. $x = \ln X$, $y = \ln Y$, and $\alpha = \ln k$. A useful property of (7.1b) is that the calculated coefficient ϵ directly yields the point elasticity of the dependent variable in the expression.

The first-order stochastic difference equation of the theoretical relationship in (7.1b) is expressed as

$$x_t = \alpha_0 + \alpha_1 x_{t-1} + \alpha_2 y_t + \alpha_3 y_{t-1} + v_t, \qquad (7.2)$$

where $0 < \alpha_1 < 1$ for the system to be stable, $\alpha_2, \alpha_3 > 0$ for the purpose of exposition, and all variables are measured in logarithmic terms.

Specification of the dynamic process by means of equations of this class has two important advantages. First, as pointed out by Harvey (1981: ch. 8), the stochastic difference equation lends itself to a specification procedure that moves from a general unrestricted dynamic model to a specific restricted model. At the outset all the explanatory variables postulated by economic theory and the lags of a relatively higher order than appear in (7.2) are deliberately included. Whether or not a particular explanatory variable should be retained and which lags are important are decided by the results obtained. An example of the usefulness of this approach as a test of variable significance occurs in the export demand relationship. The relative prices are initially included in the equation specification in order to determine whether product differentiation is reflected in the ability of an exporting country to alter the price of the product relative to that of competitors. The usefulness of this approach for the identification of important lags is particularly exemplified in the export supply relationship, where the nature of the response to price changes is central to the dynamic specification.

The second advantage of the use of the stochastic difference equation lies in the estimation procedure. Mizon (1983) has noted that, given sufficient lags in the dependent and explanatory variables, the stochastic difference equation can be so defined as to have a white noise process in its disturbance term. As a result, the ordinary least squares estimator for the coefficients will be fully efficient.

However, even if the correct estimation procedure is adopted, the coefficient estimates of the stochastic difference equation can be imprecise if the dynamics are of a relatively high order, the reason being multicollinearity between lagged values of an explanatory variable. Dynamics of a relatively high order often characterize export supply relationships because of the complex structures of the lag response to price variations. A solution to the problem of multicollinearity is the transformation of the equation in such a way that 'differences' formulations of the variables are nested in the levels form of the equation.

As a first step in the transformation, add $(-x_{t-1})$ to both sides of (7.2):

$$\Delta x_t = \alpha_0 + (\alpha_1 - 1)x_{t-1} + \alpha_2 y_t + \alpha_3 y_{t-1} + v_t.$$

Then subtract $(\alpha_2 y_{t-1})$ from the third term and add it to the fourth term:

$$\Delta x_t = \alpha_0 + (\alpha_1 - 1)x_{t-1} + \alpha_2 \Delta y_t + (\alpha_2 + \alpha_3)y_{t-1} + v_t$$

or

$$\Delta x_t = \alpha_0 + \beta_1 x_{t-1} + \alpha_2 \Delta y_t + \beta_3 y_{t-1} + v_t, \qquad (7.3)$$

where $\beta_1 = (\alpha_1 - 1)$ and $\beta_3 = (\alpha_2 + \alpha_3)$, and the expected signs are $-1 < \beta_1 < 0$ and $\alpha_2, \beta_3 > 0$. The formulation in (7.3) avoids the problem of indeterminate long-run dynamic equilibrium solutions in equations with only first-differenced variables whose relationship would then be considered to be jointly stationary. At the same time, it solves the problem of spurious correlations associated with regressions of trending variables in levels (see Granger and Newbold 1974, and, in a lighter vein, Hendry 1980).

7.2 Restricted Specification

In many instances, international commodity trade series have a long-term relationship with one or more other series after transient effects from all other series have disappeared. That part of the response of a trade variable that never decays to zero is the steady-state response, while that part that decays to zero in the long run is the transient response. Examples in international commodity trade of relationships in which steady-state responses occur are those between the export demand of a country and imports by its foreign market, between the imports of a foreign market and economic activity in that market, and between the import price of a commodity in a particular geographic market and the world market price of that commodity. An example of a transient response is the relative-price movements in the export demand function. If relative-price changes were not transient, the disparity between the price of an exporter and those of its competitors would continuously widen, with the result that the importer would eventually switch entirely to the supplier with the lower priced product.[1] Hence it is important to distinguish the short-run adjustment components from the long-run equilibrium components in commodity trade relationships.

Commodity trade series related to the long-run adjustment processes of other variables have been designated to be cointegrated series by Granger and Weiss (1983). As applied to the first-order stochastic difference equation (7.2), the theory of cointegration is as follows. Suppose that the first differences of the dependent trade variable x and the other variable y are stationary (in the sense that their observations fluctuate around a mean of zero); if x and y grow over time in such a way that the linear combination of these two variables, given by $d_t = x_t - \alpha y_t$, is stationary, and if α is unique, then x and y are said to be cointegrated. The series d_t measures the disequilibrium at period t when the long-run relationship between the two variables is $x_t = \alpha y_t$. The theory of cointegration states that movements in variables are related in a predictable way to the discrepancy between observed and equilibrium states. The sequence of this discrepancy tends to decay to its mean, which is zero.

Engle and Granger (1987) have demonstrated that a data-generating process of the form known as the 'error correction mechanism' (ECM) adjusts for any disequilibrium between variables that are cointegrated. (For an extension of Engle and Granger's work on the relationship between cointegration theory and the ECM, see Hylleberg and Mizon 1989.) The ECM specification thus provides the means by which the short-run observed behaviour of variables is associated with their long-run equilibrium growth paths. This specification was initially adapted to the time-series analysis of economic relationships by Phillips (1954) and Sargan (1964), and has been applied to a wide range of economic problems since the work of Davidson *et al.* (1978).[2]

The ECM specification can be derived by an appropriate transformation of the stochastic difference equation that describes the dynamics of the postulated relationship in (7.1). Suppose the response of the dependent trade variable X to its determinant Y in steady-state growth is proportional. The restriction $(\alpha_1 + \alpha_2 + \alpha_3) = 1$ in (7.2) imposes on the dependent trade variable a long-run proportional response to changes in the explanatory variable. To demonstrate this proportionality, we first substitute $(1 - \alpha_1 - \alpha_2)$ for α_3 in (7.2). Then we combine the lagged x and y terms in (7.3), such that $\beta_1 = (\alpha_1 - 1)$, the result of which yields

$$\Delta x_t = \alpha_0 + \beta_1(x - y)_{t-1} + \alpha_2 \Delta y_t + v_t, \qquad (7.4a)$$

with expected sign $-1 < \beta_1 < 0$. The restriction $\Sigma_1^3 \alpha_i = 1$ in (7.4a) ensures that changes in the explanatory variable produce proportional changes in the dependent trade variable over the long run. In commodity trade this result is consonant with the long-run relationships between the export demand of a country and the imports of its trading partners, and between the import price of a commodity in a particular market and the world market price of the commodity.

Where the long-run response is not necessarily proportional, as between the import growth of a geographic market and its income growth, an additional term is introduced. In this term the explanatory variable is lagged one period and the coefficient $\gamma_3 = (\alpha_1 + \alpha_2 + \alpha_3 - 1)$:

$$\Delta x_t = \alpha_0 + \beta_1(x - y)_{t-1} + \alpha_2 \Delta y_t + \gamma_3 y_{t-1} + v_t, \qquad (7.4b)$$

where $\gamma_3 > \beta_1$ when $\alpha_2, \alpha_3 > 0$ in (7.2). Hence when the coefficient γ_3 has non-zero significance, the steady-state response of the dependent trade variable to that explanatory variable is non-proportional.

The second term, $\beta_1(x - y)_{t-1}$, in (7.4a) and (7.4b) is the mechanism for adjusting any disequilibrium in the previous period. To understand how it functions, note two characteristics of the term. First, the complex variable in the term is actually the logarithm of the ratio of the dependent trade variable to the explanatory variable; i.e. $(x - y) = (\ln X - \ln Y) = \ln(X/Y)$, since all the variables are measured in logarithms. Second, the coefficient of this second term is negative; i.e. $-1 < \beta_1 < 0$. Now consider the adjustment to a deviation of the dependent variable from its steady-state path, arising for instance from some transient disturbance in Y. If the rate of growth of the dependent trade variable were to fall below its steady-state path, the value of the ratio of variables in the second term would decrease in the subsequent period. A decrease in the value, combined with the negative coefficient of the term, would have a positive influence on the growth rate of the dependent trade variable. As the variable approached its steady-state path, the actual rate would move closer to its long-term rate until eventually they converged. Conversely, if the growth rate of the dependent trade variable were to increase above its steady-state path, the adjustment mechanism embodied in the second term would generate downward pressure on the growth rate of the variable until it reached that of its steady-state path. The speed with which the system approaches its steady-state path depends on the proximity of the coefficient to minus one. If the coefficient is close to minus one, the system converges to its steady-state path quickly; if it is near to zero, the approach of the system to the steady-state path is slow.

Since the variables are measured in logarithms, Δx_t and Δy_t can be interpreted as the rate of change of the variables. Thus, the third term, $\alpha_2 \Delta y_t$, in (7.4a) and (7.4b) expresses the steady-state growth in X associated with Y.

7.3 Long-Run Equilibrium Relationships

The long-run equilibrium solutions of the dynamic relationships derived in the foregoing analysis can replicate static relationships such as those for the theory of commodity trade presented in Part II. However, the long-run solutions of dynamic relationships encompass the static equilibrium solutions of trade theories. They also contain other possible solutions in which the dependent trade variable is related not only to changes in the level of an explanatory variable, but also to changes in the rate of growth of that explanatory variable.

7.3.1 Static Equilibrium

The long-run solution of a single-equation system in a static state is a constant value if there is convergence. Since the solution is unrelated to time, the rate of change over time of the dependent trade variable (given by Δx_t, where $x = \ln X$ as before) and the rate of change of the explanatory variable (given by Δy_t) are equal to zero in the specifications of both the transformed stochastic difference equation and the ECM. In a timeless state, $x = x_{t-n}$ and $y = y_{t-n}$ when $n = 0, 1$. Hence the transformed stochastic difference equation in (7.3) has as its solution

$$0 = \alpha_0 + \beta_1 x + \alpha_2(0) + \beta_3 y.$$

Rearrangement of terms yields

$$x = -(\alpha_0/\beta_1) - (\beta_3/\beta_1)y$$

or, in terms of the original (anti-logarithmic) values of the variable,

$$X = k_0 Y^\epsilon, \qquad (7.5)$$

where $k_0 = \exp(-\alpha_0/\beta_1)$ and $\epsilon = -\beta_3/\beta_1$. Since $\beta_3 > 0$ and $-1 < \beta_1 < 0$, the expected sign of ϵ is positive.

The same static equilibrium solution can be derived from the specification of the ECM. When unitary elasticity is imposed on the specification such that $(\alpha_1 + \alpha_2 + \alpha_3) = 1$ in (7.4a), then $\epsilon = 1$ in (7.5). On the other hand, when there is a non-proportional response of X to any change in Y, as in (7.4b) if $(\alpha_1 + \alpha_2 + \alpha_3) \neq 1$, then $\epsilon > 1$ when $|\gamma_3| > |\beta_1|$ and $0 < \epsilon < 1$ when $|\gamma_3| < |\beta_1|$. Thus both the transformed stochastic difference equation and the ECM reproduce the long-run equilibrium relationship postulated in (7.1).

7.3.2 Dynamic Equilibrium

The long-run dynamic solution of a single-equation system generates a steady-state response in which growth occurs at a constant rate, say g, and all transient responses have disappeared. For the dynamic specification of the relationship described by (7.1), if g_1 is defined as the steady-state growth rate of the dependent variable X, and g_2 corresponds to the steady-state growth rate of the explanatory variable Y, then, since lower-case letters denote the logarithms of variables, $g_1 = \Delta x$ and $g_2 = \Delta y$ in dynamic equilibrium. Both Currie (1981) and Patterson and Ryding (1984) have derived the long-run dynamic equilibrium solution for the more general dynamic specification of such a relationship. The approach used by Currie will be adopted here in order to derive the long-run dynamic equilibrium properties of the simple relationship between a trade variable and its explanatory variable.

The first step is to find the relationship between the rate of change of the relevant trade variable Δx_t and that of its explanatory variable Δy_t. Given the systematic dynamics of the general stochastic difference equation (7.2),

$$x_t = \alpha_0 + \alpha_1 x_{t-1} + \alpha_2 y_t + \alpha_3 y_{t-1},$$

the first difference of that equation is

$$\Delta x_t = \alpha_1 \Delta x_{t-1} + \alpha_2 \Delta y_t + \alpha_3 \Delta y_{t-1}.$$

Since in dynamic equilibrium $\Delta x_t = g_1$ and $\Delta y_t = g_2$, it follows that

$$(1 - \alpha_1) g_1 = (\alpha_2 + \alpha_3) g_2.$$

Hence

$$g_1 = \left(\frac{\alpha_2 + \alpha_3}{1 - \alpha_1} \right) g_2. \tag{7.6}$$

Then for the systematic dynamics of the transformed stochastic difference equation (7.3),

$$g_1 = \alpha_0 + \beta_1 x + \alpha_2 g_2 + \beta_3 y,$$

the substitution of g_1 for its expression in terms of g_2,[3] and rearrangement of terms yields the long-run dynamic relationship

$$x = \frac{-\alpha_0}{\beta_1} - \left(\frac{\beta_3 + \beta_1\alpha_2}{\beta_1^2}\right)g_2 + \frac{\beta_3}{\beta_1}y,$$

or, in terms of the original values of the variables,

$$X = k_0' Y^\epsilon, \qquad (7.7)$$

where $\epsilon = -(\beta_3/\beta_1)$ and $k_0' = \exp\{(-\alpha_0/\beta_1) - [(\beta_1\alpha_2 + \beta_3)/\beta_1^2]g_2\}$. Equation (7.7) encompasses the static equilibrium solution when $g_2 = 0$.

A similar long-run dynamic equilibrium solution can easily be derived for the ECM with a non-proportional steady-state response in (7.4b). In equilibrium, its systematic dynamics are expressed as

$$g_1 = \alpha_0 + \beta_1(x - y) + \alpha_2 g_2 + \gamma_3 y.$$

Substitution of g_1 for its expression in terms of g_2,[4] and rearrangement of terms yields

$$X = k_0' Y^\epsilon, \qquad (7.7')$$

where $\epsilon = 1 - (\gamma_3/\beta_1)$ and $k_0' = \exp\{(-\alpha_0/\beta_1) + [(\beta_1 - \alpha_2\beta_1 - \gamma_3)/\beta_1^2]g_2\}$, which is the same solution as in (7.7) since $\gamma_3 = (\beta_1 + \beta_3)$.

The ECM with a proportional steady-state response in (7.4a) has the equilibrium solution

$$g_1 = \alpha_0 + \beta_1(x - y) + \alpha_2 g_2,$$

which yields

$$X = k_0^* Y, \qquad (7.8)$$

where $k_0^* = \exp\{(-\alpha_0/\beta_1) + [(1 - \alpha_2)/\beta_1]g_2\}$ since $\gamma_3 = 0$. The long-run unitary elasticity of X with respect to Y for the dynamic specification in (7.4a) is therefore consistent with the proportionality restriction imposed on the steady-state response of the dependent trade variable.

The steady-state solution, unlike the static equilibrium solution, shows X to be influenced by changes in the rate of growth of Y—despite the same long-run elasticity of X with respect to Y in both static and dynamic solutions. In particular, the dynamic solutions in (7.7) and (7.8) indicate that, were the rate of growth of the explanatory variable to accelerate, say from g_2 to g_2', the value of the trade variable X would increase.[5]

Although in our simple example X depends solely on Y and the response is one that generates steady-state growth, in practice more than one explanatory variable is often used to characterize the underlying process of the relevant trade variable. Moreover, the response to each explanatory variable can be either transient or steady-state. For example, as indicated earlier, the export demand of a country has a steady-state response to the growth of imports in its foreign markets,

Specification Strategy

whereas it has a transient response to changes in the price of its exports relative to that of its competitors in those markets. Where theoretical considerations suggest that an explanatory variable generates a transient, rather than a steady-state, response, it is appropriate to constrain its long-run effect to zero.[6]

7.4 Illustration

To illustrate the adjustment process in (7.4a), whereby the explanatory variable generates a proportional response in the dependent trade variable, let us consider a simple relationship for export demand. Let the variable X^d represent the demand for exports of Colombian coffee by the USA and M represent the total import demand for coffee by the USA. The resulting estimate of the relationship between these two variables in 1961–85 is as follows:

$$\Delta x_t^d = -0.54 - 0.28\,(x^d - m)_{t-1} + 1.56\,\Delta m_t, \qquad (7.9)$$
$$(2.9)\phantom{\,(x^d - m)_{t-1} + \,}(8.3)$$

$$\bar{R}^2 = 0.85 \qquad \text{dw} = 1.4 \qquad \hat{\sigma} = 0.084$$

where lower-case letters denote the logarithms of the corresponding capitals, the t-statistics are shown in parentheses, $\bar{R}^2$ is the corrected squared multiple correlation coefficient, dw is the Durbin–Watson statistic, and $\hat{\sigma}$ is the standard deviation of the residuals. Dummy variables were included, but are not reported in the equation in order to simplify presentation of the results. A 1981 dummy variable (1 in 1981, 0 otherwise) was used to account for the introduction of export quotas under the terms of the International Coffee Agreement; a dummy variable for 1976 was used to account for a shift in tastes and preferences for lower-quality coffee, as well as for other types of beverages, when prices rose sharply in that year; and a dummy variable for 1967 was used to account for increased US demand for Colombian coffee following disruptions in Brazilian exportable production.

Despite the statistical significance of the Δm variable, the equation failed to predict the changes in exports in 1978–9 and 1982; it seriously underestimated the movements that occurred in 1978–9, and it overestimated the movements that occurred in 1982. Notwithstanding these limitations (which will be shown in the next chapter to arise from a failure to consider the effect of relative-price movements), some long-run dynamic properties of the estimated relationship in an error correction form can be examined.

In long-run dynamic equilibrium it is reasonable to suppose that the demand by the USA for Colombia's exports of coffee has a steady-state response to the total coffee import demand of the USA. The growth rate of the US import demand in (7.9) is given by Δm, whose steady-state path will be denoted g_3. Since the long-run elasticity of the export demand of Colombia with respect to the import demand of the USA is unity on a steady-state growth path, the rate of growth of the export demand, Δx_t^d, must be equivalent to that of the import demand. Hence $\Delta x_t^d = g_3 = \Delta m_t$. Substitution of g_3 for Δx_t^d and Δm_t in (7.9) and rearrangement of terms yields the long-run dynamic relationship

$$X^d = k_0^* M, \qquad (7.10)$$

where $k_0^* = \exp(-1.96 + 2.0 g_3)$.

The estimated equation (7.9) yields a long-run unitary export demand elasticity with respect to the total coffee import demand of the USA. It also defines a conventional market shares function, (X^d/M), in which Colombia maintains a constant share of the US market equal to 15 per cent when $g_3 = 0.04$, as it did in the period over which (7.9) was estimated.

In dynamic equilibrium with non-zero, but constant, growth of the US import demand, it is possible to examine the substitution of one or more foreign suppliers by others as the import growth rate varies. The parameter of the growth rate g_3 in the long-run relationship is positive, an indication that an increase in the US import growth rate will cause the USA to expand its demand for the coffee exports of Colombia relative to those of other foreign suppliers. This decline in the level of export demand follows directly from the estimated coefficient, 1.56, that relates the growth in export demand, Δx_t^d, to the growth in foreign market import demand, Δm_t. Equation (7.10) indicates that the acceleration in the rate of growth of imports of the USA from $g_3 = 0.04$ in 1960–75 to $g_3 = 0.08$ in 1976–85 would have increased the share of Colombia in the US market from 15 to 16.5 per cent. The expansion in the export market share of Colombia when the growth of import demand in the USA accelerates suggests that importers seek out more coffee from that supply source.

Two examples will serve to demonstrate the adjustment process of export demand when its dynamic relationship is specified in an error correction form. In the first case the adjustment is to a one-time disturbance in export demand, so that exports deviate from their steady-state response to import growth in the geographic market. Export demand is assumed to be in dynamic equilibrium ($X^d/M = 15$ per cent) until 1976, when it dropped to 10 per cent of total imports by the USA. Had no other disturbances occurred, Fig. 7.1(a) shows what the time path of adjustment back to steady-state growth would have been. The

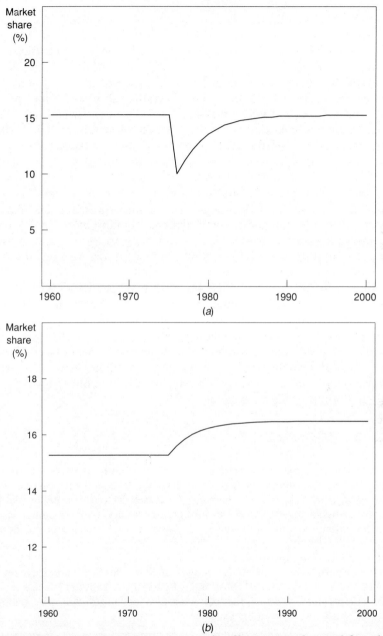

FIG. 7.1 Dynamic adjustment process of coffee export demand function
  (a) Adjustment to one-time disturbance in export demand for Colombian coffee
  (b) Adjustment of Colombian exports to increase in long-term growth of US coffee imports

deviation from equilibrium is corrected relatively quickly because the coefficient of the error correction term in (7.9) is equal to -0.28. It will be recalled that the speed of adjustment varies with the coefficient of the error correction term, $(x^d - m)_{t-1}$. If the absolute value of the coefficient had been nearer to zero, the adjustment process would have been slower. On the other hand, had the absolute value of the coefficient approached unity, the speed of adjustment would have been faster. Colombia makes up one-half of its market share reduction within three years; three-fourths after five years. But it takes seven years before the country regains 90 per cent of its previous export market share.

In the second example the adjustment of export demand is to a change in the long-term rate of growth of imports in the US market. Export demand is assumed to be in dynamic equilibrium during 1960–75 and the steady-state growth of US import demand to be equal to 4 per cent. During this period the corresponding export market share of Colombia would be 15 per cent according to (7.9). If the long-term growth rate of US imports then doubled to 8 per cent, the share of Colombia in that market would increase to 16.5 per cent at the new dynamic equilibrium, as explained earlier. However, the adjustment to the new equilibrium would not be immediate. Figure 7.1(b) shows that, according to (7.9), the adjustment would be well under way after three years. Exports would attain three-fourths of their new steady-state equilibrium within a four-year period. Thereafter, convergence to the new equilibrium would be gradual.

Notes

1 For example, suppose the marginal rate of substitution between an exporter of interest, X_i, and competing exporters, X_k, to a particular market were $(\partial U/\partial X_i)/(\partial U/\partial X_k) = 2/1$, where U is the satisfaction derived by the importer. Suppose further that the rates of change of the corresponding export prices differ over the long run, and that eventually the economic rate of substitution becomes $P_i/P_k = 3/1$. Since $2/1 < 3/1$, the importer could give up one unit of X_i's product to purchase two units of X_k's product and have \$1 left over while still remaining on the same indifference curve. Therefore, once the economic rate of substitution between the exporter of interest and its competitors exceeded the marginal rate of substitution, the importer would cease to buy from the exporter of interest.
2 It has been used to estimate the behavioural relationships for wage determination (Sargan 1964; Henry and Ormerod 1979), for the consumption function (Davidson *et al.* 1978; Hendry and von Ungern-Sternberg 1980; Hendry 1983), and for the demand for money (Hendry and Mizon 1978; Hendry 1979; Domowitz and Elbadawi 1987) and the supply of money (Davidson and Keil 1981).

3 $g_1 = -(\beta_3/\beta_1)g_2$ in terms of the coefficients in (7.6) since $\beta_1 = (\alpha_1 - 1)$ and $\beta_3 = (\alpha_2 + \alpha_3)$.
4 $g_1 = [1 - (\gamma_3/\beta_1)]g_2$ in terms of the coefficients in (7.4b) since $\gamma_3 - \beta_1 = \alpha_2 + \alpha_3$ and $\beta_1 = \alpha_1 - 1$.
5 Because of the dependence of the long-run dynamic solution on both the growth rates and the levels of the explanatory variables that determine the steady-state response, Currie (1981) has warned against policy simulations and forecasts that rely on unstable growth rates for those explanatory variables.
6 Patterson and Ryding (1984) have demonstrated that the imposition of such a constraint on one explanatory variable can induce substantial changes in the lag distribution not only of that variable, but of all other variables as well. Consequently, they warn against routine applications of such constraints.

8
The Demand Functions

The purpose of this chapter is to derive the dynamic specifications of the demand relationships in the system of equations used to characterize international commodity trade. The first section derives the import demand relationship, which uses the ECM for its dynamic specification. The application of the ECM to the import demand relationship is straightforward, and provides an opportunity to examine the ECM's characterization of the data-generating process in the context of international trade. The export demand relationship is derived in the second part of the chapter. The relationship for export demand extends the ECM to a more complex specification which imposes additional restrictions in the form of logit transformations. Thus, the procedure is to move from a general specification of the import demand equation for the foreign markets of an exporter in the first part of the chapter to a more restricted dynamic specification of the export demand equation in the second part.

8.1 Import Demand

The demand for imports of a commodity in a geographic market is postulated to have a steady-state response to the domestic economic activity, and a transient response to the constant dollar price of imports, in that market. The life-cycle approach to consumption emphasizes income as a determinant of intertemporal consumption planning and provides theoretical justification for the existence of the dynamic effect on import demand of changes in the rate of growth of domestic income (see Deaton and Muellbauer 1980: ch. 12). In contrast, there is no logical explanation for any dynamic effect of the price of imports. Were the import price of a commodity to change continually relative to the general price deflator, consumers would soon cease to purchase the commodity as the spread between the commodity price and the general price level in the importing country widened.

For the purposes of dynamic specification, an important characteristic of import demand is that its steady-state response to the growth in domestic national income is not necessarily proportional. It should be recalled that in Chapter 3 the theoretical relationship for import demand in (3.8a) had a unitary elasticity with respect to income. However, in general the share of income spent on foreign goods has historically exceeded unity, although the size of this share has changed in the post-1973 period from what it was in the period following the Second World War (Goldstein and Khan 1982); among individual countries the marginal propensity to import has varied greatly (see Houthakker and Magee 1969). This characteristic implies that the dynamic specification of the equation for import demand in (3.8a) should test whether or not its elasticity with respect to domestic income is actually unity in the long run.

Another characteristic is that in annual time-series data the dynamics for demand relationships are of a relatively small order, and therefore can be conveniently restricted to cases where the lagged values of the variables are of one period. Accordingly, in terms of the general stochastic difference specification, the expression for import demand M of a geographic market j in terms of its economic activity Y_j and the import price P_j, relative to the general price level D_j, is given by

$$m_{jt} = \alpha_{10} + \alpha_{11} y_{jt} + \alpha_{12} y_{j,t-1} + \alpha_{13}(p_j - d_j)_t + \alpha_{14}(p_j - d_j)_{t-1}$$
$$+ \alpha_{15} m_{j,t-1} + v_{1t}, \tag{8.11}$$

where lower-case letters denote the logarithms of the corresponding capitals and the expected signs are $\alpha_{11}, \alpha_{12} > 0$; $\alpha_{13}, \alpha_{14} < 0$; and $0 < \alpha_{15} < 1$.

Since the growth rate of imports depends on the expansion path of economic activity, it is appropriate to apply the ECM to the relationship between imports and domestic economic activity. The disequilibrium adjustment term in the ECM will then rectify any previous disequilibrium between the two variables. The following transformation of (8.11) incorporates an ECM driven by economic activity (see Annex D for the derivation):

$$\Delta m_{jt} = \alpha_{10} + \alpha_{11} \Delta y_{jt} + \beta_{12}(m_j - y_j)_{t-1} + \beta_{13} y_{j,t-1} + \beta_{14} \Delta(p_j - d_j)_t$$
$$+ \beta_{15}(p_j - d_j)_{t-1} + v_{1t}, \tag{8.12}$$

where $\beta_{12} = (\alpha_{15} - 1)$, $\beta_{13} = (\alpha_{11} + \alpha_{12} + \alpha_{15} - 1)$, $\beta_{14} = \alpha_{13}$, $\beta_{15} = (\alpha_{13} + \alpha_{14})$, and the expected signs of the coefficients are $\alpha_{11} > 0$, $-1 < \beta_{12} < 0$, $\beta_{12} < \beta_{13}$, and $\beta_{14}, \beta_{15} < 0$.

The import price terms in the foregoing specification have been so transformed as to nest the 'differences' formulation of the variable in the levels form of the equation. This transformation reduces the possibility of the occurrence of the spurious correlation typically associated with time-series data when the relationship between import demand and import price is estimated.

On a steady-state growth path, $\Delta m_{jt} = g_3$ and $\Delta y_{jt} = g_4$ are the growth rates of import demand and domestic economic activity, respectively. Since import price would not be expected to have any long-run dynamic influence, its effect is constrained to zero so that $\Delta(p_j - d_j)_t = 0$ in the long run. Hence the long-run dynamic equilibrium relationship implicit in (8.12) is

$$M_j = k_1^* Y_j^{1-(\beta_{13}/\beta_{12})} \left(\frac{P_j}{D_j}\right)^{-\beta_{15}/\beta_{12}}, \qquad (8.13)$$

where $k_1^* = \exp\{[-\alpha_{10} + (1 - \alpha_{11})g_4]/\beta_{12}\}$.

The elasticities in (8.13) have the expected signs. The income elasticity of import demand is expressed as

$$\epsilon_m^y = 1 - (\beta_{13}/\beta_{12}). \qquad (8.14a)$$

Its value is positive since the expected sign of β_{12} is negative and $\beta_{13} > \beta_{12}$. When $\beta_{12} < \beta_{13} < 0$, import demand is inelastic with respect to income; when $\beta_{13} = 0$, it has a unitary elasticity; and when $\beta_{13} > 0$, it is income-elastic.

The price elasticity of import demand is expressed as

$$\epsilon_m^p = -\beta_{15}/\beta_{12}. \qquad (8.14b)$$

It has a negative value since the expected signs of both β_{12} and β_{15} are negative.

The term k_1^* in (8.13) shows that, on a steady-state growth path, the level of import demand depends on the rate of growth of economic activity, g_4, as well as on the levels of economic activity and the price of the imported commodity. The income growth elasticity, denoted κ_1, is defined as the percentage change in import demand brought about by a 1 per cent change in the growth rate of economic activity g_4, and is expressed as

$$\kappa_1 = \frac{\partial M_j}{\partial g_4} \frac{1}{M_j} = \frac{1 - \alpha_{11}}{\beta_{12}}. \qquad (8.14c)$$

The value of κ_1 can be positive, negative, or zero. Consider an acceleration in the rate of economic growth in a commodity-importing country. The expression in (8.14c) states that such an acceleration will increase the demand for imports when the rate of growth of import demand has a more-than-proportional response to variations in the economic growth rate of that country, i.e. when $\alpha_{11} > 1$ in (8.12). If the response is less than proportional, i.e. when $0 < \alpha_{11} < 1$, the demand for imports will decrease in response to an acceleration in the rate of economic growth. Finally, if the growth rate of import demand has a proportional response to a change in economic growth, i.e. when $\alpha_{11} = 1$, import demand will be unaffected by variations in the economic growth rate of the country.

The steady-state solution of the import demand relationship can now be expressed as

$$M_j = k_1^* Y_j^{\epsilon_m^y} \left(\frac{P_j}{D_j}\right)^{\epsilon_m^p}, \qquad (8.13')$$

where $k_1^* = \exp(-\alpha_{10}/\beta_{12} + \kappa_1 g_4)$. This relationship shows that the effects on import demand of changes in the import price and in the level and rate of change of economic activity are directly dependent on the price elasticity ϵ_m^p, the income elasticity ϵ_m^y, and the income growth elasticity κ_1, respectively, of import demand, as defined in (8.14a)–(8.14c).

8.2 Export Demand

The demand for the exports of a country has a steady-state response to the import demand of geographic markets, and a transient response to the relative export price of the country. The justification for these long-run dynamic properties is similar to that for the import demand of the geographic markets. The demand for exports of a commodity from all foreign suppliers is equivalent to the import demand for the commodity in that market. Thus, the life-cycle model of consumption provides the same theoretical justification for the existence of a long-run dynamic effect associated with import demand in foreign markets as it did for the import demand function. In contrast, unless relative-price movements generate only transient responses, a continuous change in the price of exports from one country relative to that of exports from competing suppliers would eventually cause importers to purchase the commodity from the lower-priced supplier(s). Thus it is appropriate to constrain the long-run dynamic effect from relative prices to zero.

An important difference between the export demand relationship and the import demand relationship considered in the previous section is that a change in the demand for imports in a foreign market will bring about a proportional change in the demand for exports of each country over the long run. This characteristic appears in the export demand schedule derived in Chapter 3. Consequently, the long-run equilibrium solution of the relationship for export demand should yield a unitary elasticity with respect to import demand in the foreign market.

Consider the general first-order stochastic difference expression for export demand, X^d, of a country i by a geographic market j as a function of import demand M of the geographic market and relative price R:

$$x_{ijt}^d = \alpha_{20} + \alpha_{21} m_{jt} + \alpha_{22} m_{j,t-1} + \alpha_{23} r_{ijt} + \alpha_{24} r_{ij,t-1} + \alpha_{25} x_{ij,t-1}^d + v_{2t},$$
(8.15)

where the relative price $r_{ij} = \ln(P_{ij}/P_j)$. The expected values of the coefficients are α_{21}, $\alpha_{22} > 0$; α_{23}, $\alpha_{24} < 0$; and $0 < \alpha_{25} < 1$. It will be recalled from Chapter 7 that a long-run unitary elasticity of export demand with respect to the import demand of a foreign market implies that $(\alpha_{21} + \alpha_{22} + \alpha_{25}) = 1$. Transformation of (8.15) as described in Section 7.2 results in an export demand specification with an ECM driven by import demand and with a 'differences' formulation of the current relative price term nested in the levels form of the equation (see Annex D):

$$\Delta x_{ijt}^d = \alpha_{20} + \alpha_{21} \Delta m_{jt} + \beta_{22} (x_{ij}^d - m_j)_{t-1} + \alpha_{23} \Delta r_{ijt} + \beta_{24} r_{ij,t-1} + v_{2t},$$
(8.16)

where $\beta_{22} = (\alpha_{25} - 1)$ and $\beta_{24} = (\alpha_{23} + \alpha_{24})$, their expected signs being $-1 < \beta_{22} < 0$ and $\beta_{24} < 0$.

The disequilibrium adjustment mechanism in the third term of (8.16) corrects for previous non-proportional responses in the long-run dynamic growth of export demand. If the market share of the country were to fall below its long-run equilibrium level, the negative coefficient in the disequilibrium adjustment term would induce an increase in the demand for exports. Conversely, if the share of the country were to increase above its long-run equilibrium level, that coefficient would generate downward pressure on the export demand until the growth rate returned to its steady-state path.

When the estimated level of exports exceeds that of imports in a market, the results are logically inconsistent with the requirement that the exports to a market can never exceed the total imports of that market. Consequently it is necessary that $0 < X_{ij}^d/M_j < 1$ for all t in the regressand, as well as in the disequilibrium adjustment mechanism of the third term in (8.16). This constraint will be automatically satisfied if, first, Δm_{jt} is subtracted from both sides of (8.16) and, second, a logit transformation is applied both to the regressand and to the disequilibrium adjustment term (see Annex D):

$$\Delta z_{ijt} = \alpha_{20} + \beta_{21} \Delta m_{jt} + \beta_{22}^* z_{ij,t-1}^* + \alpha_{23} \Delta r_{ijt} + \beta_{24} r_{ij,t-1} + v_{2t}^* \quad (8.17)$$

where

$$z_{ij} = \ln\left[\frac{X_{ij}^d/M_j}{1 - (X_{ij}^d/M_j)}\right] = \ln\left(\frac{X_{ij}^d}{X_{kj}^d}\right),$$

since $X_{kj}^d = (M_j - X_{ij}^d)$, and where $\beta_{21} = (\alpha_{21} - 1)$ with expected sign $\beta_{21} > -1$ (in particular, $\beta_{21} > 0$ if $\alpha_{21} > 1$; $\beta_{21} = 0$ if $\alpha_{21} = 1$; and

$-1 < \beta_{21} < 0$ if $0 < \alpha_{21} < 1$). The expected signs of the other coefficients are $-1 < \beta_{22}^* < 0$, and α_{23}, $\beta_{24} < 0$. Then estimates of the demand for exports will automatically satisfy the constraint and the error term can be normally distributed.

In long-run dynamic equilibrium, with constant import growth $g_3 = \Delta m_{jt}$ and $\Delta r_{ijt} = 0 = \Delta z_{ijt}$, the solution of (8.17) is

$$z_{ij} = -\alpha_{20}/\beta_{22}^* - (\beta_{21}/\beta_{22}^*)g_3 - (\beta_{24}/\beta_{22}^*)r_{ij}. \quad (8.18)$$

Rearrangement of terms yields the steady-state solution for export demand in terms of a logistic function:

$$X_{ij}^d = \frac{M_j}{1 + k_2 R_{ij}^{\beta_{24}/\beta_{22}^*}} \quad (8.19)$$

where $k_2 = \exp[(\alpha_{20} + \beta_{21}g_3)/\beta_{22}^*]$.

There are three responses of export demand in the relationship described by (8.19). The first is its proportional response to a change in the level of imports. The second is its response to a change in export price, which is measured by the price elasticity of export demand:

$$\epsilon_x^p = \frac{\partial X_{ij}^d}{\partial P_{ij}} \frac{P_{ij}}{X_{ij}^d} = -\frac{\beta_{24}}{\beta_{22}^*}(1 - s), \quad (8.20)$$

where $s = X_{ij}^d/M_j$ is the market share of the country. This result shows that the price elasticity of demand for exports varies with the market share of the country. As the market share of the country increases, the absolute value of the price elasticity of export demand becomes smaller. Conversely, as the market share of the country decreases, the absolute value of the price elasticity of export demand becomes larger. The real value is bounded from above by zero and from below by $-(\beta_{24}/\beta_{22}^*)$.

The third response is stimulated by a change in the rate of growth of import demand in the foreign market. It reflects the import growth elasticity, denoted κ_2, and is defined as the percentage change in export demand brought about by a 1 per cent change in the rate of growth of import demand, g_3:

$$\kappa_2 = \frac{\partial X_{ij}^d}{\partial g_3} \frac{1}{X_{ij}^d} = \frac{\gamma e^z}{(1 + e^z)^2} \frac{1}{s}$$
$$= \gamma(1 - s), \quad (8.21)$$

where $\gamma = -(\beta_{21}/\beta_{22}^*)$. The value of κ_2 is positive when $\beta_{21} > 0$, negative when $-1 < \beta_{21} < 0$, and zero when $\beta_{21} = 0$. A positive value of κ_2 implies that the demand for exports of the country will expand, and thereby lead to a larger market share, when the import growth rate in the geographic market accelerates. The converse occurs for a negative value of κ_2, whereas the demand for exports will be unaffected by import growth rate variations when the value of κ_2 is zero.

The import growth elasticity κ_2 can be interpreted with respect to an acceleration in the rate of growth of import demand g_3, as follows.

1. The demand for exports will increase if a change in the import demand of a foreign market leads to a more-than-proportional change in export demand (i.e. when $\beta_{21} > 0$ in (8.17)).
2. The demand for exports will decrease if a change in the import demand of the foreign market leads to a less-than-proportional change in export demand (i.e. when $\beta_{21} < 0$).
3. The demand for exports will remain unchanged if a change in the import demand of the foreign market leads to a proportional change in export demand (i.e. when $\beta_{21} = 0$).

Thus, the steady-state solution, unlike the static equilibrium solution, demonstrates how export demand can be influenced by changes in the rate of growth of imports g_3 in the geographic market, as well as by changes in the level of imports M_j.

8.3 Price Competitiveness

It is also possible to obtain information about the effect of changes in price competitiveness on relative export performances from the relationship described in (8.17). The effect is measured by the elasticity of substitution, which is defined as the percentage change in the ratio of the export demand of one country to that of other suppliers to a market that is brought about by a change in the ratio of their corresponding prices.

The concept of elasticity of substitution was originally defined by Hicks (1932) in production theory and was first applied to international trade by Tinbergen (1946).[1] Armington (1969) used the basic Hicksian approach to develop a model in which the importer's allocation of the quantity demanded of goods among competing supply sources depended on the elasticity of substitution. Armington's model has been applied to aggregate trade flow data by Hickman and Lau (1973) and to agricultural trade by Collins (1977), Grennes et al. (1978), and Sarris (1981). Unfortunately, parsimony in the model was obtained at the cost of making the elasticity of substitution identical in all markets and among all suppliers to each market. In order to relax the assumption of identical elasticities of substitution, Resnick and Truman (1973) and Wells and Johnson (1979) extended the allocation of the quantity demanded by each importer to a multi-decision process for successively smaller supply sources. However, their approach excluded supply estimates, without which prices became predetermined variables. More recently, Pagoulatos and Lopez (1983) have not only relaxed Armington's constant elasticity of substitution assumption but have also incorporated a supply schedule into their model. Likewise, Geraci and Prewo (1982) have included a supply schedule in their extension of Armington's model.

Despite the strong intuitive appeal and widespread application of the concept, controversy has arisen over whether the assumptions that underlie the elasticity of substitution are realistic. In particular, its measurement requires that the income elasticities and the sum of the direct- and cross-price elasticities be equal for the competing exporters. In our model of commodity trade the first condition is indirectly satisfied, while the second does in fact impose a stringent requirement.

For the sake of simplicity, let us consider a static equilibrium framework, irrespective of the geographic market, in which the export demand for a country i and that for its competitors k are given by

$$X_i^d = k_2' M P_i^{\epsilon_i} P_k^{\epsilon_{ik}} \tag{8.22a}$$

$$X_k^d = k_2'' M P_k^{\epsilon_k} P_i^{\epsilon_{ki}} \tag{8.22b}$$

where ϵ_i and ϵ_k denote direct-price elasticities, and ϵ_{ik} and ϵ_{ki} are the cross-price elasticities of country i and competitors k, respectively. The elasticity of substitution is then derived from the ratio of the export demand equations for the two competing suppliers:

$$\frac{X_i^d}{X_k^d} = \frac{k_2'}{k_2''} \frac{P_i^{\epsilon_i - \epsilon_{ik}}}{P_k^{\epsilon_k - \epsilon_{ki}}} = k_2 \left(\frac{P_i}{P_k}\right)^{\epsilon_x^r}. \tag{8.23}$$

Hence unitary import elasticity ensures that the responses of all non-price variables are equal (from which it follows that there is a weakly imposed equivalency between the income elasticity of export demand for competing exporters). However, the ratio of export demand for competing suppliers will be functionally related to their relative price only if $\epsilon_x^r = \epsilon_i - \epsilon_{ik} = \epsilon_k - \epsilon_{ki}$. This requirement, as well as the need for the income elasticities to be identical, underlie the criticism of the elasticity of substitution concept by Prais (1962), Stern and Zupnick (1962), and Leamer and Stern (1970: 57-63). However, Richardson's (1973) empirical tests of the underlying assumptions have renewed support for the use of the elasticity of substitution in internationally traded goods by demonstrating that the assumptions do tend to hold, especially at the more disaggregated level of trade.

The exports of a country represent a negligible amount of the total world exports in monopolistic competition, so that they have no impact on the world market price and changes in their level do not produce a reaction by other exporters. Thus the price of exports of competitors to a market, denoted P_{kj}, approximates the import price of that market, P_j. Therefore $R_{ij} \simeq P_{ij}/P_{kj}$ in (8.17) and we obtain the following results.

1 *The elasticity of substitution (denoted ϵ_x^r) is given by $-\beta_{24}/\beta_{22}^*$.* The elasticity ϵ_x^r is defined as the percentage change in relative quantities demanded from competing sources $(X_i^d/X_k^d)_j$, denoted Z_{ij}, brought about by a 1 per cent change in relative price $R_{ij} = P_{ij}/P_{kj}$:

$$\epsilon_x^r = \frac{\partial Z_{ij}}{\partial R_{ij}} \frac{R_{ij}}{Z_{ij}} = -\frac{\beta_{24}}{\beta_{22}^*}. \qquad (8.24)$$

2 *The elasticity of substitution for market shares (denoted ϵ_x^s) is measured as*

$$\epsilon_x^s = \frac{\partial (X_{ij}^d/M_j)}{\partial R_{ij}} \frac{R_{ij}}{X_{ij}^d/M_j} = \frac{-\beta_{24}/\beta_{22}^*}{1 + k_1^{*-1} R_{ij}^{-(\beta_{22}^*/\beta_{24})}}$$

$$= \frac{\epsilon_x^r}{1 + Z_{ij}} = \epsilon_x^r \left(1 - \frac{X_{ij}^d}{M_j}\right) \qquad (8.25)$$

since

$$Z_{ij} = \frac{X_{ij}^d/M_j}{1 - (X_{ij}^d/M_j)}.$$

The elasticity ϵ_x^s refers to the percentage change in the export market share $(X_i^d/M)_j$, brought about by a 1 per cent change in the relative price of exports R_{ij}. Since this elasticity varies with the market share of a country, it provides a solution to the traditional problem associated with the usual way in which the elasticity of substitution is empirically measured, namely that the elasticity of substitution remains constant (Leamer and Stern 1970).

3 *The price elasticity of demand ϵ_x^p is equivalent to the elasticity of substitution for market shares ϵ_x^s.* Equations (8.21) and (8.25) are identical.

4 *The price elasticity of demand ϵ_x^p, the elasticity of substitution ϵ_x^r, and the elasticity of substitution for market shares ϵ_x^s are related to one another as follows:*

$$\epsilon_x^p = \epsilon_x^s = \epsilon_x^r (1 - s), \qquad (8.26)$$

where $s = (X_i^d/M)_j$. Then ϵ_x^s and ϵ_x^p vary with the export market share s, while ϵ_x^r remains constant, a characteristic noted by Leamer and Stern (1970: 178–9, n. 9) but often overlooked in the empirical trade literature (e.g. Kravis and Lipsey 1972). Moreover, since $\lim_{s \to 0} \epsilon_x^p = \epsilon_x^r$, the elasticity of substitution provides the lower (negative) bound for the elasticity of demand, a result first demonstrated by Harberger (1957).

8.4 Illustration

Let us return to the case of export demand for coffee from Colombia by the USA given in Section 7.4 and re-examine the relationship in the form given by equation (8.17). The specification differs from the earlier one in two respects. First, a logit constraint has been imposed on the dependent variable to ensure logical consistency with the requirement that the estimated demand for the exports of a country does not exceed the total import demand of its geographic market. Second, relative prices have been introduced to test the hypothesis that importers differentiate between sources of supply of commodities, even when they are physically homogeneous.

The empirical estimate of (8.17) relating the export demand for Colombian coffee by the USA, X^d, to the total coffee import demand of the USA, M, and to the relative price of Colombian exports to the USA, R, is as follows:

$$\Delta z_t = -0.34 + 0.51\,\Delta m_t - 0.21\,z_{t-1} - 0.77\,\Delta r_t - 0.41\,r_{t-1},$$
$$(2.9) \qquad (2.5) \qquad (3.6) \qquad (2.5) \quad (8.27a)$$
$$\bar{R}^2 = 0.82 \quad \text{dw} = 1.9 \quad \hat{\sigma} = 0.076$$

where lower-case letters denote logarithms of corresponding upper-case letters, and $Z = (X^d/M)/[1 - (X^d/M)]$. As before, dummy variables were included in the estimated relationship so as to account for unforeseen changes in export demand, although they have not been reported upon here in order to simplify presentation of the results.

The alternative hypothesis is that relative prices do not affect the demand for exports. Re-estimation of (8.17) without the relative price variable yields

$$\Delta z_t = -0.5 + 0.65\,\Delta m_t - 0.28\,z_{t-1}.$$
$$(3.0) \qquad (2.9) \qquad (8.27b)$$
$$\bar{R}^2 = 0.70 \quad \text{dw} = 1.4 \quad \hat{\sigma} = 0.098$$

Equation (8.27a) provides a better overall goodness of fit than does (8.27b), once account is taken of the difference in the degrees of freedom.[2] The relative price coefficients are statistically significant, and their inclusion in the relationship helps to explain major movements in export demand, such as occurred in 1978–9 and in 1982.

In long-run dynamic equilibrium, export demand has a steady-state response to imports in the geographic market and a transient response to relative prices. Constant growth rate $g_3 = \Delta m_t$, when $\Delta r_t = 0$ and $\Delta z_t = 0$, yields the long-run dynamic solution for the export demand function in (8.27a):

$$X^d = \frac{M}{1 + k_3 R^{1.95}}, \qquad (8.28)$$

where $k_3 = \exp(1.62 - 2.43 g_3)$.

In steady-state growth when $g_3 = 0.04$ and $R = 1.1$, Colombia's share in the US market is 15 per cent, the same as was calculated in the earlier example in (7.10). Similarly, an acceleration in the rate of growth of imports from 4 per cent in 1960–75 to 8 per cent thereafter would, as before, lead to an increase in Colombia's market share from 15 to 16.5 per cent. The adjustment of export demand to a one-time disturbance in the long-term growth of US import demand is shown in Fig. 8.1, where the market share of Colombia drops from its dynamic equilibrium level of 15 per cent to 10 per cent in 1976. Colombia regains one-half of its pre-1976 market share within four years of the disturbance and three-fourths of its share within seven years.

The introduction of relative prices in a logistic function gives rise to a data-generating process that is more consistent with expectations and more germane to observed outcomes. The logistic curve is shown in Fig. 8.2. It has an upper asymptote of 100 per cent and a lower asymptote of 0 per cent. As may be seen, the effect on Colombia's market share of a change in the relative export price of that country depends in an inverse

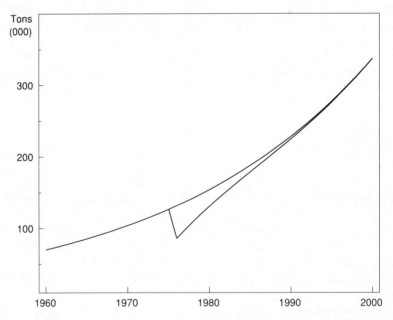

FIG. 8.1 Dynamic adjustment process of coffee export demand function with relative prices

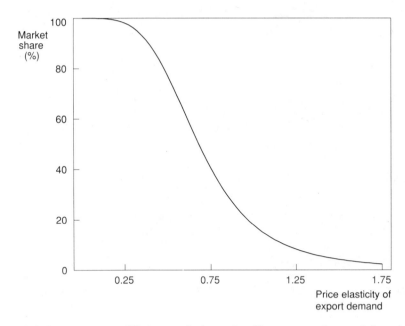

FIG. 8.2 Long-run equilibrium solution of coffee export demand function

manner on the size of its market share. The price elasticity of the US export demand, ϵ_x^p, is $-1.95(1-s)$, where s represents the Colombian market share. With an average share equal to about 15 per cent of the US market in the period over which (8.28) was estimated, the price elasticity of the export demand of Colombia is -1.66.

At the midpoint of the logit curve, where the US importer is indifferent to purchases from alternative supply sources, a change in the relative export price of Colombia in either direction will have the greatest impact on that country's market share. Thus, Colombian exporters could expect higher gains from a reduction in their relative export price if their market share were larger than 15 per cent, but smaller losses from an increase in their relative price. Conversely, Colombian exporters could not expect to penetrate the market significantly through a decline in their relative price if the country's share were smaller than 15 per cent, although they would be less vulnerable to market share losses owing to an increase in their relative export price.

The slope of the logit curve is determined by the price competitiveness of the exporting country, measured by the elasticity of substitution. For example, were the elasticity of substitution of Colombia to increase in absolute terms from -1.95 to -4.0, the curve would become steeper. Under these circumstances, a change in the relative export price of

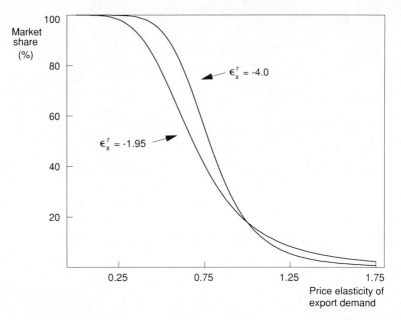

FIG. 8.3 Long-run equilibrium solution of coffee export demand function with different relative price elasticities

Colombia would induce a greater response in export demand, and hence a larger change in its market share. This effect is shown in Fig. 8.3. The response of export demand to relative-price changes is smaller at the lower range of the logit curve. A reduction in the relative export price of an exporter with a small share of the market would produce fewer gains than before, while an exporter with a larger market share would have less difficulty in penetrating the market.

8.5 World Consumption

Having derived the specification for import demand for a commodity by the principal geographic markets of each exporting country, we need to obtain the specification for all the other consumption of the commodity in the world market. The specification for rest-of-world consumption C_k in terms of the constant dollar price of the commodity P/D and constant dollar income Y_k is analogous to that for the import demand of the principal export markets given in (8.12). Its final form incorporates an ECM driven by economic activity, the price terms being transformed so that the 'differences' formulation of the variable is nested in the levels form of the stochastic difference equation:

$$\Delta c_{kt} = \alpha_{30} + \alpha_{31}\Delta y_{kt} + \beta_{32}(c_k - y_k)_{t-1} + \beta_{33}y_{k,t-1}$$
$$+ \beta_{34}\Delta(p - d)_t + \beta_{35}(p - d)_{t-1} + v_{3t}, \quad (8.29)$$

where the expected signs are $-1 < \beta_{32} < 0$, $\beta_{33} > \beta_{32}$, and β_{34}, $\beta_{35} < 0$. When $\beta_{33} < 0$, demand is inelastic with respect to income; when $\beta_{33} = 0$, it has a unitary elasticity; and when $\beta_{33} > 0$, it is income-elastic.

The long-run dynamic equilibrium relationship implicit in (8.29) is

$$C_k = k_4 Y_k^{\epsilon^y} (P/D)^{\epsilon^p} \quad (8.30)$$

where $\epsilon^y = 1 - (\beta_{33}/\beta_{32}) > 0$, and $\epsilon^p = -(\beta_{35}/\beta_{32}) < 0$.

8.6 Demand for Stocks

The demand for stocks, or inventories, arises from transactional, precautionary, and speculative motives. The transactional and precautionary demands for stocks originate from the desire of economic agents to hold stocks to meet future consumption. Stocks are also held in order to avoid delay and to provide continuity of supply. Commodity market studies have therefore focused on the relationship between the demand for stocks and either consumption or production of the commodity. In the first relationship agents adjust their stock levels to consumption; in the second they adjust their stocks to the level of output. In either case, the demand for stocks has a proportional response to changes in consumption or production of the commodity in the long run.

The speculative demand for stocks is motivated by the desire of economic agents to profit from future price changes. When prices are expected to rise, producers and processors increase their inventories in anticipation of being able to sell their products at a higher price in the future; alternatively, stocks are reduced when prices are expected to fall, since lower profits are anticipated from stocks released in the future. Inventory holdings are therefore related to expected prices, as well as to consumption or production of the commodity.

Accordingly, the demand for stocks K^d was shown in equation (5.34) to depend on consumption C or production Q for transactional and precautionary purposes, and on expected prices P^e for speculative purposes. In static equilibrium, this relationship can be expressed as

$$k_t^d = \alpha_{70} + \alpha_{71}q_t + \alpha_{72}p_t^e + v_{7t}, \quad (8.31)$$

where, as before, lower-case letters denote logarithms of corresponding capital letters; and where the expected signs are α_{71}, $\alpha_{72} > 0$.

The expected price is assumed to be formed through a rational process. To support his rational expectations hypothesis, Muth (1961) uses conditional probability theory to argue that expectations about a variable such as the price of a commodity should be treated as a problem of optimal forecasting whereby the minimum mean squared error forecast of that variable is produced by all information available at the time of making the forecast. Thus, at period t the expected price P^e of a commodity is the expected value of the price, based on all the information I available at period $t - m$ to agents concerned with stocks in the commodity market:

$$P_t^e = E_{t-m}(P_t|I), \tag{8.32}$$

where E_{t-m} is the expectation operator for expectations formed at $t - m$.

Although the concept of the rational expectations hypothesis is simple, its application has proven to be difficult because it is general about the information that rational economic agents have available in forming expectations. In practice, the mechanism for the formation of expectations follows the standard approach of specifying a time-series process for market prices, such as a univariate autoregressive moving average (ARMA) process, and of using the estimated parameter values to generate forecasts that can be revised as new information becomes available. That is, it is assumed that P_t^e is formed as the projection of P_t against lagged prices P_{t-1}, P_{t-2}, ..., P_{t-n}; hence the expected price P^e at time t is based on information available up to time $t - 1$.

The characterization of the data-generating process for expected prices in the stock demand relationship in (8.31) is based on the work of Wallis (1980). It is assumed that the price of the commodity follows a general ARMA (p, q) process:

$$\phi(L)p_t = \theta(L)u_t \tag{8.33}$$

where $\phi(L)$ and $\theta(L)$ are polynomials in the lag operator L, which are defined as

$$\phi(L) = 1 - \phi_1 L - \phi_2 L^2 - \ldots - \phi_p L^p$$

and

$$\theta(L) = 1 - \theta_1 L - \theta_2 L^2 - \ldots - \theta_q L^q.$$

Then P_t can be represented by an autoregressive process:

$$p_t = \omega(L)p_{t-1} + u_t \tag{8.34}$$

where $\omega(L) = [\theta(L) - \phi(L)]/[L\theta(L)]$. The minimum mean square error forecast based on information up to period $t - 1$ is

$$p_t^e = \omega(L)p_{t-1}, \tag{8.35}$$

which states that the forecast error is zero. Substitution of (8.35) in (8.31) gives

$$k_t^d = \alpha_{70} + \alpha_{71}q_t + \alpha_{72}\omega(L)p_{t-1} + v_{7t},$$

and multiplication of each side of the expression by $\theta(L)$ gives

$$\theta(L)k_t^d = \alpha'_{70} + \alpha_{71}\theta(L)q_t + \alpha_{72}\psi(L)p_{t-1} + v'_{7t} \qquad (8.36)$$

where

$$\psi(L) = \frac{\theta(L) - \phi(L)}{L}.$$

In practice, the $\theta(L)$ polynomial associated with the terms that contain K^d and Q is of a relatively low order. In accordance with our earlier specifications for demand equations, we consider only the first-order polynomial in the lag operator. The resulting general dynamic specification for the stock demand relationship is

$$k_t^d = \alpha'_{70} + \theta_1 k_{t-1}^d + \alpha_{71}q_t + \beta_{72}q_{t-1} + \beta_{73}\psi(L)p_{t-1} + v'_{7t}, \qquad (8.37)$$

where $\beta_{72} = -\alpha_{71}\theta_1$, and $\beta_{73} = \alpha_{72}$.

Finally, transformation of the general stochastic difference equation so that demand for stocks has an ECM driven by production or consumption with a 'differences' formulation of the expected price level gives

$$\Delta k_t^d = \alpha'_{70} + \alpha_{71}\Delta q_t + \gamma_{72}(k^d - q)_{t-1} + \beta_{73}\psi(L)p_{t-1} + v'_{7t}, \qquad (8.38)$$

where $\gamma_{72} = (\theta_1 - 1)$ and the expected signs are $\alpha_{71} > 0$, $-1 < \gamma_{72} < 0$, and $\beta_{73} > 0$. Hence the dynamic specification of the relationship for the demand for stocks yields a unitary elasticity of demand for stocks with respect to consumption or production of the commodity in its long-run dynamic equilibrium solution.

The steady-state equilibrium solution for (8.38) is

$$K^d = k_7^* Q P^{e^{\epsilon_k}}, \qquad (8.39)$$

where $k_7^* = \exp(-\alpha'_{70}/\gamma_{72} + \kappa_4 g_5)$, such that $\kappa_4 = (1 - \alpha_{71})/\gamma_{72}$ is the production or consumption growth elasticity, and $\epsilon_k = -\beta_{73}/\gamma_{72}$. Thus, the demand for stocks of the commodity is related to both the level and the rate of growth of its production or consumption, g_5, as well as to the expected price level of the commodity.

As Wallis (1980) has pointed out, direct estimation of (8.38) in its 'reduced form' generates a greater forecast error variance than does the separate estimation of the parameter values $\psi(L)$ for P_t and the use of those parameter values in the relationship for K^d. The reason is that the rational economic agent knows the data-generating process of prices, and then uses that information to obtain minimum mean square error forecasts of stock requirements. Accordingly, the relationship for the commodity price P_t is first estimated as an ARMA process; this relationship is then used to estimate the stock demand relationship in

(8.38). In this way, forecasts of the commodity price P, based on its own past behaviour, are introduced into the dynamic specification used to generate forecasts of the demand for stocks K^d.[3]

Notes

1 In subsequent applications to international trade it has been used almost exclusively to examine the export price competitiveness of the industrialized countries; see the multi-country studies by Fleming and Tsiang (1956); Maizels (1963); Junz and Rhomberg (1965); Kreinin (1967); Kravis and Lipsey (1971, 1972, 1974, 1982); Sato (1977); Hickman and Lau (1973); and Artus and Sosa (1978); and the individual country studies on Japan by Narvekar (1960) and Shinkai (1968). It has also been applied in studies that have attempted to explain the comparative advantage of the USA and the UK; see MacDougall (1952); Zelder (1958); Ginsburg and Stern (1965); and Ginsburg (1969). Attempts to determine how the comparative advantage of countries has been influenced by exchange rate realignments in industrialized countries have been made by Junz and Rhomberg (1973); Page (1975); and Kreinin (1977). An exception to the applications of the model to the analysis of industrialized countries' trade is the study by Thomas (1988), which uses the approach to analyse trade between the developing countries.
2 Equation (7.9) also shows the relationship between the US export demand for Colombian coffee and the total US coffee import demand without the relative export price variable. However, comparison between alternative specifications containing different dependent variables on the basis of goodness of fit can be misleading.
3 A similar procedure, also based on the work of Wallis, has been used by Goodwin and Sheffrin (1982) to estimate the supply relationship in the market for the broiler chicken industry, except that expected prices were estimated in terms of the expected values of the exogenous variables in the model, generated as an autoregressive integrated moving average (ARIMA) process.

9
The Supply Functions

The first part of this chapter derives the dynamic specification of the export supply relationship in the system of equations used to characterize international commodity trade. Since the lag structure is central to the relationship between supply and prices, I have adopted a general dynamic specification that allows for flexibility in the characterization of the lag structures of different commodities and of different suppliers of the same commodity. In particular, in the absence of well articulated dynamic adjustment theories, the transformed stochastic difference equation is used to characterize the export supply relationship.

In the final part of the chapter we return to the more restricted specification for the import price function. The specification, in the form of the ECM, imposes proportionality on the long-run response of import prices to movements in the world market price of a commodity.

9.1 Export Supply

For a particular product the export supply of a country is related primarily to the price received by exporters. The relationship between export supply and export price in primary commodity trade is often characterized by long lags, since the effects of price changes usually take a long time to work themselves through to export supply. Moreover, transmission of the price effects can be complex. For example, a price rise can induce a short-term response (new plantings, higher yields, or stock depletion), a medium-term response (expansion of installed capacity), or a long-term response (new capacity initiation). Consequently, the nature of the response to price changes is central to the dynamic specification of export supply.

The nature of the response can be roughly estimated in a general unrestricted equation that relates export supply to a sequence of lagged export prices. However, since unrestricted estimation is inefficient when responses are slow, I have introduced lag structures that suitably represent the underlying nature of the response once the response has been approximated in the unrestricted equation estimate. In accordance with the general approach used to represent the data-generating process outlined in Chapter 7, the lag structure in the export supply relationship will be represented by a stochastic difference equation.

Accordingly, the expression for export supply X^s of country i to its geographic market j in terms of its export price P relative to the general price level D, as well as of major disturbances W and of a secular trend T, is given by

$$x_{ijt}^s = \alpha_{40} + \alpha_{41} x_{ij,t-1}^s + \alpha_{42} x_{ij,t-2}^s + \sum_{k=0}^{n} \alpha_{43+k}(p_{ij} - d_i)_{t-k}$$
$$+ \beta_{45} T + \beta_{46} W_{it} + v_{4t}, \qquad (9.40)$$

where lower-case letters denote the logarithms of the corresponding capital letters, e.g. $(p_{ij} - d_i) = \ln(P_{ij}/D_i)$, and where the expected signs are $\alpha_{43+k} > 0$, $\beta_{45} \leqq 0$, and $\beta_{46} < 0$.

When there is only a one-period lag in the dependent variable X^s, the restriction $0 < \alpha_{41} < 1$ must be met. When there is both a one- and a two-period lag in the dependent variable X^s, the expected signs of the coefficients α_{41} and α_{42} must satisfy the following constraints if the equation is to imply a non-negative and convergent lag distribution for X^s (see Griliches 1967: 27):

(a) $0 < \alpha_{41} < 2$;
(b) $-1 < \alpha_{42} < 1$;
(c) $(1 - \alpha_{41} - \alpha_{42}) > 0$;
(d) $\alpha_{41}^2 \geq -4\alpha_{42}$.

Note that restrictions (a) and (c) imply (b).

The advantage of the stochastic equation, which by construction has white noise disturbances, diminishes when serial correlation appears in relationships formulated with variables in their levels form. Yet correction of residual autocorrelation by an autoregressive process is invalid in the absence of a common factor assumption test for dynamic misspecification (Sargan 1980). In contrast, reformulations with only variables in their differences form, designed to avoid serial correlation, will, when the true relationship is in terms of levels, introduce an additional moving average term into the disturbance (Plosser and Schwert 1977). The alternative is to transform the stochastic difference equation so that the differences formulation of appropriate variables is nested in the levels form of the equation. In accordance with the procedure described in Chapter 7 for the general dynamic specification of commodity trade relationships, the dynamic specification for the export supply relationship becomes (see Annex D for the derivation)

$$\Delta x_{ijt}^s = \alpha_{40} + \beta_{41} x_{ij,t-1}^s + \alpha_{42} x_{ij,t-2}^s + \sum_{k=0}^{n-1} \alpha_{43+k} \Delta(p_{ij} - d_i)_{t-k}$$
$$+ \sum_{k=1}^{n} \gamma_{44+k}(p_{ij} - d_i)_{t-k} + \beta_{45} T + \beta_{46} W_{it} + v_{4t}, \qquad (9.41)$$

where

$$\beta_{41} = (\alpha_{41} - 1), \sum_{k=1}^{n-1} \gamma_{44+k} = \sum_{k=0}^{n-2} \alpha_{43+k} \text{ and } \gamma_{44+n} = \alpha_{43+n-1} + \alpha_{43+n}.$$

The expected signs of the coefficients of the price variable are $\alpha_{43+k} \leq 0$ and $\gamma_{44+k} > 0$, such that $(\alpha_{43+k} + \gamma_{44+k}) > 0$. Hence $\Delta \ln(P_{ij}/D_i)$ can have a negative coefficient, which does not preclude the derived coefficients for the levels formulation of the price variables from being all positive. For the transformed equation to imply a non-negative and convergent lag distribution for X^s, the restrictions on the coefficients of the lagged dependent variable become

(a) $-1 < \beta_{41} < 1$;
(b) $-1 < \alpha_{42} < 1$;
(c) $\beta_{41} + \alpha_{42} < 0$;
(d) $(1 + \beta_{41})^2 \geq \alpha_{42}$.

The supply of exports has a transient response to the rate of change of the constant dollar price of exports. The reason is that a country's export price of a product cannot rise faster than the general price level. Were this to occur, the terms-of-trade of the exporting country would be in a continuous state of change. Accordingly, the long-run solution to the dynamic specification is the same as that for the static solution. In accordance with the previous procedure (Section 7.3), since $\Delta \ln(P_{ij}/D_i) = 0$ implies $\Delta x_{ij} = 0$, the long-run relationship in (9.41) is

$$X_{ij}^s = k_3(P_{ij}/D_i)^{\gamma_1} \exp(\sigma_1 T + \sigma_2 W), \tag{9.42}$$

where $k_3 = \exp[-\alpha_{40}/(\beta_{41} + \alpha_{42})]$, $\gamma_1 = -\Sigma_1^n \gamma_{44+k}/(\beta_{41} + \alpha_{42})$, $\sigma_1 = -\beta_{45}/(\beta_{41} + \alpha_{42})$, and $\sigma_2 = -\beta_{46}/(\beta_{41} + \alpha_{42})$, the expected signs being $\gamma_1 > 0$, $\sigma_1 \geq 0$, and $\sigma_2 < 0$. Hence the long-run equilibrium solution of export supply depends solely on the level of its explanatory variables; it is independent of the rate of growth of any one of them.

The lag coefficients that emerge from the dynamic specification of the export supply relationship in (9.41) can be obtained by the method set forth in Griliches (1967: 23). For the general stochastic difference equation,

$$x_t^s = \underbrace{\alpha_1 x_{t-1}^s + \alpha_2 x_{t-r}^s}_{\text{lagged dependent variable}} + \underbrace{\beta_0 p_t + \ldots + \beta_n p_{t-n}}_{\text{lagged explanatory variable}}, \tag{9.43}$$

the lag coefficient δ_d for the price variable is given by the formula

$$\delta_d = \sum_{i=1}^{\min(d,r)} \alpha_i \delta_{d-i} + \beta_d, \tag{9.44}$$

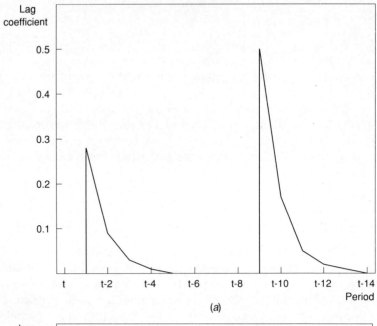
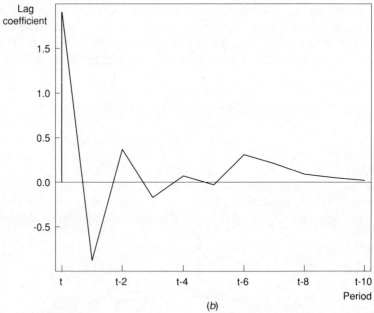

FIG. 9.1 Illustration of lag coefficients for price variable in export supply functions
(*a*) Dampened smooth response
(*b*) Dampened cyclical response

where d is the lag and the price p is measured in constant dollars. The coefficient δ_0 of the price variable in the current period t is equal to β_0.

Figure 9.1 illustrates the lag coefficients that emerge from lags of one and two periods in the dependent trade variable. A one-period lag of the dependent trade variable produces a *dampened smooth approach* to the new equilibrium solution, while a two-period lag produces a *dampened cyclical approach* to it. More than two-period lags of the dependent variable are seldom used in empirical econometrics because, as noted by Griliches (1967), they produce lag structures that are very sensitive to the parameter values.

The dampened smooth approach is demonstrated for Colombia's coffee export supply to West Germany. The results of the estimated stochastic difference equation are

$$\Delta x_t^s = 0.20 x_{t-1}^s + 0.26 \Delta p_{t-4} + 0.24 p_{t-5} + 1.62, \quad (9.45a)$$

where lower-case letters denote logarithms of upper-case letters. Transformation of the equation back to its original levels form yields the solved coefficients for the export supply equation:

$$x_t^s = 0.80 x_{t-1}^s + 0.26 p_{t-4} - 0.02 p_{t-5} + 1.62. \quad (9.45b)$$

The formula given in (9.44) is applied to derive the lag coefficients:

$\delta_0 = 0$
$\delta_1 = 0$
$\delta_2 = 0$
$\delta_3 = 0$
$\delta_4 = 0.26$
$\delta_5 = 0.80(0.26) - 0.02 = 0.19$
$\delta_6 = 0.80(0.19) = 0.15$
$\delta_7 = 0.80(0.15) = 0.12$
$\delta_8 = 0.80(0.12) = 0.10$

$\delta_9 = 0.80(0.10) = 0.08$
$\delta_{10} = 0.80(0.08) = 0.06$
$\delta_{11} = 0.80(0.06) = 0.05$
$\delta_{12} = 0.80(0.05) = 0.04$
$\delta_{13} = 0.80(0.04) = 0.03$
$\delta_{14} = 0.80(0.03) = 0.03$
$\delta_{15} = 0.80(0.02) = 0.02$
$\delta_{16} = 0.80(0.02) = 0.02$
$\delta_{17} = 0.80(0.02) = 0.01$

This equation demonstrates that the effect of a price rise is not immediate. Producers initially respond to a higher price by increasing cultivation. After lags of four and five years, new production is introduced into the market as new plantings begin to bear fruit, and thereafter there is an exponential decline in the response. The sum of the lagged coefficients is 1.2, which is the total, or long-term, price elasticity of export supply.

The dampened cyclical approach (see Griliches 1967: 27–9) is demonstrated for the solved coefficients of the estimated equation of Costa Rica's beef exports to the USA:

$$x_t^s = 0.32 x_{t-1}^s - 0.33 x_{t-2}^s + 0.63 p_{t-1} + 0.68 p_{t-2} + 0.76 p_{t-7}$$
$$+ 0.72 p_{t-8} + 2.56. \quad (9.46)$$

Application of the formula given in (9.44) yields the solved lag coefficients:

$$\begin{aligned}
\delta_0 &= 0.0 \\
\delta_1 &= 0.63 \\
\delta_2 &= 0.32(0.63) + 0.68 & &= 0.88 \\
\delta_3 &= 0.32(0.88) - 0.33(0.63) & &= 0.07 \\
\delta_4 &= 0.32(0.07) - 0.33(0.88) & &= -0.27 \\
\delta_5 &= 0.32(-0.27) - 0.33(0.07) & &= -0.11 \\
\delta_6 &= 0.32(-0.11) - 0.33(-0.27) & &= 0.05 \\
\delta_7 &= 0.32(0.05) - 0.33(-0.11) + 0.76 & &= 0.81 \\
\delta_8 &= 0.32(0.81) - 0.33(0.05) + 0.72 & &= 0.96 \\
\delta_9 &= 0.32(0.96) - 0.33(0.81) & &= 0.04 \\
\delta_{10} &= 0.32(0.04) - 0.33(0.96) & &= -0.30 \\
\delta_{11} &= 0.32(-0.30) - 0.33(0.04) & &= -0.11 \\
\delta_{12} &= 0.32(-0.11) - 0.33(-0.30) & &= 0.06 \\
\delta_{13} &= 0.32(0.06) - 0.33(-0.11) & &= 0.06
\end{aligned}$$

The explanation for the dampened cyclical response pattern in beef has been given by Jarvis (1974). A rise in beef prices creates an incentive for Costa Rican exporters to increase production. In the short run increased production is achieved through higher slaughter rates. In the long run it is accomplished by the retention of animals to increase the size of the breeding herd and to fatten them further. Retention causes a decline in the slaughter rate for about two years, after which exports increase as the expanded cattle herd is brought to market. During the period of reduction in the slaughter rate, the price of beef increases and gives rise to another cycle, which tends to be shorter than the initial one.

9.2 World Production

Total world supply of a commodity can be obtained from the sum of exports of all the countries of interest X_i (for $i = 1, \ldots, n$) and of rest-of-world production. The specification of rest-of-world production, denoted Q_k, in terms of the constant dollar market price of the commodity, P/D, is analogous to that for the export supply of the exporters of interest given in (9.41):

$$\Delta q_{kt} = \alpha_{50} + \beta_{51} q_{k,t-1} + \alpha_{52} q_{k,t-2} + \sum_{m=0}^{n-1} \alpha_{53+m} \Delta (p - d)_{t-m}$$
$$+ \sum_{m=1}^{n} \gamma_{54+m} (p - d)_{t-m} + \beta_{55} T + \beta_{56} W_{kt} + v_{5t}, \quad (9.47)$$

where

$$\beta_{51} = (\alpha_{51} - 1), \quad \sum_{m=1}^{n-1} \gamma_{54+m} = \sum_{m=0}^{n-2} \alpha_{53+m} \text{ and } \gamma_{54+n} = \alpha_{53+n-1} + \alpha_{53+n}.$$

The expected signs of the coefficients of the price variable are $\alpha_{53+m} \leq 0$ and $\gamma_{54+m} > 0$, such that $(\alpha_{53+m} + \gamma_{54+m}) > 0$. For (9.47) to imply a non-negative and convergent lag distribution for Q_k, the restrictions on the coefficients of the lagged dependent variable become

(a) $-1 < \beta_{51} < 1$;
(b) $-1 < \alpha_{52} < 1$;
(c) $\beta_{51} + \alpha_{52} < 0$;
(d) $(1 + \beta_{51})^2 \geq \alpha_{52}$.

The production of a commodity has a transient response to the rate of change of its constant dollar price. As a result, the long-run dynamic solution to the dynamic specification is the same as the static solution. Then, since $\Delta \ln(P/D) = 0$ implies $\Delta q_k = 0$, the long-run relationship in (9.47) is

$$Q_k = k_3 (P/D)^{\gamma_2} \exp(\sigma_1 T + \sigma_2 W), \qquad (9.48)$$

where $k_3 = \exp[-\alpha_{50}/(\beta_{51} + \alpha_{52})]$, $\gamma_2 = -\Sigma_1^n \gamma_{54+m}/(\beta_{51} + \alpha_{52})$, $\sigma_1 = -\beta_{55}/(\beta_{51} + \alpha_{52})$, and $\sigma_2 = -\beta_{56}/(\beta_{51} + \alpha_{52})$, the expected signs being $\gamma_2 > 0$, $\sigma_1 \leq 0$, and $\sigma_2 < 0$.

9.3 Import Price

The import price of each geographic market was derived from the inverse import supply schedule in Chapter 4. It was shown that a change in the world market price of a commodity will induce a proportional change in the import prices of individual geographic markets. In a dynamic framework, the price of an imported commodity also has a steady-state response to the world market price of that commodity because the rates of change of the import price of a commodity and its world market price must be the same in the long run.

These response characteristics help to determine the dynamic specification for the import price relationship. In the first place, the steady-state response makes the ECM an appropriate specification for the relationship between the import price and the world market price. Secondly, the requirement that the long-run rate of change of the import price be equal to that of the world market price implies that the long-run elasticity should be restricted to unity in the specification.

In accordance with these guidelines, the dynamic specification for the import price relationship can be derived as follows. The first-order stochastic difference equation for the relationship between the import price of a geographic market P_j and the world market price P is given by

$$p_{jt} = \alpha_{60} + \alpha_{61}p_t + \alpha_{62}p_{t-1} + \alpha_{63}p_{j,t-1} + v_{6t}, \quad (9.49)$$

where, as before, lower-case letters denote the logarithms of the corresponding capital letters and the expected signs are α_{61}, $\alpha_{62} > 0$, and $0 < \alpha_{63} < 1$.

Transformation of the equation with an ECM driven by the world market price yields the dynamic specification (see Annex D)

$$\Delta p_{jt} = \alpha_{60} + \alpha_{61}\Delta p_t + \beta_{62}(p_j - p)_{t-1} + v_{6t}, \quad (9.50)$$

where $\beta_{62} = (\alpha_{63} - 1)$ and where the expected signs are $\alpha_{61} > 0$ and $-1 < \beta_{62} < 0$.

Since the rates of change of the import price and the market price must be the same over the long term, $\Delta p_j = g_6 = \Delta p$. Hence the steady-state solution of (9.50) is

$$P_j = k_6 P, \quad (9.51)$$

where $k_6 = \exp\{-[\alpha_{60} + (\alpha_{61} - 1)g_6]/\beta_{62}\}$. The result shows that (1) the elasticity of the import price with respect to the market price is unity and (2) the long-run equilibrium solution depends not only on the level of the world market price, but also on its rate of change, both of which are characteristics derived from the dynamic specification in (9.50).

10
Summary

This part of the study has derived the dynamic specification of trade relationships for adjustment to long-run equilibrium solutions that are consonant with the equilibrium specifications of the commodity trade theory formulated in Part II. The dynamic processes of adjustment have been described by stochastic difference equations. The supply relationships for an exporter of interest and for the world market of a commodity usually have long and complicated lag structures. Since the nature of the response to price changes is central to the dynamic specification of those relationships, the stochastic difference equation framework provides a convenient means by which to move from a general to a specific lag structure.

For other relationships, the ECM offers a particularly appropriate means by which to characterize the data-generating processes within equations of this class. It has been applied to demand relationships where, on the one hand, it yields long-run unitary elasticity of export demand for a country with respect to the import demand of its geographic markets and, on the other, it does not preclude non-unitary elasticity of import demand in the geographic markets with respect to their domestic economic activity. It has also been used in the relationships for import prices, where over the long run the rate of change of these prices must be proportional to that of the world market price for the commodity.

While each equation has been shown to yield a long-run equilibrium solution that is consistent with its static solution in the system of trade equations derived in Part II, there is also a dynamic effect. This effect arises from the existence of a steady-state response of the dependent trade variables to certain explanatory variables. Thus, in dynamic equilibrium these dependent trade variables respond not only to the changes in the level of the explanatory variables, but also to the rate of change of those variables that generate a steady-state response. The relationship between trade theory and observations of trader behaviour is that observed behaviour in trade would eventually converge to steady-state equilibrium growth were all change to cease in the variables used to characterize the underlying processes of trade.

Annex B summarizes the dynamic system of equations derived in this part of the study in order to characterize data-generating processes in international commodity trade.

Part IV

Empirical Analysis of Commodity Trade

Introduction

In this part the dynamic specification of the system of equations used to characterize commodity trade is applied to Latin America's major non-fuel commodity exports. The objective is to test the empirical validity of development trade theories concerning the elasticities of supply and demand for primary commodities in international trade, as well as the theory of imperfect competition in international commodity trade set forth in Part II. To the extent that the dynamics used to characterize the data-generating process of commodity trade in Part III are appropriate, one of the questions to be addressed is whether estimates of the export demand equation reflect product differentiation, and hence whether commodity exporting countries have downward-sloping demand functions. If such differentiation existed, the exporters could exercise a significant, albeit probably limited, degree of control over their competitive position and could influence their market shares.

The commodity exports of Latin America furnish a choice sample with which to evaluate the previously formulated theory of imperfect competition in international commodity trade and the dynamic specification of relationships in the system. On the one hand, the composition of the region's exports is fairly well distributed both between foods and raw materials, and between agricultural and mineral raw materials, a feature that allows comparative analysis across different commodities and thus generalization of findings. On the other, the number of commodity exports that are significant at the regional level is relatively small, and for the most part are those most important in international commodity trade; commodity exports that are nationally, but not regionally, significant are few. Hence the coverage within countries of the region is relatively comprehensive. The analysis of regionally significant commodities permits a comparison across countries that export the same commodity. The result is a sample that lends itself to comparative analyses both across commodities and across exporters of the same commodity.

The comparative nature of the sample provides a means by which to test the validity of past generalizations about commodity trade. Thus the first chapter in this part (Chapter 11) reviews the development trade issues that are to be empirically evaluated in subsequent chapters, and Chapter 12 describes the trade structure and performance of Latin America's major commodity exports. The results of the estimated system of equations are presented in Chapters 13–16. In Chapter 13 empirical estimates are presented of the import demand function of commodities in their principal export markets—the industrialized countries—and these results are compared with other estimates of import demand functions for manufactured goods of the industrialized countries. Chapters 14 and 15 examine the results of empirical estimates of the export demand and supply of developing countries. In describing the results of the demand for exports, the aggregation procedure that needs to be used when the export demand functions of countries are obtained from calculations made at the level of bilateral trade is explained. In describing the export supply functions, estimates for short-run and long-run elasticities are examined, as is the characterization of their lag structures. Finally, Chapter 16 presents empirical estimates of the system of equations used to determine market prices in each of the commodities.

The system of equations as a whole is dealt with in Part V. The first chapter in that part assesses the stability of the system when it is subjected to external disturbances, the way the system converges back to the equilibrium path, and the response of the system to external and internal disturbances. In the analysis of these aspects, one of the main issues that has been raised in the development trade literature is investigated, namely how and to what extent changes in the level of economic activity in the industrialized countries are transmitted to the developing countries. In our system of equations, changes in the economic activity of the industrialized countries are shown to have both a direct effect on import demand and several indirect effects on domestic demand and foreign export supply and demand. As a result, the total impact of changes in the economic activity of the industrialized countries on the commodity export growth of the developing countries can be quite different from the direct effects associated with single-equation estimates of import demand functions, such as those described in Chapter 13. It is particularly important to bear this difference in mind in any examination of the above-mentioned issues.

The ultimate objective of this part of the study is to test the legitimacy of the behavioural relationships formulated as a means of characterizing international commodity trade. Once the inferences about the behavioural relationships and their system interaction are shown to be reliable, the relationships can be used to analyse trade policies in the final part of this study.

11
Testing Development Trade Theories

One of the characteristics most often noted in the development trade literature has been the tendency of world trade in primary commodities to lag behind trade in manufactured goods. Moreover, primary commodity exports from the developing countries have grown at a slower rate than exports of goods of the same type from the industrialized countries. This pattern of growth in traded goods has not altered since Kravis (1970a) made the observation two decades ago.

Table 11.1 shows the trend growth rates of trade in primary commodities and in manufactured goods, and the resulting changes in the composition of trade, from 1960 to 1987. During this period, world trade in manufactured goods expanded 1.5 times faster than trade in primary commodities. In commodity trade, exports by the industrialized countries expanded 1.6 times faster than exports by the developing countries. As a result, the export market share of the developing countries in world trade of primary commodities fell from 34 per cent in the first part of the 1960s to 28 per cent in the first part of the 1980s; the industrialized countries increased their market share of traded commodities from 55 to 63 per cent during the same period.

Despite an aggressive export drive by some developing countries, most of them have found it difficult to capture a larger share of the market. The Latin American countries, however, fared considerably better than did the rest of the developing countries in the last quarter-century. The commodity export growth of this region averaged 4.7 per cent a year, compared with an average of only 3.4 per cent in the other developing regions. Yet the growth rate of commodity exports from Latin America was only 0.77 times that of commodity exports from the industrialized countries. As a result, the commodity market share of Latin America remained virtually unchanged during 1960–87.

11.1 Commodity Trade Experiences and Theories

From the above empirical observations of commodity trade, it is not difficult to understand why Ian Little (1982) noted that the export

Table 11.1 Growth rates of real values of trade and economic activity, 1960–1987
(Average annual growth rates and market shares)

	Growth rates %	Market shares %	
	1960–87	1960–7	1980–7
World: Total merchandise trade	5.8	100.0[a]	100.0[a]
of which:			
Primary commodities	5.1	33.9	17.4
Manufactured goods	7.4	60.9	64.7
Developing countries: Primary commodity exports	3.8	34.0[b]	28.2[b]
of which:			
Latin America	4.7	12.7[b]	12.3[b]
Other developing regions	3.4	21.3[b]	15.9[b]
Industrialized countries: Total merchandise exports	6.8		
of which			
Primary commodity exports	6.1	55.2[b]	63.4[b]
Manufactured exports	6.7		
Imports of primary commodities	5.0	72.2[c]	64.7[c]
Gross domestic product (GDP)	3.6		

[a] The difference between total merchandise trade and the combined share of trade in primary commodities and manufactured goods is equal to trade in fuels (SITC 3) and products not elsewhere classified (SITC 9).
[b] Export market share of primary commodity trade. The difference between world trade in primary commodities and the sum of commodity exports of the developing countries and the industrialized countries is equal to the export market share of the centrally planned economies.
[c] Import market share of primary commodity trade.
Note: Primary commodities are covered by the SITC, Rev. 2, categories 0 + 1 + 2 + 4. Manufactured goods are covered by the SITC, Rev. 2, categories 5 + 6 + 7 + 8.
Sources: Derived from trade value data and unit value data in UN, *Monthly Bulletin of Statistics*, Special Tables (various issues) and IMF, *International Financial Statistics Yearbook* (various issues). Data for real GDP from UN, *National Accounts Statistics: Analysis of Main Aggregates* (various issues).

pessimism mentality of postwar thinking about the developing countries has continued to cast its shadow on development trade theories, even though empirical studies in the 1960s had questioned the validity of this view. For the most part, this pessimism reflects the overall postwar experience of the developing countries.

In the postwar era the world economy started out on a comparatively well-defined steady-state growth path. The international trade system was characterized by the dominance of the USA and the pegging of currencies to the US dollar. For the USA, one of the more serious problems that this situation created was the shortage of dollars. But for the other industrialized countries, especially Japan and the European Community, it helped hold down inflation, which in turn encouraged savings and raised domestic investment, and consequently permitted the expansion of export supplies to help satisfy the growth in these countries' intra-industry trade.

In the developing countries, however, the existing system was less conducive to economic growth and export expansion. Between 1950 and 1969 their average per capita income rose by 2.5 per cent a year, compared with 3.3 per cent in the industrialized countries as a whole. At the same time, exports rose at an average annual rate of 4.8 per cent, versus 7.9 per cent in the industrialized countries. As a result, the market share of developing countries in total world trade declined from 32 per cent in 1950 to 21 per cent in 1969. For the Latin American region the situation was even worse: the region's share of total world exports dropped from 12 to 6 per cent between 1950 and 1969.

Raul Prebisch and Hans Singer, of the United Nations, were among the first to build these events into a general thesis attacking the orthodox Ricardian trade theory of comparative advantage based on differences in production technologies. Starting from the notion that the difference observed in the growth of the developing countries and the industrialized countries had been the result of a problem of interrelations between trade and development, they asserted that technological progress had not been symmetrical for the industrialized (centre) and for the developing (periphery) countries. Improvements in production and marketing processes were transferred to the industrialized countries in the form of lower prices for primary products, and when improvements were introduced in the industrialized countries the resultant lower prices were not transferred to the developing countries. Prebisch (1959) attributed this phenomenon to technological progress, as evidenced by the bias towards substitution of often imported raw material products. Cyclical swings brought on by changes in economic activity in the industrialized countries further aggravated this situation; during an upswing wage rises parallel prices, whereas in a recession they resist downward pressure. This resistance maintains higher incomes in the industrialized countries at the expense of the developing countries.[1] Singer (1950) believed this occurrence took place in basic food imports because of Engel's Law, which states that preferences are not homothetic for goods of this type. Hence observed differences in growth between primary commodity trade and trade in manufactured goods were explained in the Prebisch–Singer thesis as arising from low income and price elasticities of foreign demand for primary commodity exports from the developing countries.

The commodity trade situation was believed to be aggravated by low price elasticities of supply for primary commodities in the developing countries, which were attributed to institutional rigidities in rural exportables and to trade barriers in the export markets. As Diaz-Alejandro (1975: 117–19) pointed out in his comprehensive review of trade and development issues, these two strands of thought generated an export pessimism mentality in the development trade literature by focusing attention on the gap between primary commodity exports by the developing countries and exports of manufactured goods by the industrialized countries.

Notwithstanding initiatives by the developing countries designed to regulate international commodity markets, beginning with the first United Nations Conference on Trade and Development (UNCTAD I) in 1964, the international trade system continued to focus almost exclusively on trade between the industrialized countries, despite the gradual erosion and eventual breakdown of the postwar trade system. During the late 1960s the relative superiority of the USA declined, and the role assigned to the dollar deteriorated as the dollar shortage (partly in response to the Vietnam War) turned into a dollar surplus. In 1971 the international monetary system established in Bretton Woods finally collapsed, when the USA was forced to devalue the dollar. But during the subsequent two-year negotiations on the reform of the international monetary system, the fundamental changes that had occurred in the international order in the preceding 20 years were overlooked. The concluding report of the Committee of Twenty (1974) gave a renewed emphasis to the maintenance of balance only among the industrialized countries as a necessary and sufficient condition for equilibrium in the world economy.

A period of growth and prosperity in commodity trade began with the commodity boom of 1973–4. Increased foreign demand, combined with supply shortages arising from natural disasters and lagged responses to demand changes, caused commodity market prices to rise by an average of more than 40 per cent a year in that period. However, the commodity boom was followed by a secular improvement in commodity trade, and the focus of development trade literature shifted from concerns about the long-term growth of commodity trade to its short-term instability. Spurred by the success of OPEC, developing countries began to clamour for the negotiation of international agreements, while the industrialized countries were forced to enter into negotiations by shortages of, and sharp price rises in, many commodities.

The first concrete proposals for the establishment of a new international economic order were adopted at UNCTAD IV in Conference

Resolution 93(IV) on the Integrated Programme for Commodities (IPC). The principal objective of the IPC was to achieve 'stable conditions in commodity trade, including avoidance of excessive price fluctuations', by means of 'pricing arrangements, in particular negotiated price ranges' (UNCTAD 1976: 3–4). The arrangements were to operate through supply management, principally in the form of commodity stockpiling, or through production controls and export quotas.[2] A common fund was to serve as the main financial support of individual commodity stocking agreements, but it was not until the meeting of UNCTAD VII in 1987 that the required two-thirds of the funds needed for stockpile financing was almost met. During the period 1976–87 stabilization schemes were successfully negotiated for sugar, coffee, and cocoa. It was due largely to the success of the coffee agreement that prices did not plummet in the 1980s.

Empirical work on commodity trade in the 1970s was concerned with three major issues: the magnitude of short-term instability in commodity trade, the consequences of export instability for economic growth, and the impact of price stabilization on producing and consuming countries. Coppock (1962) and MacBean (1966) had found that the degree of export instability in developing countries had not been significantly different from that in the industrialized countries. Conflicting results were reported by Massell (1970), Naya (1973), Glezakos (1973), Mathieson and McKinnon (1974), Knudsen and Parnes (1975), Stein (1977), and Sheehey (1977), all of whom found that developing countries do suffer from more instability than the industrialized countries. Conflicting results were also found for the consequences of instability after MacBean (1966) questioned intuitive assertions that had been made regarding the adverse effects of export instability on the economic growth of the developing countries. The studies of Glezakos (1973) and Massell *et al.* (1972), among others, supported the traditional view that export instability can inhibit economic development, while Rangarajan and Sundararajan (1976) found conflicting results in their analysis for individual countries. Finally, the comprehensive empirical econometric analysis by Behrman (1977) on the benefits and costs of price stabilization schemes under the IPC was followed by a series of other studies (see the survey by Labys 1980) that culminated in the work by Newbery and Stiglitz (1981) emphasizing the effects of risk from price instability in commodity markets.

In their theoretical work, Newbery and Stiglitz (1981, 1982) argued that optimal intervention rules, as first developed by Gustafson (1958), could be derived so as to achieve well-defined objectives for economic agents in commodity markets, and that these rules would provide a better approach to the design of price stabilization policies than other types of intervention rules that maintained price movements within a band width. Ghosh *et al.* (1987), using an econometric model of the world copper market, verified the practical superiority of optimal intervention rules over a wide variety of conventional band width rules. They also corroborated the finding of previous studies which showed that, although price stabilization schemes can reduce the variability of developing countries' export earnings, they do not significantly influence the long-term growth of those earnings.

The 1981–2 world-wide recession renewed interest in the long-term growth of commodity trade. Initially, it was thought that the 10 per cent average annual drop in commodity prices during this period and the 7 per cent contraction in world commodity trade in 1983 was cyclical rather than secular in nature. However, commodity markets did not recover and another contraction in trade occurred in 1985. In view of a trend growth rate in the value of commodity exports by the developing countries of only 1.4 per cent a year between 1981 and 1987 compared with a trend growth rate of 4.6 per cent a year for manufactured goods exported by the industrialized countries, empirical work on trade and development again focused on the observed divergence in the long-term growth of the developing countries and the industrialized countries.

The next two sections of this chapter examine the arguments that continue to be advanced to explain the comparatively slower expansion of commodity exports from the developing countries. Those who have put the blame on demand conditions maintain that there is a relatively low foreign income elasticity of export demand; those who blame supply conditions argue that there is a relatively low price elasticity of export supply. The assumptions underlying these arguments are examined in this chapter, and the validity of the assertions are assessed in subsequent chapters in the light of the empirical results presented.

11.2 Characterization of Demand

The export pessimism view about demand conditions is well laid out by one of its main proponents, Sir Arthur Lewis (1980), in his 1979 Nobel lecture:

We need no elaborate statistical proof that trade [of the less developed countries (LDCs)] depends on prosperity in the industrial countries ... More interesting is that the coefficient [of commodity export growth of the developing countries to economic growth of the more developed countries (MDCs)] is less than one, viz. 0.87. This means that if the engines of growth were industrial production in MDCs and exports of primary products in LDCs, then the MDC engine was beating slightly faster than the LDC engine.

Lewis's thesis that there is a low demand for the primary commodities of the developing countries can be illustrated by the data presented in Table 11.1. According to these data, without other influences at work, i.e. if supply were to be perfectly elastic with respect to price and if the price elasticity of export demand of the developing countries were infinite, then, for every 1 per cent increase in real gross domestic product (GDP) of the industrialized countries, the demand for primary commodities from the developing countries grew by 1.1 per cent from 1960 to 1987; in contrast, under the same assumptions about export supply and demand elasticities, the demand for manufactured goods from the industrialized countries rose by 1.9 per cent for every 1 per cent increase in real GDP during this period.

However, the *ceteris paribus* condition needed to arrive at this conclusion relies on three stringent assumptions: (1) that the price elasticity of the export supply of developing countries is very high, or completely elastic; (2) that the price elasticity of the import demand in the industrialized countries is very low, if not completely inelastic, so that changes in commodity market prices resulting from income changes in the industrialized countries have little or no effect on the quantity of import demand; and (3) that the price elasticity of the export demand of these countries is infinite, so that changes in the import demand bring about proportional changes in the export demand of all foreign suppliers. If we remove these assumptions and allow for less than perfectly elastic export supply and demand functions and feedback effects between market prices and import demand in the industrialized countries as a result of less than perfectly inelastic import demand functions in these countries, it is uncertain whether Lewis's argument would hold. Casual observation about the weak transmission of economic activity growth in the industrialized countries to demand for primary commodities can be misleading. The only way in which the legitimacy of the assertion regarding the low income elasticity of demand for the commodity exports of developing countries can be properly evaluated is to compute the income elasticities of demand for primary commodities in the industrialized countries and to compare these estimates with those calculated for manufactured goods. This is the approach that has been adopted in Chapter 13 for a sample of commodities.

None the less, direct estimation of the income elasticity of import demand can be misleading, since the link between economic activity in the industrialized countries and demand for commodities can be bipolar. When the general level of economic activity in all the industrialized countries expands or contracts at the same time, changes in economic activity can have opposite influences on the demand for commodities. For example, as economic activity in the industrialized countries increases, the demand for commodity imports is likely to expand. However, the increase in economic activity will simultaneously induce higher market prices, and consequently import prices, and higher prices will exert downward pressure on the quantity of imports demanded. As a result, the direct positive effect on import demand is dampened by the indirect negative effect on import demand induced by variations in market prices when there are concurrent changes in economic activity in the industrialized countries. These considerations suggest that direct equation estimates do not provide an adequate measure of the transmission of economic expansion in the industrialized countries to demand for primary commodities. In Chapter 18, multiplier analysis is used to calculate the foreign income multiplier for a sample of commodities. The difference between these calculations and those obtained from elasticity estimates in the import demand relationship in Chapter 13 is that influences arising from changes in income that are sufficiently widespread to affect market prices are also captured. In Chapter 13, estimates are obtained of influences arising from changes in income in one or more foreign markets that do not perceptibly affect market prices.

The demand for exports can also be influenced by relative-price changes in exporting countries that compete in international commodity markets. The importance of this factor has been underscored in the recent theory of trade characterized by commodity differentiation and imperfect competition. In a critique of the rigid link established by Lewis (1980) between the export performance of primary commodities and the prosperity of industrialized economies, Riedel (1984) has noted the failure of such studies to consider the capacity of exporters 'to engage in price competition by which they could expand exports despite slowdown in developed countries by claiming a larger share of the DC market'. Riedel's empirical analysis demonstrated the weakness of the link between commodity trade and economic activity in the industrialized countries. Yet he did not analyse the influence of relative-price changes on developing country exports because, as he observed,

The appropriate procedure for measuring the link between MDC export growth and LDC export growth ... is simultaneous estimation of supply and demand equations for LDC exports. The enormous complexity of demand and supply relationships, however, makes this approach methodologically infeasible and generally restricts analysis to highly reduced-form relationships.

Chapter 15 and Part V tackle the issue of the effect that relative-price changes can have on the demand for developing country exports of primary commodities by adapting the model of commodity trade developed in Parts II and III to a disaggregated sample of countries and simultaneously simulating the system of equations. The results demonstrate the extent to which price competition can be used to influence the export market shares of a country and thereby serve to stimulate its growth rate despite a possibly sluggish overall demand for its commodity exports in foreign markets.

11.3 Characterization of Supply

The structuralist explanation of the weak growth of commodity exports from the developing countries is that producers have had a weak response to real price improvements. The most notable proponent of this view is Cairncross (1961), who argued that in the short run the commodity export supply of developing countries is highly inelastic, and that this characteristic induces sharp price fluctuations whenever demand changes. In the long run, he asserted, the quantity of exports supplied is unresponsive to growth in demand because of conservative social and governmental attitudes and practices that withstand transformation and dampen external impulses. However, this inertia need not be caused by government intervention measures that distort market price signals; recent work has established that the efficiency of the private sector can be raised by government intervention when there is imperfect information on the market. (For a proof, see Greenwald and Stiglitz 1988.) The key point is that market signals are not transmitted to producers and that this isolation of producers from the market causes export supply to be unresponsive to market price movements.

That this view of the sluggish response of producers to price changes continues to influence interpretations of the export performance of the developing countries is reflected in the empirical studies by Kravis (1970*a*), Cohen and Sisler (1971), and Hanson (1977). All these studies attributed the decrease in the export market shares of developing countries to a low price elasticity of export supply relative to that of the industrialized countries. In Kravis's (1970*a*) critique of proponents of trade as an engine, rather than a handmaiden, of economic growth, he found that successful export growth by the developing countries was positively associated with increased market shares of traditional exports (see also Kravis 1970*b*). Kravis suggested that the comparatively poor overall export performance of the developing countries relative to that of the industrialized countries has been due not to weak external demand, but rather to unfavourable trade policies and other internal factors that have impeded mobility of resources. Consequently, he concluded that the developing countries were unable to increase their market shares as demand for their commodity exports grew because the industrialized countries, which compete in many of the same non-tropical commodity exports, were more successful in mobilizing their resources, and thus in increasing their market shares.

Using Kravis's approach, Hanson (1977) examined developments in

1860–1900, a period when demand expansion stemming from increased world income and population was believed to govern international trade developments. He too concluded that the main constraint on the exports of the developing countries had been internal influences. The same explanation was offered by Cohen and Sisler (1971) for the decline in the shares of the developing countries in the markets for commodity exports during the 1960s. All these studies suggested that domestic policies had caused production to be less responsive to market improvements in the form of higher prices.

The aforementioned empirical analyses have rested on the fundamental assumption that the demand for exports of individual developing countries is infinitely elastic with respect to price. In all the studies, any increase in the export market share of a country has been explained by a higher price elasticity of export supply relative to that of competing suppliers; conversely, a decrease in a country's market share has been ascribed to a lower price elasticity of export supply compared with that of other producing countries. Thus, any change in the market share of a country originates from a relatively greater increase in its export supply than in that of other countries as a result of an overall expansion in demand. For example, suppose there are two suppliers of a commodity, one a developing country producer and the other an industrialized country producer, and that the price elasticity of the export supply of the developing country is lower than that of the industrialized country. Then, as the trend growth of demand increases, the market share of the industrialized country will rise, while that of the developing country will fall.

However, when the price elasticity of the export demand is less than infinite, changes in market shares cannot be ascribed solely to differences in the price elasticity of the export supply. In such cases, changes in market shares can be either supply-induced or demand-induced. They are demand-induced when export supply shifts and relative-price variations cause a change in the quantity of the exports demanded. They are supply-induced when demand shifts produce different responses in the quantity of exports of a commodity supplied by alternative supplying countries. Consequently, product differentiation in international commodity trade invalidates the use of market share analysis for determining whether differences in import demand or export supply responses account for changes in the market shares of countries.

Instead, empirical econometric analysis offers an appropriate means by which to infer whether the supply response of commodity exports from the developing countries has been responsible for observed differences in their export performances, and in that of the developing countries as compared with the industrialized countries. In this approach, the price elasticity of export supply is calculated and the estimates are compared, first, at a disaggregated level for the developing countries in order to identify possible sources of differential export growth, and second, at an aggregate level for the developing countries, with comparable estimates for the export supply of manufactured goods of the industrialized countries, in order to ascertain whether an overall growth of foreign demand in the industrialized countries leads to a greater or smaller increase in commodity exports than in exports of manufactured goods from the industrialized countries. This is the approach adopted in the analysis of the supply of commodity exports in Chapter 14.

11.4 Testing Development Trade Theories

Although the specific nature of the factors governing the composition and direction of exports remained largely undefined until the recent emergence of trade theories based on product differentiation and imperfect competition, considerable progress was made in identifying and measuring the major determinants of the comparative export performance of countries by linking empirical theories explaining the structure of trade to those explaining the level of trade.

Yet such progress has been almost wholly limited to the analysis of the exports of the industrialized countries. This type of analysis initially emerged from efforts to explain the changing comparative advantage in manufacturing exports of the USA and the UK (see, *inter alia*, Baldwin 1962; Ginsburg 1969), the postwar emergence of Japan in the world economy (Narvekar 1960), and, more recently, the export-oriented industrialization of the East Asian newly industrialized countries (NICs) (see the survey by Meier 1989). Attempts to analyse empirically the export performance of the developing countries have been scarce. While the cited studies pointed to supply constraints as the main impediment to export growth, those of Naya (1968), Banerji (1974), and Jepma (1986) attributed most of the slow growth of the developing countries to demand-related factors.

However, these studies have relied on CMS analysis, which Richardson (1971*a*, 1971*b*) has shown to have a number of severe theoretical and empirical limitations. They include the following: (1) the empirical results depend on the order in which the 'commodity composition effect' and the 'geographic distribution effect' are calculated; (2) the identity is unable to determine whether changes in the 'competitive effect' are supply- or demand-related; and (3) the results of the identity, which are based on calculations in two discrete time periods, are arbitrarily dependent on the periods selected.

128   *Empirical Analysis*

Empirical econometric analysis is used in this part of the study to identify the major influences on commodity trade. The next chapter delimits the commodity and country coverage of the empirical analysis, and examines the patterns of trade of the commodities and those characteristics that have a bearing on their supply and demand. The results of the calculations of the import demand functions and export demand and supply functions are presented in the following three chapters; those of the other relationships that determine market prices are analysed in Chapter 16. In the final chapter of this part of the study the development trade theories examined in this chapter are assessed in the light of these empirical results. In Part V, the empirical econometric models are used to assess the viability of alternative trade policies.

Notes

1   A closely related hypothesis has recently been put forward by Kaldor (1976) and termed the 'ratchet-effect' by Cooper and Lawrence (1975). It suggests that primary commodity price rises are passed on to the final product and have a deflationary impact on the effective demand for goods in the industrialized countries. Since market imperfections in these countries prevent declines in prices of basic products from having a symmetrically downward effect on final products, any significant change in primary commodity prices eventually has a dampening effect on industrial activity. However, the validity of this hypothesis has been questioned by Finger and DeRosa (1978) on the grounds that empirical support for downward price inflexibility of final products is unsubstantiated.
2   In their historical review and comprehensive documentation of the events leading up to the IPC, Erb and Fisher (1977: 482–3) maintained that the major impediment to the success of single-commodity agreements appeared to be the dominance of the USA in overall consumption. This dominance led the US government to resist international commodity agreements. According to Vastine (1977: 460), the eventual commitment of the USA to the IPC was, in fact, the result of a misunderstanding. During the final days of negotiations, the chief US spokesperson at the conference had led the intensive consultation groups to expect from his government a more forthcoming stance on the resolution than was, in fact, acceptable to the various foreign policy agencies in Washington, so that when the language of the draft resolution went beyond the US position, the US negotiator was forced to join in the consensus, though with reservation.

12
Commodity Exports of Latin America

The empirical analysis of international commodity trade will be conducted with a sample of data that cover Latin America's major commodity exports. The Latin American region depends on a relatively small number of primary commodity exports for its foreign exchange earnings. Of them, ten non-fuel commodities are important,[1] since they each contribute at least 1 per cent to the total export earnings of the region:[2]

Coffee	10.0%
Soybeans	5.9%
Copper	3.5%
Sugar	2.9%
Iron ore	2.8%
Beef	1.8%
Cotton	1.6%
Bananas	1.4%
Cocoa	1.3%
Maize	1.2%

These primary commodities provide the basis for a comparison of cross-country experiences in each commodity, and of experiences in different types of primary commodities, since they are well distributed both between foods and raw materials, and between agricultural and mineral raw materials.[3] (For a comprehensive description of early patterns of trade in Latin America's primary commodity exports, see Lord and Boye 1991, and Grunwald and Musgrove 1970.)

The contribution of these commodities to the total exports of each country in the region is shown in Table 12.1. The countries with a high degree of reliance on the commodities (deriving at least half of their total export earnings from them) are the Central American countries, Chile, Colombia, the Dominican Republic, and Paraguay. The countries with a moderate degree of reliance (deriving between 30 and 49 per cent of their total export earnings from them) are Brazil, Ecuador, Guyana, Haiti, Panama, and Peru. The remaining countries in the region often rely on another primary commodity for much of their export earnings (for example, Jamaica on bauxite, and Mexico on petroleum).

Table 12.1 Percentage contribution of major commodity exports of Latin America[a] to total exports, 1970–1987 average (Percentages and indices)

Country	Beef	Maize	Bananas	Sugar	Coffee	Cocoa[b]	Soy-beans[b]	Cotton	Iron ore	Copper[b]	Total of 10 goods	Total merch. exports	Index of commodity concentration[c]
Argentina	7.2	10.3	–	1.5	–	–	7.2	1.0	–	–	27.2	100.0	50.2
Barbados	–	–	–	21.3	–	–	–	–	–	–	21.3	100.0	100.0
Brazil	0.7	0.5	0.1	4.3	13.5	3.4	10.0	0.8	7.6	0.1	41.0	100.0	40.2
Chile	–	–	–	–	–	–	–	–	4.0	46.9	50.9	100.0	86.3
Colombia	1.1	–	3.1	2.0	54.2	–	–	2.4	–	–	62.8	100.0	86.5
Costa Rica	6.6	–	21.7	3.0	29.6	1.2	–	–	–	–	62.2	100.0	60.2
Dom. Rep.	1.0	–	0.2	35.0	11.0	7.3	–	–	–	–	54.6	100.0	68.6
Ecuador	–	–	11.3	0.9	10.6	7.7	–	0.1	–	–	30.5	100.0	55.3
El Salvador	0.5	0.2	–	4.2	54.5	–	–	8.8	–	–	68.3	100.0	81.4
Guatemala	2.4	–	4.2	6.6	36.3	0.4	–	9.9	–	–	59.9	100.0	64.6
Guyana	–	–	–	34.2	–	–	–	–	–	–	34.2	100.0	100.0
Haiti	0.9	–	–	2.5	34.2	2.7	–	–	–	–	40.2	100.0	85.7
Honduras	5.5	–	28.5	2.1	24.7	–	–	1.1	–	–	62.1	100.0	48.2
Jamaica	–	–	2.0	9.6	0.6	0.6	–	–	–	–	12.9	100.0	76.7
Mexico	0.6	0.1	–	1.2	6.0	–	–	3.1	–	0.6	11.6	100.0	58.7
Nicaragua	8.6	0.1	2.5	4.7	26.4	–	–	23.3	–	0.2	65.7	100.0	56.3
Panama	0.9	–	27.0	10.0	3.2	0.5	–	–	–	–	41.7	100.0	69.4
Paraguay	3.5	0.1	–	1.1	1.4	–	18.2	25.8	–	–	50.1	100.0	62.0
Peru	–	–	–	3.9	5.8	2.6	–	2.8	4.4	20.8	40.3	100.0	56.5
Uruguay	18.6	–	–	–	–	–	–	0.1	–	–	18.7	100.0	98.2
LATIN AMERICA	1.8	1.2	1.4	2.9	10.0	1.3	5.9	1.6	2.8	3.5	32.4	100.0	40.7

| Index of exporter concentration[d] | 46.0 | 89.8 | 42.6 | 43.4 | 45.8 | — | 68.3 | 61.1 | 35.2 | 76.9 | 76.1 | 26.3 |

[a] Major commodity exports are defined as those regionally significant, non-fuel goods that during 1970–87 accounted for an average of at least 1% of the total value of merchandise exports.
[b] Cocoa consists of cocoa beans, cocoa powder, and cocoa paste; soybeans is made up of soybeans and soybean cake; copper is composed of copper concentrate and refined copper.
[c] Measured by the Gini–Hirschman coefficient of concentration

$$GH_j = \left[\sum_{i=1}^{10} (X_{ij}/X_j)^2\right]^{1/2} \times 100,$$

where X_{ij} is the export value of the ith country in the jth commodity and X_j is the total export value of the 10 commodities by country. Calculated for countries in which the commodity accounted for at least 1% of merchandise exports; regional total includes all exporters of the commodity.
[d] Measured by the Gini–Hirschman coefficient of concentration

$$GH_i = \left[\sum_{j=1}^{20} (X_{ij}/X_i)^2\right]^{1/2} \times 100,$$

where X_{ij} is the export value of the ith country in the jth commodity and X_i is the total export value of the 10 commodities by region. Calculated for commodities that accounted for at least 1% of total merchandise exports.

Note: Product contributions to total exports of over 5% have their values in italics.

Sources: Commodity data from UN Food and Agriculture Organization (FAO) *Trade Yearbook*, and UN Conference on Trade and Development (UNCTAD) *Yearbook of Commodity Trade Statistics*; data for total value of merchandise exports from International Monetary Fund (IMF) *International Financial Statistics*.

132 *Empirical Analysis*

Most of the Latin American countries tend to specialize in only a few of the ten commodities. The commodity concentration indices, shown in the last column of the table, have a mean average value of 70 for the countries in the region (the standard deviation is ±18). However, there is no correlation between a country's dependence on the region's major commodity exports and the degree of its commodity concentration within this group of goods. Thus, the exports of the nine countries that rely the most on these ten commodities are, on the whole, as diversified as those of the other countries in the region. Furthermore, as Kravis (1970*a*) found for a sample of 21 countries from various regions in the world, there is no correlation between the level of commodity concentration and the growth performance of total exports in the Latin American countries.

12.1 The Principal Export Markets

Just as Latin America's exports are concentrated in a few commodities, the destination of these exports is concentrated in a small number of markets. The industrialized countries are the principal trading partners of the countries in the region and absorb more than two-thirds of their commodity exports. Of the combined exports of these goods, the USA is the principal market of the region, followed by the EEC and Japan. Although the destinations of Latin America's commodity exports have not changed radically over the years, there has none the less been a gradual shift since the 1960s to Western European countries that are not members of the EEC, and to the centrally planned economies.

Shifts have also taken place within the pattern of the traditional markets themselves (see Table 12.2). Latin American exporters of iron ore and bananas have increasingly diverted their supplies from the Japanese market to US and EEC markets. In contrast, an increased proportion of copper shipments has been switched from the USA and the EEC to Japan. The composition of the export markets for the region's beverages has also changed. The centrally planned economies absorbed over 25 per cent of Latin America's cocoa exports in 1975–85, compared with less than 17 per cent in 1960–74. The preference of coffee consumers in the USA, the world's leading importer of coffee, for the robusta-type coffee produced in Africa caused a drop in Latin America's market share by 15 percentage points between 1960–74 and 1975–85. In contrast, the Latin American countries increased their share of the coffee market in the EEC, as well as in the non-traditional markets of non-EEC European countries and Japan. The geographic distribution of other commodity exports of the region, e.g. cotton, has remained virtually unchanged, and the EEC has continued to be the

Table 12.2 Distribution of Latin America's major commodity exports by principal geographic market[a]
(Percentages)

	USA	Western Europe EEC	Western Europe Other	Japan	Centrally planned economies
TOTAL					
1975–85	52.1	41.8	0.5	4.6	0.9
1960–74	60.7	31.6	0.3	7.1	0.3
Beef					
1975–85	52.7	47.3	—	—	—
1960–74	41.1	58.9	—	—	—
Maize					
1975–85	—	100.0	—	—	—
1960–74	—	100.0	—	—	—
Bananas					
1975–85	79.9	19.5	—	0.6	—
1960–74	73.2	19.3	—	7.5	—
Sugar					
1975–85	84.6	15.4	—	—	—
1960–74	83.6	16.4	—	—	—
Coffee					
1975–85	58.7	39.3	1.1	0.8	—
1960–74	73.5	25.7	0.7	—	—
Cocoa					
1975–85	43.7	25.7	—	3.0	27.6
1960–74	59.8	20.1	—	3.3	16.8
Soybeans					
1975–85	—	100.0	—	—	—
1960–74	—	100.0	—	—	—
Cotton					
1975–85	16.1	41.2	—	42.7	—
1960–74	18.4	38.7	—	42.8	—
Iron ore					
1975–85	13.6	79.4	—	7.1	—
1960–74	21.2	51.8	—	26.9	—
Copper					
1975–85	50.6	33.4	—	16.0	—
1960–74	63.2	25.9	—	10.9	—

[a] Principal geographic markets are defined as those country destinations that absorbed at least 5% of a country's exports in 1970–85.

Source: National statistical offices (see Annex A for details).

dominant market for both maize and soybeans. Trade policies in the industrialized countries often limit market access, and in Chapter 20 we shall examine in detail the various ways in which these markets have regulated their imports of the commodities covered by this study.

12.2 Comparative Performance of the Region's Exports

Notwithstanding the attempts the Latin American countries have made to diversify their markets (see Labys and Lord 1990), the industrialized countries continue to absorb most of the region's major non-fuel commodity exports. As a result, the export performance of the region is closely associated with the growth in imports of this group of countries. Historically, the growth in imports of the industrialized countries has exceeded that of their overall economic activity. As Table 12.3 shows, the import growth rate of the region's major commodities in these countries has been substantially below their 3.6 per cent average annual rate of economic growth in 1961–87. The trade-weighted average rate of import growth of these commodities was only 2.2 per cent between 1961 and 1987, although the unweighted average was significantly higher (3.3 per cent), but still below the real GDP growth rate. The import growth rate of Latin America's commodities was considerably below the industrialized countries' overall 6.5 per cent average annual rate of import growth, and even further below the 7.8 per cent average annual growth rate of imports of manufactured goods.

These overall trends obscure differences in the performance of individual commodities, however. In some cases the differences have been dramatic. The growth of imports of beef and soybeans by the industrialized countries has been as high as the growth of their imports of manufactured goods. The reason for the overall sluggish growth of imports of Latin America's commodities is that the real value of sugar, which has had a stagnant growth, and that of coffee, which averaged only 2 per cent a year between 1961 and 1987, have accounted for nearly one-third of the real value of imports of the region's major commodities. The industrialized countries also recorded above-average import growth rates for iron ore, copper, bananas, and cocoa.

There have been even greater differences in the growth rates of Latin America's exports of individual commodities than in those of the imports of these goods by their principal markets. It is the large real value of sugar and coffee exports in the overall total of the commodities—their combined total amounting to more than one-half of the overall total—and the low rates of export growth of these two commodities that caused the region's overall export growth rate to be so low in 1961–87. Indeed, the growth rate of soybeans, maize, and iron ore exports from the region has greatly surpassed that of imports of all types of goods by the industrialized countries. In fact, the other commodities of the region, except cotton, had growth rates exceeding the trade-weighted average of the commodities.

Notwithstanding these differences, the performance of Latin America's commodity exports has been closely related to economic growth in foreign markets. In the 1960s, a period in which the developing countries recorded unexpectedly large gains in their export earnings, according to Cohen and Sisler (1971), the unweighted average growth rate of Latin America's major commodity exports was about 2.5 times that of real GDP growth in the industrialized countries. In the 1970s, and again in the 1980s, the region's growth rate of these exports decelerated. Yet the unweighted average annual growth rate of Latin America's major commodity exports remained 2.5 times that of real GDP growth in the industrialized countries.

Even though Latin America has generally maintained a high ratio of export growth to economic growth in its foreign markets for many commodities, the region's market shares fell between the 1960s and the 1980s. Latin America increased its market shares only of soybeans, iron ore, and bananas between the 1960s and the 1980s; the region's share of the beef, maize, coffee, and cotton markets eroded dramatically. There was also a decline in its share of the copper market, although by a somewhat smaller amount. Accordingly, the growth in exports of these commodities did not keep up with that of total world trade in them. In temperate zone agricultural commodities, as well as in copper, the decline in the region's market shares was accompanied by increased penetration into these markets by the industrialized countries.

These observations concerning Latin America's comparative export growth in the above-mentioned commodities obscure differences in the export performance of countries within the region. For instance, the loss of the region's market share of total world trade in copper was due to the drop in Chile's market share from 24 per cent in the first part of the 1960s to 19 per cent in the 1980s; Peru maintained its share of the copper market. Thus, there have also been considerable differences in the performances of countries exporting the same commodity.

The foregoing analysis of the major commodity exports of Latin America underscores the diversity of experiences of commodities. Maizels's (1968: 68) observation, made over two decades ago, is still germane to generalizations about trade patterns: 'Once we get away from the overall totals, an outstanding feature [of commodity trade] is the remarkable diversity of experience, between commodities and between countries.' For the commodities discussed in this chapter, the principal behavioural relationships in trade will be examined in the remainder of this part of the book. This analysis will help to explain the

Table 12.3 Growth rates of real values of imports and economic activity of the industrialized countries and of major commodity exports of Latin America, 1961–1987
(Average annual growth rates and market shares)

	1961–87	1961–69	1970–79	1980–87
Industrialized countries				
Total imports	6.5	9.4	6.0	3.2
Manufactured goods	7.8	11.6	7.0	3.5
Latin America's major commodity exports	*2.2*	*3.5*	*1.9*	*1.2*
of which:				
Beef	7.4	11.8	5.9	3.2
Maize	2.2	19.8	−1.3	−6.1
Bananas	3.6	5.4	2.6	3.3
Sugar	−0.6	0.8	−0.5	−2.0
Coffee	2.0	1.5	2.3	2.1
Cocoa	2.6	4.1	0.5	5.2
Soybeans	9.5	12.6	11.5	4.2
Cotton	−0.7	−1.3	−1.1	1.5
Iron ore	3.9	6.8	3.1	0.2
Copper	3.5	7.3	2.2	2.5
Real GDP	3.6	4.9	3.4	2.3

Commodity Exports of Latin America 137

					World market shares		
					1960–69	1970–79	1980–87
Latin America							
Exports of major commodities	2.5	3.4	1.8	2.3	23.2	20.1	18.3
of which:							
Beef	2.3	7.8	6.1	−12.3	22.4	12.6	8.1
Maize	12.6	10.6	10.3	19.4	11.2	7.3	6.2
Bananas	3.2	7.7	2.1	−1.9	30.3	31.4	30.8
Sugar	2.5	1.1	4.7	0.9	18.3	20.5	15.5
Coffee	1.7	1.7	0.7	3.5	63.1	51.8	52.4
Cocoa	2.8	0.8	3.8	4.1	14.1	16.7	15.5
Soybeans	51.1	68.8	42.3	39.3	1.6	9.9	6.8
Cotton	0.7	9.1	−3.8	−4.5	16.9	12.3	9.1
Iron ore	10.4	13.4	11.6	4.3	6.5	12.8	17.3
Copper	6.4	8.2	6.1	4.1	23.3	19.4	19.2

Sources: Total import data from IMF, *International Financial Statistics Yearbook* (1988), and IMF, *International Financial Statistics* (April 1989). Data for manufactured goods from UN, *Monthly Bulletin of Statistics* (various issues). Data for imports of Latin America's major commodities from FAO trade tapes, and UNCTAD, *Commodity Yearbook* (1988). Data for real GDP from UN, *National Accounts Statistics: Analysis of Main Aggregates* (various issues). Data for Latin America's major commodity exports from national statistical offices.

138   *Empirical Analysis*

comparative export performance of commodities and of countries exporting the same commodities, as well as empirically to assess the development trade theories discussed in the previous chapter.

Notes

1 An additional 13 commodities are significant for some countries in the region in that they contribute 5% or more to their total export revenues. The nationally significant, but not regionally significant, non-fuel exports of Latin America are:

> Bauxite and alumina (in Guyana and Jamaica)
> Ferronickel (in the Dominican Republic)
> Fishmeal (in Peru)
> Hides and skins (in Argentina and Uruguay)
> Lead (in Peru)
> Rice (in Guyana)
> Shrimp (in Mexico and Panama)
> Silver (in Peru)
> Timber (in Paraguay)
> Tobacco (in the Dominican Republic)
> Wheat (in Argentina)
> Wool (in Argentina and Uruguay)
> Zinc (in Peru)

2 Average contribution to the total regional value of merchandise exports in 1970–87. A relatively long time span was used to measure the degree of importance so as to avoid transient influences on the commodity selection.

3 Annex A describes the criteria used to specify the four parameters of the data: the commodities, the exporting countries, the geographic markets, and the period of analysis. It also describes the data and specifies their sources.

13
The Demand for Commodity Imports

This chapter examines the response of primary commodity imports to changes in incomes and prices in the industrialized countries. The analysis is based on the import demand schedule derived in Chapter 3, which was shown to have a unitary elasticity with respect to income and a negative elasticity with respect to price. These results are consistent with expectations for price and, to a lesser extent, income. Murray and Ginman (1976) found the income elasticity of demand for aggregate imports of the industrialized countries to be near unity, a conclusion that was consistent with the fact that the share of aggregate imports in real GDP had remained relatively stable during most of their sample period. The same observation holds for imports of primary commodities in the industrialized countries, which, in 1960–87, increased at about the same rate as real GDP.

Nevertheless, generalizations about aggregate imports do not necessarily apply to their components. For instance, the demand for imports of manufactured goods in the industrialized countries has generally increased at a much faster rate than real GDP growth in those countries. (For a recent survey of the literature, see Goldstein and Khan 1985.) For this reason, the dynamic specification of the import demand relationship was formulated in Chapter 8 in such a way as to test whether or not the demand for imports of particular commodities has, in fact, had a long-run unitary elasticity with respect to the economic activity of their markets.

The first part of this chapter analyses the income elasticity of demand for commodity imports, as well as the dynamics underlying the response of import demand to income and price changes. Import prices were shown in Chapter 4 to have a unitary elasticity with respect to the world market price of the commodities. The second part of this chapter examines the dynamics associated with the adjustment of import prices to changes in world market prices.

13.1 Import Demand

The import demand functions of the principal geographic markets of Latin America have been derived from estimates of the dynamic specification of the relationship in equation (8.12). As was to be expected, income has always been found to be statistically significant in explaining the level of demand for commodity imports. In most cases, estimates of the income coefficients have 99 per cent confidence intervals. Prices are statistically different from zero in 40 of the 52 markets, and, of these, one-half are significant at the 1 per cent level, nearly one-fourth are significant at the 5 per cent level, and the remainder are significant at the 10 per cent level.

The coefficients of the error-correcting terms in the import demand relationships are, in general, close to unity in absolute terms. This fact reflects the relatively quick response of importers to changes in income and prices. Most beef, coffee, cocoa, soybeans, iron ore, and copper importers respond fairly quickly to temporary disequilibria in their markets. Several importers of maize, sugar, and cotton adjust to income and price changes very slowly, a characteristic that is reflected in near-zero coefficients of the error-correcting term. In these cases, Granger (1986) has noted that 'error-correcting' can be a somewhat too optimistic title for the term, since it can take a great deal of time for import demand to resume its long-run equilibrium growth path when a short-run disequilibrium arises between import demand and income.

Major disturbances in import demand do not occur as often as they do in the export supply of primary commodities. When they did occur in the period for which the equations were estimated, the frequency of disturbances in the markets was the same for all commodities, and the disturbances were fairly evenly distributed over the sample period. The exceptions were the periods 1974–5 and 1980–1, when generally large movements in commodity prices induced consumer shifts to substitutes. However, these were transitory shifts. A test of parameter constancy based on the Chow test showed the coefficients to be stable at the 5 per cent level of significance in all the estimated relationships.

A comparison of income and price elasticity estimates of import demand for the ten major exports of Latin America with one another and with similar estimates for other types of goods reveals some important differences. These are described below. However, such comparisons are possible only if the aggregation of income elasticities, and in certain cases of price elasticities, is adjusted by so-called 'distribution elasticities', since otherwise possible biases may arise. The following section describes the aggregation problem and its remedy in the import demand function. The next two sections examine the empirical findings on the income and price elasticities of import demand.

13.1.1 Aggregation

Trade-weighted average income elasticities of import demand have been adjusted for non-uniform changes that occur in the incomes of geographic markets. These adjustments apply to both the aggregate income elasticity of a commodity for all importing countries and the aggregate income elasticity of all commodities imported by a particular country. They also apply to trade-weighted averages of price elasticities for different goods, but not to those for the same good, since changes in import prices in response to world market price changes are uniform across countries.

Consider the aggregate income elasticity for a commodity. When real income variations differ among importing countries, the total change in imports calculated as the trade-weighted product of income elasticities and percentage changes in incomes will not necessarily be equal to the change in imports calculated as the sum of individual import changes in response to income variations. As a result, for the aggregate income elasticity of import demand ϵ^y,

$$\epsilon^y \neq \sum_{j=1}^{m} w_j \epsilon_j^y$$

(w_j being the share of the commodity imported by geographic market j) if the percentage changes in income of the geographic markets are not the same.[1]

The appropriate procedure for the aggregation of income elasticities in import demand functions is one that takes into account the differences between changes in real incomes of individual markets and changes in aggregate income. The general procedure has been derived by Roy (1952), and Barker (1969) has derived the formula for the income elasticity of import demand. His formula for the total elasticity of all principal importers, $j = 1, \ldots, m$, of a commodity is given by

$$\epsilon^y = \sum_{j=1}^{m} w_j \delta_j \epsilon_j^y. \tag{13.1}$$

The second term, δ_j, on the right-hand side of the equation is the so-called 'distribution elasticity'. It is defined as the percentage change in real income Y in the jth market associated with a change in aggregate real income:

$$\delta_j = \frac{\partial Y_j / Y_j}{\partial Y / Y}. \tag{13.2}$$

In estimating the income distribution elasticity, I used the same sample period as that used to calculate the import demand function, and the following equation

$$y_j = \lambda + \delta_j y + \epsilon, \tag{13.3}$$

where lower-case letters denote the logarithms of upper-case letters. Table 13.1 shows estimates of these distribution elasticities, together with estimates of their t-statistics.

Table 13.1 Income distribution elasticities

Importer	Distribution elasticity	Importer	Distribution elasticity
EEC members		Other European countries	
Belgium	0.99	Finland	1.09
	(42.8)		(37.7)
France	1.10	Sweden	0.69
	(60.8)		(39.6)
W. Germany	0.89	Switzerland	0.58
	(61.3)		(14.5)
Greece	1.44	Other countries	
	(48.1)	USA	0.85
Italy	0.98		(57.1)
	(58.4)	Japan	1.88
Netherlands	1.06		(56.8)
	(50.3)	Soviet Union	1.52
Spain	1.25		(68.6)
	(35.3)		
UK	0.58		
	(45.9)		

Note: Numbers in parentheses refer to t-statistics.

13.1.2 Income Elasticities

Table 13.2 shows the import demand functions of the principal geographic markets for the major commodity exports of Latin America. The average income elasticities of import demand, adjusted by the distribution elasticities, range from around 2 for soybeans and iron ore to significantly less than unity for cotton, coffee, and cocoa. The income elasticities for beef, copper, sugar, maize, and bananas are near unity.

The variance of the income elasticities of the importing countries is generally greater in commodities with average elasticities that are

Table 13.2 Import demand functions of principal export markets

Commodity Importer	Income elasticity	Price elasticity	Income growth elasticity[a]
Beef	1.0	−0.7	3.9
USA	0.2	−1.0	−3.6
Japan	3.4	−1.5	7.1
France	4.1	—	8.4
W. Germany	1.0	−0.3	5.4
Greece	1.8	−0.6	3.6
Maize	1.0	—	−2.4
Italy	1.0	—	−2.4
Bananas	1.0	−0.9	−0.3
USA	1.0	—	−0.7
Japan	1.0	−1.3	1.2
W. Germany	1.0	−2.7	−3.3
Italy	1.0	−0.5	0.9
Sugar	0.9	−0.7	9.4
USA	0.6	−0.3	1.1
Japan	1.0	−1.6	20.5
UK	1.0	−0.8	−7.7
Coffee	0.7	−0.2	−0.4
USA	0.2	−0.2	−0.4
Japan	2.2	−0.2	3.3
Belgium	1.1	−0.3	0.7
France	0.7	−0.1	−1.3
W. Germany	1.0	−0.2	−3.7
Italy	1.2	−0.2	0.2
Netherlands	1.1	−0.2	2.4
Finland	0.7	−0.2	1.9
Sweden	0.1	−0.3	−2.2
Switzerland	0.5	−0.2	−1.5
Cocoa	0.8	−0.3	−0.4
USA	1.0	−0.5	3.3
Japan	0.3	−0.4	1.1
W. Germany	0.5	−0.1	−1.3
Italy	0.2	−0.3	2.2
Netherlands	1.0	−0.4	−3.1
Spain	0.4	−0.1	−1.0
Soviet Union	0.6	−0.2	−1.5
Soybeans	2.4	−0.4	3.9
W. Germany	2.0	—	−0.6
Netherlands	3.1	−0.7	13.8

Table 13.2 (*cont.*)

Commodity Importer	Income elasticity	Price elasticity	Income growth elasticity[a]
Soybeans (cont.)			
Spain	1.9	−0.5	−0.1
Switzerland	1.0	−1.1	4.6
Cotton	*0.4*	*−1.0*	*5.1*
USA	1.0	—	−3.7
Japan	0.1	−0.6	−1.8
Belgium	1.0	−4.3	40.8
France	1.0	−1.4	18.1
W. Germany	1.0	−2.4	−11.1
Italy	0.2	−0.5	2.5
Netherlands	1.0	−0.3	−0.3
UK	1.0	—	91.6
Iron ore	*1.8*	*−1.2*	*4.3*
USA	1.0	−5.9	19.8
Japan	1.4	−0.6	−0.2
France	2.6	—	10.7
W. Germany	0.8	—	5.9
Italy	1.3	—	0.6
UK	0.3	—	2.2
Copper	*1.1*	*−0.5*	*4.9*
USA	1.0	−1.5	6.0
Japan	1.9	—	10.3
Belgium	1.0	−0.3	1.6
W. Germany	0.8	—	1.1

[a] The income growth elasticity is the percentage change in import demand brought about by a 1% change in the rate of growth of economic activity. It is defined by the coefficient for g_4, the rate of growth of economic activity, in equation (8.13).

Notes:
— Not significant at the 5% level.
Aggregate elasticities are trade-weighted averages, adjusted by distribution elasticities in the case of income and income growth elasticities.

Source: Derived from Table 13.1 and Annex C.

greater than unity. The income elasticities of import demand for iron ore range from significantly less than unity in the UK to more than 2.5 in France. Similarly, the income elasticities of import demand for soybeans range from unity in Switzerland to over 3 in the Netherlands. These large differences have important implications for sales by exporters. Markets with high income elasticities of import demand have a considerably stronger growth potential than others, either because of a strong response of buyers in those countries to long-term improvements in their real income or because of lower import restrictions. Of course, exports to these markets will also be susceptible to large swings of demand during business cycles. The income elasticities vary much less among importers of cotton, cocoa, and (to a lesser extent) coffee, all of which have average income elasticities that are less than unity. These results suggest a consistently weak growth of the markets for such goods. The selection of markets based on differences in growth is consequently less important for exporters of these commodities.

A comparison of the estimates of income elasticities for imports of primary commodities with other estimates of import demand functions for the industrialized countries reveals three important points.[2] First, the average income elasticity of import demand for primary commodities by the industrialized countries estimated in the present study is about three-quarters of that for all goods imported into the industrialized countries. The trade-weighted average elasticity for the ten commodities covered in this study is 1.2. In comparison, the average of Houthakker and Magee's (1969) estimated income elasticity of demand for all goods imported into the industrialized countries is 1.6. Subsequent studies have yielded very similar results.[3] Goldstein and Khan (1976) have estimated the income elasticities of several industrialized countries, the average of which is 1.5. More recently, Warner and Kreinin (1983) have calculated import demand relationships for the industrialized countries, and the average income elasticity of those whose coefficients are statistically significant equals 1.8; Thursby and Thursby (1984) have estimated the average income elasticity of import demand in five industrialized countries to be 1.4.[4]

Second, disaggregated estimates of import demand relationships have yielded income elasticities of a much larger magnitude than those found in aggregate estimates. The elasticities calculated by Kreinin (1973) for a cross-section of goods imported by the USA ranged from 1 to 9, most of them being concentrated between 2 and 4. Recently, Grossman (1982) also estimated demand relationships for a cross-section of goods imported into the USA and found income elasticities that ranged from less than unity to over 7, the overall mean average being around 2.5.

This point is also evidenced by the differences between our income elasticities for individual commodities and aggregate elasticities for goods of this class. All the commodities in our sample for which the USA is an important market other than coffee, sugar, and beef have unitary elasticities with respect to income. Those three products have much lower elasticities and, since they represent two-thirds of the value of US imports of the commodities in the sample, the total elasticity is only 0.4 (see Table 13.3). This total is the same as that found by

Table 13.3 Aggregate import demand functions of selected commodities in principal markets[a]

Importer	Income elasticity	Price elasticity	Income growth elasticity[b]
USA	0.4	−0.9	3.1
Japan	2.3	−1.3	6.6
EEC members	*1.2*	*−0.3*	*2.4*
Belgium	1.0	−0.5	3.8
France	2.2	−0.3	7.3
W. Germany	0.9	−0.3	0.0
Greece	2.6	−0.9	5.2
Italy	0.9	−0.2	0.1
Netherlands	2.0	−0.5	6.1
Spain	2.2	−0.6	−0.3
UK	0.4	−0.2	6.6
Other Europe	*0.3*	*−0.2*	*0.1*
Finland	0.8	−0.2	2.1
Sweden	0.1	−0.2	−1.5
Switzerland	0.3	−0.2	−0.6
Soviet Union	0.9	−0.2	−2.2
TOTAL	*1.2*	*−0.5*	*4.4*

[a] Trade-weighted averages, adjusted by distribution elasticities.
[b] The income growth elasticity is the percentage change in import demand brought about by a 1% change in the rate of growth of economic activity. It is defined by the coefficient for g_4, the rate of growth of economic activity, in equation (8.13).
Source: Table 13.2.

Houthakker and Magee (1969) for raw materials, including fuels, and by Marquez and McNeilly (1988) for foods originating in the developing countries, but it is larger than the Houthakker–Magee estimate of 0.3 for foods and the Marquez–McNeilly estimate of −0.3 for raw materials. Similarly, the total elasticity of 0.4 for UK imports of the commodities is the same as that calculated by Humphrey (1976) for foods and raw material aggregates, while that of 0.9 for West German imports of the commodities is somewhat larger. A much greater difference occurs in the estimates for France, where my own calculations yield a total elasticity for the sample of commodities of 2.2 and those of Humphrey yield only 0.9 for foods and 0.6 for raw materials. This difference is explained by the large income elasticities for several commodities in the present sample, particularly beef and iron ore. Accordingly, estimates of the import demand relationship for individual commodities can yield income elasticities of considerably different magnitudes from those that have been calculated by aggregate estimates of goods of this class.[5]

A third point to emerge from the empirical estimates of import demand relationships is that the income elasticities for manufactured goods in the industrialized countries are, in general, considerably higher than the income elasticities for the ten primary commodities. The US income elasticity for manufactured goods calculated by Kreinin (1967), Houthakker and Magee (1969), Price and Thornblade (1972), and Goldstein and Khan (1976) ranges from 2.3 to 2.9, while that of the UK is 2.0 according to Barker (1969), and 3.3 according to Humphrey (1976). Humphrey also calculated the income elasticity of both France and West Germany to be 2.3; and the average elasticity of eight industrialized countries for which Goldstein and Khan (1976) estimated import demand relationships was equal to 2. As Table 13.3 shows, the total income elasticities for the ten primary commodities are smaller in most of these countries. These results help to explain the larger expansion that has taken place in trade in manufactured goods than in trade in primary commodities since 1960.

Nevertheless, the total income elasticities for the commodities are as high as those estimated for manufactured goods in some of the industrialized countries. Several EEC countries, namely France, Greece, the Netherlands, and Spain, have income elasticities that are equal to or greater than 2. In fact, the income elasticity of demand for beef in Greece, which has been estimated to be 1.8, is greater than the income elasticity of demand for manufactured goods, which Hitiris and Petoussis (1984) estimated to be less than unity. Thus, while generalizations about income elasticities being larger for manufactured goods than for primary commodities are supported by empirical evidence, particular geographic markets constitute exceptions.

The dynamic specification of the import demand relationship also allows us to calculate the income growth elasticity. In the long-run dynamic equilibrium relationships this ranges from 6.6 in Japan to −2.2 in the Soviet Union, the overall average for the ten commodities being 4.4 (see Table 13.3). This average elasticity has the following interpretation. If the long-term rate of growth of income of all importing countries were to rise by 1 per cent, e. g. from 4 to 5 per cent, the level of import demand for all the commodities would increase by 4.4 per cent. Hence changes in the *rate of growth* of income in the export markets for these commodities can bring about significant changes in the *level* of import demand in those countries. Accordingly, the total change in import demand arising from a change in both the rate of growth and the level of economic activity is substantially greater in a dynamic framework in which there is steady-state growth than in a static framework.[6] As Table

13.2 shows, this result is particularly applicable to sugar, copper, cotton, iron ore, soybeans, and beef imports, whose income growth elasticities are, on average, high in the principal importing countries.

13.1.3 Price Elasticities

The trade-weighted average price elasticity of import demand, adjusted by distribution elasticities across commodities, is equal to -0.5 in the principal markets of Latin America. This average is similar to estimates by others for total merchandise imports of the industrialized countries, which range from -0.5 to -1.0, according to the survey by Goldstein and Khan (1985).

Although the average price elasticity for the ten commodities lies within the range of price elasticities estimated by others for the total merchandise imports of the industrialized countries, there is conflicting evidence about its similarity to aggregate estimates of primary commodity imports by others. For foods and raw materials Houthakker and Magee (1969) estimated relatively low price elasticities of -0.2 in the USA. In contrast, Humphrey (1976) found a relatively high aggregate price elasticity of -0.8 for the food imports of West Germany, and Kreinin (1973) calculated a price elasticity of -0.7 for raw materials in that country.

The reason for the difference between the estimated price elasticity of demand for the ten commodities and the aggregated estimates for primary commodity imports of the industrialized countries lies in the commodity coverage. An examination of elasticity estimates for individual commodities in Table 13.2 reveals that those that have large price elasticities represent a larger share of the value of imports in markets with large aggregate elasticities. Iron ore accounts for almost one-half of the total value of the commodity imports for which Japan is a major market, and has a relatively low price elasticity compared with that of some of the other commodities imported by Japan. In contrast, the price elasticity is small, or insignificant, for coffee in France and Italy, and for iron ore in the UK, and each of these commodities accounts for at least one-fourth of the total value of the commodity imports for which these countries are major markets. This observation is analogous to that made by Goldstein and Khan (1985: 1085, n. 64) to the effect that differences in the commodity group composition of imports are the most important explanation of inter-country differences in price elasticities.

Inter-country differences in price elasticities for the commodities are consistent with differences that have been found by others for price elasticities of total imports. As Table 13.3 shows, the price elasticity of -1.3 in Japan is larger than that of -0.9 in the USA and is considerably larger than the -0.3 total for the EEC, as well as for other Western

European countries. These lower price elasticities for the European countries are consistent with the elasticities calculated by Boylan et al. (1980) for three members of the EEC. Exceptionally, Thursby and Thursby's (1984) elasticity estimates for West Germany were of nearly the same magnitude as for Japan. In line with the results presented here, Khan and Ross (1977) and Thursby and Thursby (1984) found the price responsiveness for aggregate imports to be much greater in Japan than in the USA. The results also show that, whereas the EEC countries and the Soviet Union have relatively small aggregate price elasticities for the commodities covered in this study, they have significantly larger aggregate income elasticities than does the USA.

As would be expected, the price elasticities for the primary commodities tend to be significantly lower than those for manufactured goods. The average of price elasticities of manufactured goods in the studies surveyed by Goldstein and Khan (1985: Table 4.4) is −2, compared with an average of −0.5 for the price elasticity of the ten primary commodities. Nevertheless, the range of elasticities varies greatly among both primary commodities and manufactured goods. The average elasticity of iron ore is −1.2 and that of cotton and bananas is equal to, or near, unity; beef and sugar have lower elasticities; and the remaining five commodities have elasticities of less than −0.5 in absolute terms.

Stone (1979) has found an equally wide range of elasticities for 34 manufactured goods. His estimates range from −0.5 to −3.0 or more in each of the markets of the USA, the EEC, and Japan. None the less, the average of elasticity estimates by Stone for individual manufactured goods, trade-weighted and adjusted by distribution elasticities, is −1.5 for the USA, −1.3 for Japan, and −0.8 for the EEC, whereas my own comparable average elasticities for individual primary commodities are −0.9 for the USA, −1.3 for Japan, and −0.3 for the EEC. Thus, while generalizations about differences between price elasticities for the two types of goods are supported by my estimates, the elasticities for individual products sometimes vary greatly in each class of goods.

The policy implication of the empirical estimates of price elasticities of import demand for the ten commodities is that exchange rate policies and commercial policy intervention measures in the form of tariff and non-tariff barriers to trade and subsidies to domestic producers can have a strong impact on the demand for imports of most of these goods. A devaluation would be effective in bringing about a reduction in the quantity of imports demanded, and the adjustment to the devaluation would, in most cases, be quick. The imposition of a tariff or a domestic subsidy that effectively raised the price of foreign goods would also lower the demand for imports in a similar manner.[7] Overall, exchange rate and tariff policies have a greater impact on the demand for imports in the USA and Japan than they do in the Western European countries.

However, the EEC countries, as well as Japan, have a greater response to incomes policies than does the USA. These policies are examined in greater detail in later chapters.

13.2 Import Prices

The pattern of the response of import prices to changes in the world market price of a commodity is generally characterized by a dampened smooth response. After an initial impact on import prices, a change in the world market price quickly becomes ineffective. Once a change in market price has been transmitted by traders, the total change in import prices, as mentioned earlier, will always be proportional to the change in the world market price. A unitary elasticity of import prices with respect to the world market price of a commodity is consistent with theory, and the dynamic specification of the import price relationship in equation (9.50) was therefore constrained to yield a long-term proportional response to world market price changes. My principal concern in examining the results of the import price function is the speed with which import prices adjust to changes in world market prices of commodities.

Figure 13.1 shows the average lag response of import prices to a one-time change in the world market price of each commodity, and

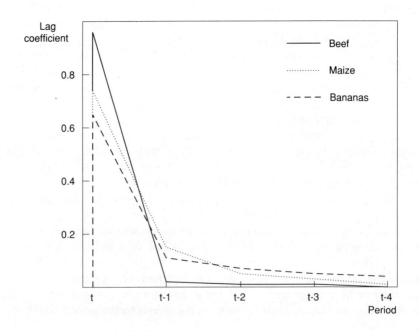

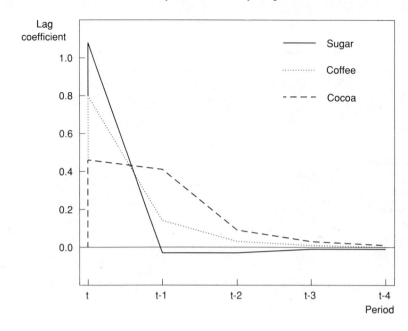

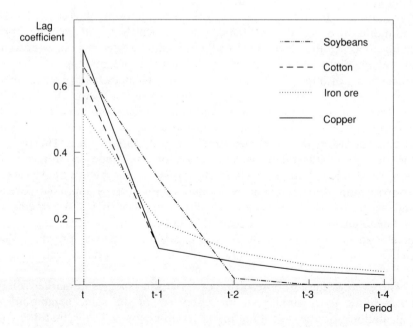

FIG. 13.1 Average normalized lag distribution of import prices in major markets

Table 13.4 Average lag response of import prices to a one-time change in market price

Product	No. of markets	No. of periods for % of response		
		50%	75%	90%
Beef	5	0.0	0.0	0.0
Maize	1	0.0	0.7	2.3
Bananas	4	0.0	0.9	3.4
Sugar	3	0.0	0.0	0.0
Coffee	10	0.0	0.0	0.7
Cocoa	7	0.1	0.7	1.0
Soybeans	4	0.0	0.3	0.8
Cotton	8	0.0	0.4	2.5
Iron ore	6	0.0	1.3	3.5
Copper	4	0.0	0.3	2.2

Source: Derived from Annex C.

Table 13.4 presents summary statistics for the solved lag distributions of equation (9.50). As expected, the response profiles have a J-shaped distribution. One-half of the import price adjustments of nearly all commodities occur in the same year in which world market price changes occur, and 75 per cent of the adjustments occur within one year of a change in world market prices. Among the commodities, the fastest adjustments of import prices to market price changes take place in beef, sugar, coffee, cocoa, and soybeans; the slowest adjustments occur in iron ore, bananas, cotton, maize, and copper. Only the import prices of iron ore take almost as long as a year and a half to complete 75 per cent of their adjustment to a change in the world market price. The reason for this slow adjustment is that most of the trade in iron ore is conducted under long-term contracts. Western European and Japanese iron ore importers negotiate with their major foreign suppliers (Brazil and Australia, respectively) after a year or more of their last agreement; the negotiated price then serves as a benchmark for other traders in their own negotiations, with the result that the transmission of a price change among traders throughout the iron ore industry can take a relatively long time.

The average lag response of import prices to changes in market prices tends to be fairly similar among importers of the same commodities. Calculations of the variation in lag coefficients indicate that the lag distributions are generally very similar. None the less, as Fig. 13.2 illustrates for cocoa, the lag distributions of importing countries can be

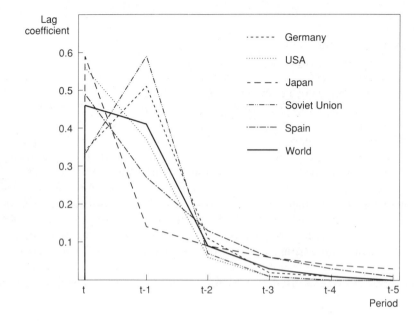

FIG. 13.2 Lag distribution of import prices in some major markets for cocoa

significantly different from the average. Such differences are important when the demand for imports of different geographic markets is considered, since the speed of adjustment of import prices to changes in world market prices determines how quickly import demand in particular geographic markets will adjust to improved conditions in world commodity markets.

These results are based on coefficient estimates of the import price relationship, which in nearly all cases are significant at the 1 per cent level. Half of the estimates for the growth coefficient of cotton are significant at the 5 per cent level, and all other coefficients are significant at the 1 per cent level. The test for stability of coefficients does not reject the null hypothesis of parameter constancy in any of the estimated relationships.

The import prices of other goods in the industrialized countries appear to adjust more quickly to international price movements than do primary commodities. Llewellyn and Pesaran (1976) found that more than two-thirds of the average price adjustment in all types of goods imported into the USA occurs within the same period as changes in foreign prices of the goods. Stern *et al.* (1979) estimated a similar lag structure for US import prices. In contrast, about one-half of the adjustment in import prices of most commodities in our sample occurs in the same period as market price changes. This difference in lag structure is consistent with other empirical results. In the LINK models, three-quarters of the average response of import prices of manufactured goods in the industrialized countries occurs in the same year as changes in the international prices of those goods (Sawyer 1979).

Brorsen *et al.* (1984) suggested that the somewhat slower observed reaction of some import prices to changes in market prices reflected inefficiency on the part of agents in processing information. This may be the case for some commodities. The absence of major commodity exchanges, or trade conducted outside existing commodity exchanges, can delay the transmission of information to traders. But information-processing inefficiency does not always explain observed delays in import price adjustments. Marketing arrangements, for example long-term contracts in the iron ore trade, can also give rise to slow import price adjustments. Consequently, the somewhat slower adjustment of import prices of primary commodities may be due, for the most part, to the types of systems used to transmit international price changes in the markets.

13.3 Conclusions

Three key features of import demand functions for commodities were examined in this chapter: (1) the total response to income and price changes, (2) the length of time required for the response to occur, and (3) the uniformity of response of importers of the same commodity to income and price changes. The coefficient estimates of the import demand functions show that, in the industrialized countries, there is a relatively strong demand response for Latin America's commodity exports. Indeed, demand for primary commodity imports by the industrialized countries was, in general, found to be elastic with respect to income. The trade-weighted average income elasticity of import demand for these commodities has been calculated to be equal to 1.2, which is nearly two-thirds of that for the manufactured goods of the industrialized countries.

Moreover, changes in the economic growth of the industrialized countries were found to have, on balance, a positive dynamic effect on the demand for imports. Hence an increase in the steady-state growth of income not only leads to a more than proportional increase in the demand for commodity imports; it also induces a further expansion in the demand for imports as a result of a change in the rate of economic growth in the foreign markets. The trade-weighted average price elasticity of import demand by the industrialized countries for the commodities was found to equal −0.5, which is about the same as that for the average of all types of goods imported by the industrialized countries, but substantially smaller in absolute terms than that for imports of manufactured goods by the industrialized countries.

Notes

1 Magee (1975: app. A) provides the following illustration of aggregation error, expressed here in terms of imports of a commodity in two geographic markets when there are three sets of income changes whose trade-weighted averages are none the less the same:

Geographic market	Market trade weight	Income elasticity	Changes in income (%)		
			Case A	Case B	Case C
USA	0.5	3	15	10	20
Japan	0.5	1	15	20	10
Totals of trade-weighted averages	1.0	2	15	15	15
Changes in aggregate imports			30	25	35

2 The comparison of elasticities estimated for different sample periods is valid if the estimated parameters in the import demand relationship are stable. Our test for stability of coefficients did not reject the null hypothesis of parameter constancy in any of the estimated relationships. Volker (1982) has also confirmed the stability of elasticity estimates for total imports into the USA, and Warner and Kreinin (1983) have found that 9 out of 15 industrialized countries had stable coefficients in their import demand relationships between 1957–70 and 1972–80.
3 For surveys of early empirical estimates of import demand relationships, see Magee (1975) and Stern et al. (1976). Comparison of my own elasticity estimates for particular goods with those of others is limited by the dearth of previous empirical analyses of the import demand relationship for individual commodities in the industrialized countries. A recent paper by Husted and Kollintzas (1984) has estimated the import demand relationships for coffee and cocoa, as well as for bauxite and petroleum, of the USA; however, elasticity estimates were not obtained from the type of equation that was fitted to the data. In another study, Chang (1977) has calculated the income elasticity of demand for meat in the USA to be 0.5 in a static model. This result is similar to my own estimate of 0.2 for the income elasticity of the demand for beef imports in the USA.

4 The pattern of variation of income elasticities among countries has been an important finding of these studies. The elasticities of the USA and the UK have been consistently shown to be much smaller than that of Japan. Thus, if prices, real incomes, and exports all rose at the same rate, there would be a secular worsening of the trade balance of Japan and an improvement of the trade balances of the USA and the UK. For a discussion of this issue, see the recent survey of empirical estimates of trade relationships by Goldstein and Khan (1985).

5 Any further comparison between my own estimates and aggregate estimates for primary commodity imports in the industrialized countries is limited by differences in commodity coverages. Nevertheless, the total income elasticities for individual countries reported in Table 13.3 do, with the exceptions noted above, generally replicate the pattern of elasticities found in studies that have estimated aggregate income elasticities for groups of commodities.

6 The dynamic properties of the import demand function have been considered elsewhere by Hitiris and Petoussis (1984). For manufactured imports of Greece, the implied income growth elasticity of their estimate of the demand relationship averages 0.3. In contrast, the income growth elasticity of demand for beef imports in that country has been estimated to be 3.6. This result is consistent with the more dynamic growth of beef imports than of overall manufactured imports in Greece.

7 Non-tariff barriers to trade eliminate the effect that changes in prices of foreign goods have on import demand. For example, sugar imports into the USA are subject to quotas, and the import price is subject to a support system when prices fall below a predetermined level; consequently, the price elasticity of import demand for sugar is not statistically different from zero in the USA.

14
The Demand for Commodity Exports

One of the principal features of the system of equations developed to represent international commodity trade is that it allows for the possibility that exporters can exert some influence on the demand for their exports. The reason is that importers can perceive a commodity as being heterogeneous in terms of its country of origin because of either vertical or horizontal differentiation. Commodity differentiation is reflected in the ability of exporters to influence the demand for their exports through relative-price changes, so that each exporter confronts a downward-sloping demand schedule. This chapter presents the results of estimates of the export demand relationship for the Latin American countries. The results not only provide a test of product heterogeneity by country of origin; they also permit an examination of some of the questions about the magnitude of the demand elasticities of internationally traded commodities that have been posed in the development trade literature and were raised at the beginning of this part of the study.

Disaggregated estimates of export demand by geographic markets have been combined in order to synthesize information about Latin American exports. However, as with the import demand functions in the previous chapter, a simple trade-weighted averaging procedure can lead to incorrect aggregate statistics when movements in disaggregated explanatory variables are different from those of their aggregates.[1] The first section reports the empirical findings on the elasticity of substitution and compares them with estimates by others for manufactured exports of the industrialized countries. Section 14.2 presents the results of coefficient estimates of the export demand function and also compares the results with those estimated by others. The chapter concludes with a summary of the findings.

14.1 Elasticity of Substitution

To assess whether the commodity differentiation hypothesis embodied in a country's export demand function is valid, equation (8.17) was tested

Table 14.1 Elasticities of substitution for commodity exports

Commodity	Elasticity of substitution[a]		Commodity	Elasticity of substitution[a]	
Exporter	Short-run	Long-run	Exporter	Short-run	Long-run
Beef			*Cotton*		
Argentina	−0.82	−3.13	Brazil	−0.48	−1.26
Costa Rica	−1.02	−4.33	El Salvador	—	—
Honduras	−1.11	−5.03	Guatemala	−1.52[b]	−3.64
Nicaragua	—	—	Mexico	−0.69	−1.81
Uruguay	—	—	Nicaragua	−0.43	−0.72
Maize	−0.49	−0.54	Paraguay	−0.38	−0.95
Argentina	−1.06	−1.53	Peru	—	—
Bananas	−1.06	−1.53	*Coffee*	−1.30	−5.24
Costa Rica	−0.45	−1.20	Brazil	−0.76	−2.46
Ecuador	−0.46	−1.07	Colombia	−0.43	−2.07
Honduras	−0.42[b]	−1.67	Costa Rica	−0.83	−3.03
Panama	−0.54	−0.58	Dom. Republic	−2.81	−4.45
Cocoa	−0.36[b]	−1.14	Ecuador	—	—
Brazil	−0.28	−0.40	El Salvador	−0.66	−1.47
Dom. Republic	−0.37	−0.51	Guatemala	−1.51	−3.48
Ecuador	—	—	Haiti	−1.03	−2.32
Soybeans	−0.23	−0.34	Honduras	−0.56	−1.77
Brazil	−0.74	−1.11	Mexico	−2.09	−5.69
Paraguay	−0.76	−1.17	Nicaragua	−0.37	−0.82
	−0.49	−0.49		−0.78	−3.01

	Sugar	Iron ore	
Barbados	—	Brazil	−0.20 / −0.85
Brazil	—	Peru	−0.22 / −0.77
Costa Rica	—		−0.13[b] / −1.11
Dom. Republic	—	Copper	−0.63 / −2.01
El Salvador	—	Chile	−0.68 / −2.20
Guatemala	—	Peru	−0.47 / −1.43
Guyana	—		
Jamaica	—		
Panama	—		
Peru	—		

[a] Calculated from separate estimates for each country's exports to its principal geographic markets. See text for aggregation method.
[b] One-period lag.

Note: — Not significant at the 5% level.

Source: Annex C.

for the significance of the relative price variables. Significant coefficients for these variables imply that the exporter's market share has been influenced by price competitiveness. The empirical results show the coefficients to be significant in most cases, which supports the hypothesis that these commodities have been differentiated by importers on the basis of country of origin.

The extent to which price competitiveness has influenced the market shares of commodity exports is shown in Table 14.1, where short-run and long-run elasticities of substitution are presented for the total product exports of each Latin American country. The elasticities of substitution are measured by the estimated coefficients for the relative-price variables in (8.17). The short-run and long-run elasticities are derived from the solved coefficients for the estimated stochastic difference equations. The solved coefficients are associated with the variables of the originally specified levels form of the equations. Hence the short-run elasticity is simply the impact of, or first-period response to, a change in the relative-price variable. The long-run elasticity was shown in (8.24) to be the negative of the ratio of the coefficient of the lagged relative-price variable to that of the lagged dependent variable (i.e. $-\beta_{24}/\beta_{22}^*$).[2]

The results presented in Table 14.1 indicate that in nearly all commodities relative-price movements have influenced the exports of the Latin American countries. The one exception is sugar. Under preferential trade arrangements with the region's two principal markets, the USA and the UK, the price received by exporters of sugar to these two markets has been fixed (usually at a substantial premium when world market prices have been depressed). Consequently, exporters to these markets have been unable to vary their relative prices. For this reason, the price elasticities of Latin American exporters to the US and the UK markets are $-\infty$, where own export price equals the import price in the geographic market (in terms of earlier notation, where $P_{ij} = P_j$); elsewhere they are zero.

For all other major commodities of Latin America, estimates of elasticities of substitution were often found to be statistically significant at the 5 per cent level. In particular, relative prices were found to be statistically significant in 71 of the 125 export demand relationships (excluding those for sugar) that were estimated. The proportion of relationships within each commodity in which prices were found to be significant ranged from 50 to 70 per cent.

The trade-weighted averages, adjusted by the distribution elasticities, of seven of the ten commodities have long-run elasticities of substitution that are greater than unity. These commodities are beef (-3.1), coffee (-2.5), copper (-2.0), maize (-1.5), cotton (-1.3), bananas (-1.2), and soybeans (-1.1). In general, exporters of these commodities have very competitive marketing practices, although important differences exist between exporters of the same commodity. In copper, for example, Chile has a more developed marketing and trading system than Peru. In beef, Argentina and Costa Rica have more developed marketing systems than other Latin American countries. In bananas, Ecuador is more flexible in its trade practices than the Central American countries (see Kawata 1975). And in coffee, Brazil, Colombia, and the Central American countries have well-established trade practices. These differences give rise to the wide range of elasticities of exporters of the same commodity that can be observed in Table 14.1.

These results can be contrasted with those for manufactured exports of the industrialized countries. As far as is known, no comparable econometric estimates exist for primary commodity exports of developing countries.[3] Until now, empirical studies on elasticities of substitution in international trade have focused on the exports of the industrialized countries (see the survey by Stern et al. 1976). Among these studies, that of Junz and Rhomberg (1965) found the long-run average elasticity of substitution for manufactures of the industrialized countries to equal -5.1. Hickman and Lau's (1973) results were very similar; they estimated an average long-run elasticity of substitution of -4.8 for aggregate exports of the industrialized countries. (Their calculations included an estimate for the centrally planned economies of -3.4, and another for rest-of-world exports, equal to -5.2.)

The short-run elasticity of substitution for manufactured exports of the industrialized countries has been estimated by Kreinin (1967) to be, on average, equal to -2.6. (A subsequent estimate by Kreinin (1973) for all types of goods exported by the USA yielded an elasticity of substitution equal to -1.8.) Kravis and Lipsey (1982) have also calculated short-run elasticities of substitution for West Germany and Japan in an effort to explain the rise in the shares of these countries in the world markets for manufactured goods. For West German exports of machinery and equipment (SITC 7) relative to those of the USA, they calculated a short-run elasticity equal to between -1.7 and -1.9; in this same commodity category, they estimated the elasticity of substitution for Japanese exports relative to US exports to be -2.8.

These various estimates of the elasticity of substitution for manufactured exports from the industrialized countries appear to be fairly consistent with one another. In the short run the elasticity is around -2.5; in the long run it is somewhat over -5. In comparison, the estimated average elasticity of substitution for Latin America's major exports is -0.5 in the short run and -1.4 in the long run. These results are based on unweighted averages of the elasticities of substitution reported in Table 14.1. The trade-weighted average is somewhat higher: it is -0.6 in the short run and -1.6 in the long run. Hence the estimated average elasticity of substitution for Latin America's major commodity exports is about one-quarter of that for the manufactured exports of the industrialized countries, both in the short run and in the long run.

14.2 The Export Demand Function

Dynamic considerations of commodity trade in Chapter 8 suggested (1) that export demand has a proportional response to changes in the level of import demand in geographic markets, (2) that it is influenced by the level of the relative export price, and (3) that it is affected by variations in the rate of growth of imports.

The first property is a requisite constraint for the equation since, *ceteris paribus*, any increase or decrease in imports by a geographic market would be reflected in an equivalent percentage change in its demand for exports from all foreign suppliers. Thus, the export demand function in equation (8.17) was constrained to yield a long-run unitary import demand elasticity. The adjustment of export demand from one level of foreign import demand to another is determined by the error correction term. As discussed in Chapter 8, the closer the absolute value of the coefficient of the error correction term to unity, the faster the rate of adjustment of export demand to import demand changes. Alternatively, the closer the value of the coefficient to zero, the slower the rate of adjustment.

Table 14.2 shows the frequency distribution of the coefficients of the error correction terms for all the estimated equations. The total number of coefficients suggests that there is an equally frequent distribution. However, the distribution reflects superimposed distributions of individual commodities. The frequency distribution of coffee is asymmetrical, there being a maximum number of coefficients in the -0.2 to -0.3 range and then a very gradual decrease in frequency thereafter. The frequency of cotton reaches its maximum in the -0.4 to -0.5 range of coefficients and then tails off. In the distribution of the coefficients for iron ore, which is J-shaped, the maximum frequency occurs very close to the start. Cocoa and soybeans have skewed distributions, the former with its maximum frequency at -0.7 to -0.8, the latter with its maximum frequency at -0.2 to -0.3. Sugar has two maxima; near the start of the distribution there is a maximum and a rapid fall in the frequency, which then rises to another maximum, this time at -0.5 to -0.6. The coefficients for the error correction terms are also important in determining the long-run parameters of other factors that influence the export demand function.

Table 14.2 Distribution of error-correction coefficients

	Frequency per 0.1 interval									
	−0.0	−0.1	−0.2	−0.3	−0.4	−0.5	−0.6	−0.7	−0.8	−0.9
Beef	1	2	1	1	1	2	—	—	1	1
Maize	—	—	1	—	—	—	—	—	—	1
Bananas	—	—	3	1	2	1	1	2	—	2
Sugar	—	3	1	2	—	3	2	—	—	2
Coffee	—	3	10	6	5	5	4	5	3	—
Cocoa	—	—	—	1	—	1	1	4	—	3
Soybeans	—	—	3	1	1	—	1	1	—	1
Cotton	2	2	1	4	6	3	1	4	1	1
Iron ore	—	6	—	1	2	—	—	—	1	—
Copper	—	1	1	1	1	—	—	2	1	—
TOTAL	3	17	21	18	18	15	10	18	6	11

Source: Annex C.

The other factors that influence export demand, namely relative export prices and the rate of growth of imports, operate to produce non-proportional responses between changes in imports of a geographic market and changes in export demand of foreign suppliers to that market. Relative-price movements can affect the export demand of a country and thereby alter the market share of that country. In the previous section the elasticity of substitution was often found to be statistically significant—except for exports of sugar—in the estimated export demand relationships, and in Chapter 8 the price elasticity of export demand ϵ_x^p was shown to be related to the elasticity of substitution ϵ_x^r by the formula $\epsilon_x^p = (1 - s)\epsilon_x^r$, where s is the exporter's market share. Thus, the price elasticity of export demand (which is equivalent to the 'elasticity of substitution for market shares') varies with the market share of an exporter. Small or moderate-sized exporting countries can therefore influence their shares, albeit to a limited extent, through relative-price changes. Although not new to international trade literature (see Marshall 1924: app. J in the context of devaluation policies), the finding in this chapter that relative prices can influence export demand for primary commodities, which is also supported by the empirical results reported in Lord (1989a), undermines one of the conventional assumptions of trade theory and has important implications for development trade issues.

Another influence on the export demand, or market share, of an exporter is the dynamic effect originating from changes in the rate of growth of imports. Equation (8.19) states that, at given import and relative-price levels, a country's export demand can vary positively or negatively with the rate of growth of imports in its geographic market. The direction of the dynamic effect from import growth is determined by the estimated coefficient for the rate of growth of imports. When a change in the rate of growth of imports leads to a positive change in the rate of growth of exports, the import growth elasticity is positive. Alternatively, when a change in the rate of growth of imports brings about a negative change in the export growth rate, the import growth elasticity is negative.

In addition to these determinants of the demand for exports, non-price competitiveness has also played an important role in the ability of exporting countries to bring about changes in their market shares (see Ireland 1987; Bismut and Oliveira-Martins 1987). Non-price competitive influences on export demand have been empirically measured by assigning binary variables to their occurrence (1 in the year in which demand shifted; 0 otherwise). These binary variables are reported in Annex C. An examination of their frequency distribution indicates that, on the one hand, the shifts were fairly symmetrical over the period in so far as the number of negative-signed coefficients was approximately the same as the number of positive-signed coefficients; on the other, the shifts in the demand schedules were concentrated in the first part of the 1970s and in the first part of the 1980s.

The average price elasticities and the average import growth elasticities of the export demand function for Latin America's major commodity exports are reported in Table 14.3. For the combined major commodity exports of the region, the unweighted average price elasticity of export demand for Latin America is equal to −0.5 in the short run and −1.2 in the long run. The trade-weighted average price elasticity of export demand is also −0.5 in the short run, but it is −1.4 in the long run. Among individual commodities, the price elasticity of regional export demand is greater than unity in half of them. In the upper range

Table 14.3 Dynamic equilibrium solution of export demand function

Commodity	Price elasticity of export demand		Import growth elasticity[a]
Exporter	Short-run	Long-run	
Beef	−0.72	−2.75	3.05
Argentina	−0.87	−3.74	4.97
Costa Rica	−1.07	−4.87	−1.45
Honduras	−	−	0.00
Nicaragua	−	−	−0.77
Uruguay	−0.47	−0.52	0.00
Maize	−0.99	−1.43	0.00
Argentina	−0.99	−1.43	0.00
Bananas	−0.35	−0.95	−0.34
Costa Rica	−0.38	−0.90	0.00
Ecuador	−0.32	−1.27	−0.75
Honduras	−0.40	−0.43	−0.24
Panama	−0.31	−0.99	0.00
Sugar	−	−	0.01
Barbados	−	−	0.00
Brazil	−	−	0.31
Costa Rica	−	−	−0.85
Dom. Republic	−	−	0.00
El Salvador	−	−	−4.69
Guatemala	−	−	4.00
Guyana	−	−	−0.79
Jamaica	−	−	0.00
Panama	−	−	−5.40
Peru	−	−	0.00

Table 14.3 (*cont.*)

Commodity Exporter	Price elasticity of export demand		Import growth elasticity[a]
	Short-run	Long-run	
Coffee	−0.70	−2.21	1.61
Brazil	−0.38	−1.78	2.70
Colombia	−0.71	−2.63	0.87
Costa Rica	−2.76	−4.36	−0.33
Dom. Republic	—	—	−0.01
Ecuador	−0.64	−1.43	1.15
El Salvador	−1.49	−3.43	0.00
Guatemala	−1.00	−2.24	0.00
Haiti	−0.55	−1.75	0.24
Honduras	−2.05	−5.54	1.99
Mexico	−0.34	−0.76	1.28
Nicaragua	−0.76	−2.91	0.67
Cocoa	−0.26	−0.37	−0.25
Brazil	−0.34	−0.47	0.24
Dom. Republic	—	—	−2.35
Ecuador	−0.22	−0.33	−0.23
Soybeans	−0.63	−0.96	1.35
Brazil	−0.67	−1.03	1.49
Paraguay	−0.25	−0.25	0.10
Cotton	−0.47	−1.20	0.38
Brazil	—	—	−0.18
El Salvador	−1.45	−3.47	2.44
Guatemala	−0.68	−1.75	−0.75
Mexico	−0.41	−0.69	0.00
Nicaragua	−0.38	−0.94	1.79
Paraguay	—	—	2.77
Peru	−1.26	−4.90	−1.17
Iron ore	−0.18	−0.76	−0.03
Brazil	−0.20	−0.68	0.74
Peru	−0.12	−1.05	−2.78
Copper	−0.54	−1.76	−0.51
Chile	−0.56	−1.89	−0.64
Peru	−0.46	−1.39	−0.13

[a] Defined as the percentage change in export demand brought about by a 1% change in the rate of growth of import demand.

Note: — Not significant at the 5% level.

Source: Annex C.

are beef (−2.8), coffee (−2.2), and copper (−1.8); in the lower range are maize (−1.4) and cotton (−1.2). Regional exports whose demand is inelastic with respect to prices in the long run are soybeans and bananas (both just under −1.0), iron ore (−0.8), and cocoa (−0.4).

These different elasticities of export demand seem to be associated at times with the commodity itself and at other times with the geographic distribution of the commodities. Beef, cocoa, cotton, and copper have similar elasticities in the US, Japanese, and European markets. In contrast, bananas, coffee, soybeans, and iron ore have significantly different elasticities in these market areas. For instance, Japanese and European importers of coffee and bananas are much more responsive to relative-price changes by their foreign suppliers than are US importers. Thus, the capacity of exporting countries to influence their market shares depends on the geographic distribution of their exports of these commodities rather than of those commodities for which the price elasticity of export demand is fairly similar across markets.

At the regional level the short-term price elasticity of demand ranges from just under −1.0 for maize to −0.2 for iron ore. In the upper range are beef (−0.7), coffee (−0.7), and soybeans (−0.6); in the lower range are cocoa (−0.3), bananas (−0.4), and cotton (−0.5). As with the estimated elasticities of substitution, there is a fairly wide distribution of price elasticities of demand among exporters of the same commodity. Care should therefore be exercised in generalizations about the price elasticities of demand for commodity exports.

Most other estimated price elasticities of export demand with which the present results can be compared are for exports of the industrialized countries. There are two exceptions. The first is Khan's (1974) estimates of the price elasticities of export demand for a sample of developing countries, among which are eight Latin American countries. His estimates are based on the relationship between aggregate export demand of each country of interest and its total unit export value relative to some measure of the world price level, as well as to the total real GDP of the industrialized countries. Only two of the eight Latin American countries had coefficients for the relative-price variable that were significant at the 5 per cent level. In any event, the average price elasticity of export demand for the eight Latin American countries is equal to −0.24 in the short run, and the derived long-run average price elasticity is equal to −0.28.[4] Khan's results are in sharp contrast to the −0.5 short-run and −1.4 long-run price elasticities of export demand for the region's major commodity exports that have been estimated in this study.

These differences can in part be the result of Khan's use of aggregated data, as well as his inappropriate price data for competing exporters to the markets of the countries of interest. Geographic aggregation of markets, as noted by Junz and Rhomberg (1965), can introduce a significant bias towards lower elasticity estimates. This bias is evident in Khan's estimated price elasticities of demand for Latin American exports. In particular, his estimated elasticities for composite markets are lower than they would have been if they had been estimated for the component markets. For instance, in the case of Chile, which during 1960–85 derived two-thirds of its export revenue from copper, Khan's (statistically significant) long-run price elasticity of total export demand is -0.08; in contrast, the long-run elasticity for Chile's copper exports is -1.9 in Table 14.3. Another example is Khan's (statistically non-significant) estimate of the price elasticity of total export demand for Colombia, which is -0.51; in contrast, the estimated long-run elasticity of export demand for Colombia's coffee, which accounted for 62 per cent of that country's revenue in 1960–85, is shown to be equal to -2.6 in Table 14.3. So global estimates, as Junz and Rhomberg warn, 'may seriously underestimate the true effects of relative price changes' (1965: 242). Although computationally tedious, price elasticities of export demand should be calculated from geographically and commodity disaggregated data.

The second exception to the tendency of empirical research on international trade to focus on the industrialized countries also demonstrates the importance of disaggregated estimates of price elasticities. Bond (1987) has estimated the price elasticity of export demand for five developing regions, one of which is Latin America, and for five commodity groups. The commodity groups are foods; beverages and tobacco; agricultural raw materials; minerals; and energy. The results for the demand for Latin American exports of non-fuel commodities are similar to those for the other regions. Regional-level estimates of export demand for Latin America yield the following results for four aggregated groups of commodities: -0.11 for foods, -0.33 for beverages and tobacco, -0.14 for agricultural raw materials, and -0.38 for minerals. The average of these four groups is -0.24, compared with an average elasticity of -1.4 estimated in this study for individual commodity exports of Latin America within the same commodity classification. Accordingly, the conclusions of Junz and Rhomberg concerning the need for disaggregated estimates of price elasticities are applicable.

Estimates of the price elasticity of export demand, calculated from disaggregated data, are available for exports of the industrialized countries. For their total exports, Houthakker and Magee (1969) have estimated the short-run elasticity for the USA at -1.2. Goldstein and Khan (1978) have computed a similar short-run price elasticity for the USA. They have also found statistically significant price elasticities for the UK, France, Italy, and the Netherlands in the dynamic specification of the export demand relationship, the average of the estimates being

−1.1. For manufactured exports, higher price elasticities of export demand have been computed by Stone (1979) for the USA (−1.3) and Japan (−1.5), but not for the EEC (−0.7). In contrast, for agricultural exports, the short-run price elasticity of −0.8 for the USA estimated by Houthakker and Magee is considerably lower than that for manufactured goods (which is equal to −1.2). This lower price elasticity of demand for primary commodity exports is consistent with my own estimated short-run average elasticity of −0.5 for Latin America's major commodity exports.

The long-run price elasticity of demand for the total exports of the industrialized countries implied by the results of Goldstein and Khan averages −2.1 for four countries whose coefficient estimates were statistically significant. According to a recent survey by these authors (Goldstein and Khan 1985), the average estimated price elasticity of demand for exports of the industrialized countries estimated by others has also been found to be around −2.0 in the long run. This elasticity is considerably higher than the average long-run price elasticity of −1.4 computed for Latin America's commodity exports. Thus, the estimated price elasticities of demand for industrialized country exports, particularly manufactured goods, tend on average to be higher—both in the short run and in the long run—than those for Latin America's commodity exports. None the less, the present results point up the important role of relative prices in determining demand for Latin America's exports.

Finally, the results show that demand for exports varies not only with the level of imports and relative prices, but also with the rate of growth of imports on a steady-state path. Variations in the rate of growth of imports in foreign markets induce substantial changes in the export market shares of countries. In four of the ten commodities, a change in the rate of growth of imports leads to an increase in the average ratio of exports of the countries in the sample to imports in their foreign markets; in another four it induces a decrease in their average export market share; and in the remaining two it produces no change.

14.3 Conclusions

The findings presented in this chapter support the recent theory of trade dealing with commodity differentiation and imperfect competition. The empirical results indicate that, for a large sample of commodity exports from developing countries, relative-price variations affect the demand for exports, even though the average of the estimated price elasticities of export demand was not as high as that for the manufactured exports of the industrialized countries. The trade-weighted average price elasticity of export demand for the commodities is equal to -0.5 in the short run and -1.4 in the long run. By way of comparison, the average estimated price elasticity of demand for exports of the industrialized countries estimated by others is around -1.2 in the short run and -2.0 in the long run.

Nevertheless, the estimated price elasticities of export demand conceal considerable variations among commodities, among exporters of the same commodity, and among geographic markets of the same exporting country. Accordingly, caution should be exercised in generalizations about these estimates.

Notes

1. In accordance with the approach described earlier for income changes in the import demand relationship, the aggregation problem in export demand can be solved by taking into account the differences between changes in the relative price in individual markets and the changes in their weighted averages. The formula for the aggregation of price elasticities has been derived by Roy (1952), and is directly applicable to the elasticity of substitution, ϵ_x^r, and the price elasticity of export demand, ϵ_x^p:

$$\epsilon_i^r = \sum_{j=1}^{m} w_{ij} \frac{\partial P_{ij}/P_{ij}}{\partial P_i/P_i} \epsilon_{ij}^r \qquad (14.4a)$$

$$\epsilon_i^p = \sum_{j=1}^{m} w_{ij} \frac{\partial P_{ij}/P_{ij}}{\partial P_i/P_i} \epsilon_{ij}^p. \qquad (14.4b)$$

The first component on the right-hand side of each expression is the proportion of exports of country i shipped to geographic market j. The second component is the so-called 'distribution elasticity', denoted γ. It is defined for a given commodity as the percentage change in the relative price P in the jth market relative to that of the average price of the country. It has been estimated over the same time period as was used to estimate the export demand relationship. For that purpose, we used the following equation:

$$p_{ij} = \lambda + \gamma_{ij} p_i, \qquad (14.5)$$

where lower-case letters denote the logarithms of upper-case letters.

2. The long-run elasticity can be calculated as the sum of the solved coefficients for (8.17) or, more simply, as the ratio of two polynomials in the expression for the derived coefficients of the equation. In particular, given the systematic dynamics of (8.17) without the intercept,

$$\Delta z_t = \beta_{21} \Delta m_t + \beta_{22}^* z_{t-1} + \alpha_{23} \Delta r_t + \beta_{24} r_{t-1},$$

its expression in terms of the derived coefficients is

$$z_t = \beta_{21} m_t - \beta_{21} m_{t-1} + (1 + \beta_{22}^*) z_{t-1} + \alpha_{23} r_t + (\beta_{24} - \alpha_{23}) r_{t-1},$$

which can be expressed in terms of lag operators and rewritten as:

$$z_t = \frac{\beta_{21} - \beta_{21}L}{1 - (1 + \beta_{22}^*)L} m_t + \frac{\alpha_{23} + (\beta_{24} - \alpha_{23})L}{1 - (1 + \beta_{22}^*)L} r_t.$$

Then the long-term response of z equals the ratio of the polynomials in the lag operator evaluated at $L = 1$, which for imports is equal to zero and for relative prices is equal to $-\beta_{24}/\beta_{22}^*$.

3 The study on wheat by Grennes *et al.* (1978) adopts the Armington (1969) assumption of equal elasticities of substitution in a market, and assumes the elasticity of substitution to be -3.0. A few other studies, referenced in Gardiner and Dixit (1987), have used a method of calculation for the price elasticity of the foreign demand for US agricultural exports that is based on an algebraic formula derived by Yntema (1932).

4 The average of Khan's estimated price elasticities of export demand was calculated as follows:

	Trade weight	×	Distribution elasticity	×	Price elasticity Short-run	Price elasticity Long-run	=	Average elasticity Short-run	Average elasticity Long-run
Argentina	0.199		0.764		−0.40	−0.35		−0.0608	−0.0537
Brazil	0.433		0.900		−0.13	−0.10		−0.0507	−0.0401
Chile	0.100		0.654		−0.12	−0.08		−0.0078	−0.0054
Colombia	0.087		1.087		−0.23	−0.51		−0.0218	−0.0480
Costa Rica	0.026		0.694		−0.15	−0.35		−0.0027	−0.0062
Ecuador	0.050		1.066		−1.08	−0.52		−0.0576	−0.0278
Peru	0.082		1.018		−0.72	−1.32		−0.0601	−0.1103
Uruguay	0.023		0.694		+1.65	+1.05		+0.0263	+0.0168
Total:	1.000					Average:		−0.235	−0.275

(Figures may not equal sums or products because of rounding errors.)

15
Characterization of Commodity Export Supply

Two statistics are central to the characterization of the export supply relationship. The first is the price elasticity, which determines how well exporters respond to changes in conditions prevailing in commodity markets, and how well they would respond to export incentive policies. The other is the lag distribution. Since the effects of price changes usually take a long time to work themselves through to export supply, and since the transmission of the price effects can be complex, we need to provide relevant information about the shape and length of the lag. In some cases, such as beef, coffee, and cocoa, we know the general form of the lag; in others the form is less well known. In either case the lag distribution is critical. For individual producing countries it determines how the amount exported will be allocated over time; for commodity markets it determines the speed and manner in which adjustments from one steady-state solution to another take place.

The results of the estimated export supply relationship for the major commodity exports of Latin America are summarized in Fig. 15.1 and Table 15.1. As with export demand, each of the export supply relationships has been estimated for the exports of the countries to each of their principal markets. Details of these estimates are presented in Annex C. The statistics for the price elasticities and lag distributions in the figure and table have been calculated from the trade-weighted averages of those separate estimates.

The problem of aggregation, discussed in the previous chapter, also arises for export supply, since the explanatory price variable is also export prices to a geographic market. Consequently, aggregation of export demand and supply responses to price movements, which do not necessarily occur in a uniform manner across export markets, can be erroneous if simple average price elasticities are calculated.

The remainder of this chapter examines the values of the derived statistics of the estimated supply relationships. First, the results are analysed in terms of the responsiveness of exports to changes in market prices. Then lag distributions and the differences between exporters of the same commodity are examined.

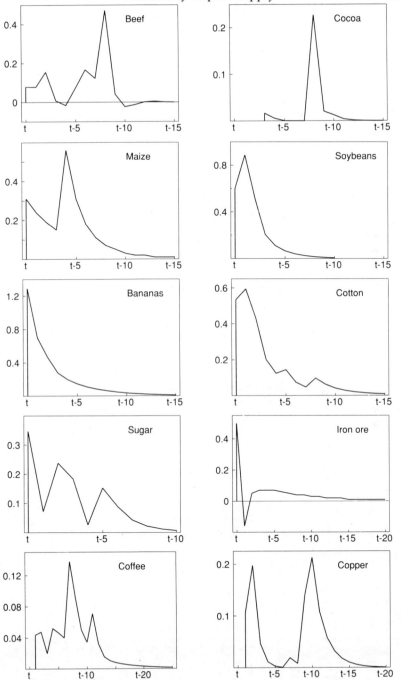

FIG. 15.1 Average lag coefficients of export supply functions
(Trade-weighted averages)

Table 15.1 Export supply response to price change[a]

Commodity Exporter	Elasticity Short-run	Long-run	Average lag Mean	Median	No. of periods for: 75% of response	90% of response
Beef						
Argentina	0.25	1.15	5.59	6.25	7.41	7.78
Costa Rica	0.12[c]	0.95	7.28	7.30	7.70	7.94
Honduras	0.63[b]	2.76	4.35	1.84	7.00	7.43
Nicaragua	2.38	5.55	3.52	0.45	6.71	7.12
Uruguay	0.99[c]	1.52	6.54	5.77	6.44	7.28
	0.07[b]	0.65	2.10	1.53	1.86	2.42
Maize						
Argentina	0.31	2.27	3.99	3.44	4.84	7.00
	0.31	2.27	3.99	3.44	4.84	7.00
Bananas						
Costa Rica	1.51	3.53	2.27	0.67	2.67	6.00
Ecuador	0.09[b]	0.70	3.34	1.78	3.41	6.75
Honduras	2.79	7.17	2.52	0.75	3.24	6.62
Panama	0.86[b]	1.61	1.90	0.93	1.86	3.18
	1.42	1.57	−0.59	0.00	0.00	0.00
Sugar						
Barbados	0.54	1.20	2.72	1.77	4.23	5.70
Brazil	3.15[c]	7.72	5.15	4.80	5.71	6.78
Costa Rica	0.29	0.46	0.59	0.00	1.11	1.54
Dom. Republic	0.15[c]	0.23	5.36	4.78	5.17	5.73
El Salvador	0.43	1.61	2.55	1.30	3.09	5.79
Guatemala	1.13[b]	1.14	1.01	0.50	0.76	0.91
Guyana	1.89[c]	2.37	2.35	1.63	1.94	3.77
	0.98	0.66	−1.10	0.00	0.00	0.00

Jamaica	0.10	0.95	4.27	3.68	4.66	6.00
Panama	0.54[c]	1.21	4.75	4.48	5.25	6.33
Peru	0.58	1.50	2.11	1.72	2.61	3.75
Coffee	*0.10*[b]	*0.74*	*7.96*	*6.88*	*9.78*	*12.00*
Brazil	0.01[c]	0.41	7.57	6.94	7.69	8.51
Colombia	0.14[b]	0.99	6.93	4.00	8.50	16.00
Costa Rica	0.07[c]	0.44	7.78	7.56	8.69	9.87
Dom. Republic	0.35[b]	0.53	2.36	0.76	3.49	4.81
Ecuador	0.26[c]	1.43	9.00	7.63	11.32	11.90
El Salvador	0.22[c]	1.17	8.77	6.82	9.43	14.35
Guatemala	0.33[c]	0.86	8.80	8.43	10.51	13.27
Haiti	0.05[c]	1.38	7.13	5.75	8.86	10.67
Honduras	0.34[b]	0.85	2.77	1.43	4.12	5.32
Mexico	0.04[c]	1.51	11.46	10.60	10.99	13.53
Nicaragua	0.18[c]	0.93	9.97	10.97	12.00	15.24
Cocoa	*0.07*[c]	*0.29*	*7.92*	*7.53*	*7.85*	*8.04*
Brazil	0.03[c]	0.30	7.25	7.35	7.58	7.72
Dom. Republic	0.10[c]	0.31	9.05	11.51	13.26	14.30
Ecuador	0.19[c]	0.23	9.21	8.59	8.89	8.49
Soybeans	*0.60*	*2.48*	*1.64*	*0.72*	*1.73*	*3.28*
Brazil	0.56	2.53	1.69	0.76	1.77	3.37
Paraguay	0.93	2.08	1.08	0.21	1.27	2.34
Cotton	*0.87*	*2.53*	*3.65*	*1.32*	*4.09*	*8.50*
Brazil	1.22[b]	3.63	4.12	1.95	7.00	8.90
El Salvador	1.35	0.98	−0.39	0.00	0.00	0.00
Guatemala	1.39	2.75	0.90	0.00	0.81	1.91

Table 15.1 (cont.)

Commodity	Elasticity		Average lag		No. of periods for:	
Exporter	Short-run	Long-run	Mean	Median	75% of response	90% of response
Cotton (cont.)						
Mexico	0.59	1.39	1.88	1.45	2.28	4.80
Nicaragua	0.61[b]	3.03	3.94	2.82	4.73	6.84
Paraguay	0.69	6.16	7.78	1.79	12.00	25.00
Peru	0.53	1.21	0.79	0.15	0.76	1.52
Iron ore						
Brazil	*0.52*	*1.00*	*5.11*	*0.00*	*7.56*	*12.63*
Peru	0.63	1.11	5.28	0.00	8.00	13.00
Peru	0.11[b]	0.59	3.99	2.43	4.62	7.83
Copper						
Copper	*0.29*[b]	*0.98*	*7.34*	*8.72*	*9.97*	*11.56*
Chile	0.25[c]	0.75	6.86	0.00	0.00	0.00
Peru	0.41[b]	1.66	7.97	8.86	10.23	12.00

[a] Calculations based on estimates of the export supply of a country to each of its principal export markets.
[b] Response begun after a one-period lag.
[c] Response begun after more than a one-period lag.

Source: Annex C.

15.1 Identification of Lag Structures

The identification procedure for the export supply relationship consists in determining the shape of the lag structure and finding a suitable specification of the dynamics. In the demand equation most of the response to changes in explanatory variables occurs within the first and second years; hence the *a priori* given identification of the lag structure. However, in the supply equation there can be multiple, complex, and delayed responses to price changes.

The shape of the lag structure in the supply relationship can be roughly estimated from the coefficients obtained when export supply, X_t^s, is regressed on a sequence of lagged prices, $P_t, P_{t-1}, \ldots, P_{t-m}$, for the product (see Harvey 1981: 244–5). For each commodity the maximum lag was set two years greater than the last response suggested by other studies. The procedure yielded very imprecise estimates because of the existence of slow lag responses, but it provided a rough approximation of the shape of the lag structure.

Once the general pattern of response was determined, a lag structure was applied that suitably represented the underlying nature of the response. The results suggested that the lag structures given in Table 15.2 be considered. The dampened cyclical response of beef was explained in Chapter 9 when we illustrated the convergence property of a two-period lag structure in the expression used to characterize the export supply relationship. A dampened cyclical response has also been observed in the supply of coffee (Wellman 1961) and has been tested in that of cocoa because of its similarities as a tree crop. Wellman has described two cycles in coffee, a short-term one associated with a biennial production cycle and a long-term one associated with a yield cycle. The biennial production cycle in coffee, mainly arabica-type, is attributed to the strain placed on the tree by a large crop, which leads to a small crop in the following year, during which it accumulates new strength. The yield cycle for a robusta-type tree is much longer. New plantings begin to bear fruit after about three years and production accelerates over the next 10–12 years, after which the peak yield can be maintained for over 30 years. A similar pattern is followed by cocoa trees. The traditional variety starts to yield cocoa beans five years after planting and reaches its peak yield at around 11 years; the hybrid variety takes about three years to mature and reaches its peak at around seven years. Both varieties can maintain their peak yields for over 20 years. In contrast, the banana tree produces fruit within 11–14 months; because of this, and because the banana tree has to be felled after it has produced, the possibilities for adjustment of production to prices are greater than usual with tree crops.

The exportable harvest of cotton, maize, soybeans, and sugar is

Table 15.2 Initial lag structures of export supply functions

Commodity	Lag(s) of dependent supply variable	Lags of explanatory price variable	Commodity	Lag(s) of dependent supply variable	Lags of explanatory price variable
Beef	1,2	1,6	Cocoa	1,2	3,8
Maize	1	0,4	Soybeans	1	0,2
Bananas	1	0,2	Cotton	1	0,2
Sugar	1	0,2	Iron ore	1	0,2
Coffee	1,2	5,7	Copper	1	2,7

generally influenced by the previous year's price effect on planting decisions. Sugar is additionally influenced by the capacity of processing factories and by transportation infrastructure, which respond to price changes with delays of between three and five years.

15.2 Price Elasticities

The price elasticities of export supply derived from the estimated equations are reported in the first two columns of Table 15.1. Export supply is almost always inelastic with respect to price in the short run, whereas it is nearly always elastic in the long run (except in the cases of coffee and cocoa). On average, the unweighted mean price elasticity of export supply is 0.5 in the short run and 1.6 in the long run. The trade-weighted average elasticity is about the same in the short run (0.4) and the long run (1.4). As we shall see in Chapter 16 and Part V, short-run price elasticities are germane to price stabilization policies, whereas long-run elasticities are relevant to issues of export promotion and trade liberalization.

The results for the short-term responsiveness of export supply are closely associated with the interval between the planting decision and harvesting in agricultural commodities, and between the investment decision and capacity expansion or initiation in minerals. For example, coffee has a low short-run elasticity because of the long lead time period (three to four years) between the planting response to a price change and the production of coffee beans. Exports of such commodities may be increased in the short run only by an expansion of output arising from yields or the release of stocks. The same factors operate in the export supply response of cocoa. Copper and iron ore exporters also take a long time to respond to price changes; capacity expansion requires three to four years and capacity initiation takes six to seven years in those industries. The short-term export response of these two products is associated mainly with changes in the utilization of existing capacity and changes in stocks. In contrast, over one-fourth of the total response of the export supply of bananas, maize, and sugar occurs within one or two years of a price change. Apart from periods of maturation, short-term supply elasticities are also affected by the ability or willingness of exporters to respond to price changes. Banana exports from Ecuador and cotton exports from Guatemala are very sensitive to changes that occur in their markets. In contrast, a number of exporters have a poor response to price changes. They include beef exporters from Argentina and Uruguay, cotton exporters from Peru and Mexico, and iron ore exporters from Peru. Hence there is considerable variation in the short-run price responsiveness of exporters of the same commodity.

The long-run price elasticity of export supply is equal to the sum of the lag coefficients. For both a particular exporter and the average of all exporters of a commodity, therefore, it can be calculated as the sum of the trade-weighted averages of the derived lag coefficients.[1]

The long-run average elasticity of export supply of countries in the Latin American region varies widely for the ten commodities, from a low of only 0.3 for cocoa exports to a high of 3.5 for banana exports. Above-average elasticities occur in cotton (2.5), soybeans (2.5), and maize (2.3); average elasticities occur in beef (1.2) and sugar (1.2); and below-average elasticities occur in iron ore (1.0) copper (1.0), and coffee (0.7). In the case of banana exports, an additional stimulus to the high supply response of Ecuador has been the low export-to-production ratio that is maintained in the Central American countries. This excess production policy allows for an above-normal response in the exportable volume of bananas when changes in prices take place. The low elasticity of the sugar export supply of nearly all Latin American exporters is largely attributable to preferential arrangements that have dominated trade with the USA and, in the case of Jamaica, with the UK. Exports of sugar to these two markets are established by quota allocations.

These estimated elasticities cannot be compared with estimated elasticities for production of the same commodities. The reason is that world trade has grown at a much faster rate than world output as economies have become more open. From 1960 to 1987, for example, the volume of world trade in goods rose, on average, one-and-a-half times faster than real GDP (see Table 11.1). Consequently, changes in export supply associated with changes in market prices of commodities have been greater than changes in output associated with the same price changes.[2] The appropriate comparison is with other elasticity estimates of export supply.

Unfortunately, there are far fewer estimates of export supply relationships than there are of export demand relationships. Haynes and Stone (1983) report that 'Stern, Francis, and Schumacher's (1976) bibliographical survey of price elasticities in international trade devotes over 350 pages to demand estimates but barely 10 pages to supply estimates.' The more recent survey by Goldstein and Khan (1985: Table 4.5) lists only seven studies that report estimates of price elasticities of export supply, and these refer to exports of the industrialized countries. Since one of the questions to be addressed in this chapter is whether generally lower price elasticities of export supply have caused the growth in exports of the developing countries to lag behind that of the industrialized countries, those estimates will be used as the basis for comparison.

As estimated by other economists, the average long-run price elasticity of the total export supply of the industrialized countries is equal to 3.2. This average represents the unweighted mean of those studies surveyed in Goldstein and Khan (1985) that provide elasticity estimates for total exports from several industrialized countries. The average price elasticity of supply for semi-manufactured and manufactured goods exported by the industrialized countries is somewhat higher, i.e. 3.5 in the survey of Goldstein and Khan. Accordingly, the unweighted average long-term price elasticity of supply for commodity exports in the present sample, which is 1.6, is equal to one-half of that which has been estimated by others for total exports of the industrialized countries.

By themselves, these results would appear to support the view that the slower growth of exports from the developing countries compared with the industrialized countries is attributable to a poor response by exporters to price improvements in the world markets. As Table 11.1 showed, commodity exports of Latin America grew at about two-thirds of the rate of growth of manufactured exports from the industrialized countries in 1960–1987. Assuming that prices of commodities relative to those of manufactured goods remained constant during this period, the slower rate of growth in exports could be explained by the lower response of commodity exporting countries in our sample to price improvements in the world markets.

This inference rests on *ceteris paribus* assumptions that were rejected in the last two chapters. Demand for commodity imports in our sample was also found in Chapter 13 to have a substantially lower response to economic growth in the industrialized countries than imports of manufactured goods, or imports of all types of goods, by the industrialized countries. Furthermore, the price elasticity of import demand for the commodities was found to be the same as that for all types of imports by the industrialized countries, although it was lower than that of imports of manufactured goods by the industrialized countries.

The conclusion we draw from these results is that relatively lower elasticities in both the demand for and the supply of commodity exports of the Latin American countries have caused their average growth rate to be lower than that of either overall exports or manufactured exports of the industrialized countries. This conclusion does not necessarily support the pessimistic view of the exports of the developing countries. Demand for exports of the commodities was found in Chapter 14 to be generally price-elastic and to average −1.4 for all the commodity exporters. Accordingly, relative prices have an important role to play in determining the demand for commodity exports, and consequently there is considerable leeway for accelerating the rate of growth of commodity exports from the developing countries through policies that increase their competitive position.

The results presented in this chapter suggest that the export supply of commodities from the Latin American countries is generally responsive to price movements. This responsiveness varies considerably among commodity exports of the region. Consequently, export incentive policies would tend to work in these countries, though the degree of their effectiveness would vary considerably across countries.

15.3 Lag Distributions

The speed and the manner in which export supply adjusts to a price change differ not only among commodities, but also among exporters of the same commodity. The lag coefficients, which determine the way in which export supply will respond to a change in price, have been derived from solved coefficients of the export supply equation (9.41).

Four statistics reported in Table 15.1 help to describe the shape of the lag distribution. The first two are the mean and median lags. The mean lag, denoted μ, is the average time for the response. It is calculated as[3]

$$\mu_d = \sum_d d\delta_d / \sum_d \delta_d, \qquad (15.6)$$

where δ is the lag coefficient and d is the lag.

The major limitation of average mean lags, as Hendry et al. (1984) point out, is their inability to describe asymmetrical lag distributions and their erroneous results when the lag coefficients are not all of the same sign. For this reason, additional summary statistics should be used to describe a response, for example the time taken for a certain portion of the total response to occur, or the amount of response that has transpired at the mean lag.

The median lag is the number of time periods it takes for one-half of the adjustment δ to be completed. In particular, it is the number of lagged periods required for the interim response, the sum of the normalized lag coefficients, to equal 0.5.[4] Normalized lag coefficients are the ratio of individual lag coefficients to the total of the lag coefficients. By construction, the normalized lag coefficients sum to unity. Consequently, the first step in identifying the median lag of the export supply response to price changes was to calculate the normalized lag coefficients. Next the interim response was computed. Finally, since in a number of instances half of the total response did not occur at one of the discrete intervals, the median lag was determined by interpolation.[5]

A comparison of the mean and median lags can give an idea of the shape of the lag distribution. If the lag distribution were symmetrical, the mean lag would equal the median lag.[6] A mean lag greater than a median lag indicates that the lag coefficients decline over time. This 'tailing off' of asymmetrical distributions is commonly observed in the response of most agents, since the more remote the occurrence of an explanatory variable, the less its behaviour affects the present behaviour of most agents. The extent of the difference between the mean and median lags can provide an indication of how rapidly the effect is abated. The more rapid the convergence of the lag coefficients to zero, the greater will be the difference between the mean and median lags. The rate at which the lag coefficients converge to zero is also reflected in the difference between the median lag and the time it takes for 75 or 90 per cent of the total response to occur.

A simple example will illustrate the point. Consider the distributions for two response profiles of export supply to a price change tabulated in Table 15.3 and graphed in Fig. 15.2. Both have the same total response, the sum of the lag coefficients (the elasticities) being identical, and both have the same immediate response, the short-term elasticities being equal to 1.25 (which is one-half of the long-term elasticity). The difference between the two distributions defines how rapidly export supply completes its adjustment to a price change. In the first distribution 90 per cent of the adjustment is completed by the end of the second period; in the second, it takes four periods for 90 per cent of the adjustment to be completed. Since the adjustment is fairly rapid in the first distribution, the mean lag (0.7) is near the median lag (0.5). In the second, the mean lag (1.3) is significantly greater than the median lag (0.5) — a reflection of the slower adjustment of export supply to a change in price.

The median lag already provides information on the number of periods that transpire before half of the adjustment in export supply occurs. The other periods reported are those needed to complete 75 and 90 per cent of the total response. In the two examples of lag distributions in Table 15.3 and Fig. 15.2, the distribution that represents a rapid adjustment achieves 75 per cent of its total response before a lag of one period, while the distribution that represents a slow adjustment requires 1.6 periods to achieve 75 per cent of its total response. Similarly, 90 per cent of the adjustment in export supply takes place after 1.6 periods in the first case, while in the second 3.8 periods are required for the same proportion of the adjustment.

For all the major commodity exports of Latin America, Fig. 15.1 shows that there is an asymmetrical distribution that 'tails off' in the export supply response. The export of bananas, soybeans, and cotton declines exponentially after the first or second year of a price change. However, the supply of other exports does not begin an exponential decline until several years after the price change because of the lagged responses described above. The distribution of these responses is closely associated with the length of the lags. Commodity exports that have

Table 15.3 Illustration of summary statistics to describe lag distribution

		Response profile I		
Lag	Lag coefficient	Lag-weighted average time	Normalized lag coefficient	Interim response
0	1.25	0.00	0.50	0.50
1	0.80	0.80	0.32	0.82
2	0.35	0.70	0.14	0.96
3	0.08	0.23	0.03	0.99
4	0.03	0.10	0.01	1.00
5	0.00	0.00	0.00	1.00
	2.50	1.83	1.00	

		No. of periods for	
Mean:	0.73	75% of response:	0.78
Median:	0.50[a]	90% of response:	1.57

		Response profile II		
Lag	Lag coefficient	Lag-weighted average time	Normalized lag coefficient	Interim response
0	1.25	0.00	0.50	0.50
1	0.48	0.48	0.19	0.69
2	0.25	0.50	0.10	0.79
3	0.18	0.53	0.07	0.86
4	0.13	0.52	0.05	0.91
5	0.09	0.44	0.04	0.95
6	0.06	0.34	0.02	0.97
7	0.04	0.30	0.02	0.99
8	0.03	0.20	0.01	1.00
	2.50	3.29	1.00	

		No. of periods for	
Mean:	1.32	75% of response:	1.60
Median:	0.50[a]	90% of response:	3.77

[a]The amount of 0.5 is usually added to the calculated value to make the median identical with the mean for symmetrical patterns of lag coefficients.

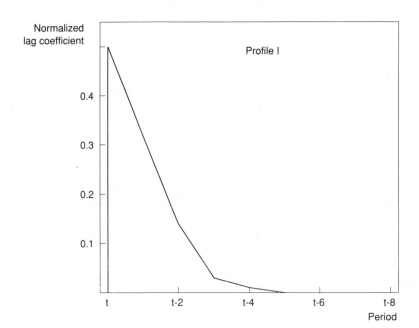
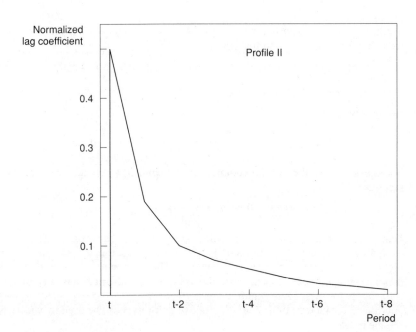

FIG. 15.2 Hypothetical response profiles of export supply functions

short-run adjustments tend to have lag distributions that quickly converge to zero. The large difference between the mean and median lags for bananas, soybeans, cotton, and iron ore indicate that the export supply of these commodities has a fast adjustment to price changes. In contrast, the supply of exports of the other commodities takes several years to adjust to price changes, and consequently the mean lags are quite similar to the median lags.

Notes

1. The long-run price elasticity can also be directly calculated, as suggested by Griliches (1967: 23), from the derived coefficients of the stochastic difference equation:

$$A(L)x_t^s = B(L)p_t + e_t$$

as the ratio of the two polynomials evaluated at $L = 1$:

$$\frac{B(1)}{A(1)} = \frac{\beta_0 + \beta_1 + \ldots + \beta_s}{1 - \alpha_1 - \alpha_2},$$

where the lag polynomials $A(L)$ and $B(L)$ are defined in the form $1 - \alpha_1 L - \alpha_2 L^2$ and $\beta_0 + \beta_1 L + \beta_2 L^2 + \ldots + \beta_s L^s$ respectively (the lag operator being defined as $Lp_t = p_{t-1}$, $L^2 p_t = p_{t-2}$, and so forth). For example, the second-order stochastic difference equation estimate for beef exports from Costa Rica to the USA,

$$x_t^s = 0.32 x_{t-1}^s - 0.33 x_{t-2}^s + 0.63 p_{t-1} + 0.68 p_{t-2} + 0.76 p_{t-7} + 0.72 p_{t-8} + 2.56,$$

can be rewritten as

$$x_t^s = \frac{0.63L + 0.68L^2 + 0.76L^7 + 0.72L^8}{1 - 0.32L + 0.33L^2} p_t.$$

Setting $L = 1$ in the lag polynomials $A(L)$ and $B(L)$ yields the long-run response,

$$\frac{0.63 + 0.68 + 0.76 + 0.72}{1 - 0.32 + 0.33} = 2.76.$$

2. Using the same specification as the export supply equation in (9.41), I estimated coefficients for the supply relationships in the world markets of eight of the commodities in our sample. The long-run price elasticity of supply averaged 0.23. In contrast, the trade-weighted price elasticity of export supply for these same commodities is 1.4.

3. As shown by Harvey (1981: 234), the following formula can be used to calculate the mean lag:

$$\mu_d = \frac{B'(1)}{B(1)} - \frac{A'(1)}{A(1)},$$

where a prime denotes differentiation of the lag polynomial. For example, given the estimated coefficients for the export supply equation of Costa Rica's beef exports to the USA expressed in terms of lag polynomials,

$$x_t^s = \frac{0.63L + 0.68L^2 + 0.76L^7 + 0.72L^8}{1 - 0.32L + 0.33L^2} p_t,$$

differentiation with respect to the lag in the polynomials gives

$$B'(L) = 0.63 + 1.36L + 5.32L^6 + 5.76L^7$$

and

$$A'(L) = -0.32 + 0.66L.$$

Hence

$$\mu_d = \frac{0.63 + 1.36 + 5.32 + 5.76}{0.63 + 0.68 + 0.76 + 0.72} - \frac{-0.32 + 0.66}{1 - 0.32 + 0.33} = 4.35.$$

4 The usual practice is to calculate the median lag directly from the derived coefficients of the estimated equation (see Pindyck and Rubinfeld 1981: 233–4, and Harvey 1981: 234). However, when the explanatory variables are lagged, the formula used will lead to erroneous results. The resulting error can easily be demonstrated. The total response in the second-order case was shown to be

$$\frac{B(1)}{A(1)} = \frac{\beta_0 + \beta_1 + \ldots + \beta_s}{1 - \alpha_1 - \alpha_2}.$$

The interim response is

$$\frac{\beta_0 + \beta_1 + \ldots + \beta_s[1 - (\alpha_1 + \alpha_2)^{d+1}]}{1 - \alpha_1 - \alpha_2}.$$

(Recall that $\alpha_1 + \alpha_2 < 0$ is a necessary restriction for a non-negative lag distribution in the export supply equation (9.41).) Hence the *normalized* interim response equals the ratio between the interim response and the total response, which yields $1 - (\alpha_1 + \alpha_2)^{d+1}$. The median lag is found at the point where $1 - (\alpha_1 + \alpha_2)^{d+1} = 0.5$. The solution for d at that point is

$$d\ln(\alpha_1 + \alpha_2) + \ln(\alpha_1 + \alpha_2) = \ln 0.5$$

or

$$\mathrm{med}_d = \frac{\ln 0.5}{\ln(\alpha_1 + \alpha_2)} - 1.$$

It follows directly that if $\alpha_2 = 0$, such that the lag distribution follows a dampened smooth path, then $\mathrm{med}_d = \ln 0.5/(\ln \alpha_1 - 1)$. When $\alpha_1 + \alpha_2 < 0.5$, the median lag will be negative, in which case Harvey (1981: 234) suggests that it should be rounded to zero. It is readily seen that the formula does not take into account the coefficients $(\beta_0 + \beta_1 + \ldots + \beta_s)$ of the explanatory variable, and is therefore only a general approximation of the total response.

5 For example, one-half of the total supply response to a unit increase in the

price of beef exports from Costa Rica to the USA is 1.84, and occurs between the second period, when the normalized interim response is 0.23, and the third period, when the normalized interim response is 0.55. The median lag therefore occurs at

$$\text{med}_d = 1 + \frac{0.50 - 0.23}{0.55 - 0.23} = 1.84.$$

If the median lag is to be identical with the mean lag for a symmetrical pattern of lag coefficients, 0.5 must be added to the median lag.

6 In a symmetrical lag distribution, the observed lags around the median are identically distributed. In particular, a distribution is symmetrical about the mean μ if for any constant, denoted c, the values $\mu - c$ and $-(\mu - c)$ have the same lag distribution.

16
Commodity Market Price Formation

Commodity market models have proliferated in the last decade. The first bibliography on commodity models by Labys (1973) lists 241 entries; in the ten years that followed that first survey an additional 684 commodity models were constructed (Labys 1987). Concurrent with the proliferation of commodity market models has been the growth in diversity of methodologies used to analyse and forecast price movements. (For a taxonomic classification of methodologies, see Labys and Pollak 1984: 38-47.) Methodologies have become so innovative, and motivation to score accurate forecasts so great, that even astrology is being used to forecast the apparently unpredictable movement, or 'random walk', of commodity prices. (See Rotton's (1985) article, 'Astrological Forecasts and the Commodity Market: Random Walks As a Source of Illusory Correlation', in *The Skeptical Inquirer*, published by the Committee for the Scientific Investigation of Claims of the Paranormal.) The random walk nature of commodity prices has led to the use of pure time-series models, whereby an attempt is made to identify regularities in movements of time series of prices that might be obscured by noise. (For an example of this type of approach see Labys and Granger 1970; Leuthold *et al.* 1970; and Chu 1978.)

However, these methods of analysing and forecasting commodity markets ignore the fundamentals of supply and demand in the determination of price. Consequently, they cannot be used to examine policies in which intervention in a market occurs either through supply, be they in the form of production, export, or stockpiling policies, or through demand, for example in the form of protectionism and subsidies in importing countries, or in the form of sectoral or macroeconomic policies in the domestic economies. For this reason, in the approach adopted in the present study the supply and demand components in the market are separately estimated and the price is determined by its steady-state equilibrium solution. This approach is certainly not new. Multi-commodity market models based on parsimonious specifications of the underlying data-generating processes have been constructed by Adams and Behrman (1976), Hwa (1985), and Gilbert and Palaskas (1990), among others.

These multi-commodity market models are motivated by the desire to apply a common modelling framework to the characterization of the underlying data-generating processes in commodity markets and, at the same time, to retain the specific features inherent in each commodity market. The features that differentiate commodity markets from one another are the structure of the market, which defines the parameters to be included in the model, and the model specification, which determines the values to be assigned to the parameters. The structure of commodity markets differs in the extent to which there is fragmentation; for example, the US and EEC sugar markets are considered to be separate from those of the rest of the world because of price discrimination. The structure of a market is also determined by the degree of substitution of commodities, as is the fats and oils market. In contrast, the model specification determines differences in the magnitude of the price and income elasticities and in the lag structures of economic agents, which are particularly important in commodity markets because differences between lag structures in supply and demand functions are what give rise to observed price cycles in many of these markets.

The guidelines for applying the common modelling framework to price formation in commodity markets have already been set out in earlier chapters. The theory of commodity markets was described in Part II, and the dynamics underlying the data-generating processes of the relationships in those markets were specified in Part III. The postulated theory provided a parsimonious interpretation of the process of price formation. It described demand and supply, as well as equilibrium conditions, in commodity markets, and it made explicit the constraints that need to be imposed in order to formulate and estimate a complete market system. The price of a commodity that can be stored for a long time period was shown to be determined by the adjustment of actual stock levels to desired stock levels. Figure 16.1 provides a visual representation of the equations of the system that have been derived in order to describe the underlying features of commodity markets. (The variables are enclosed in a box if they are endogenous and in a circle if they are exogenous.) In contrast, the price of a commodity that is perishable was shown to be determined by the equilibrium condition between production and consumption. The dynamic specification of the relationships in the system used to determine the market price of a commodity both encompasses previous specifications and reproduces the postulated theory of market price formation. It reflects the findings of recent studies on dynamic time-series models that explain observed disequilibrium in the context of long-term, or steady-state, solutions of behavioural relationships in a market.

The plan of this chapter is as follows. The first two sections describe the results for the consumption and production functions respectively;

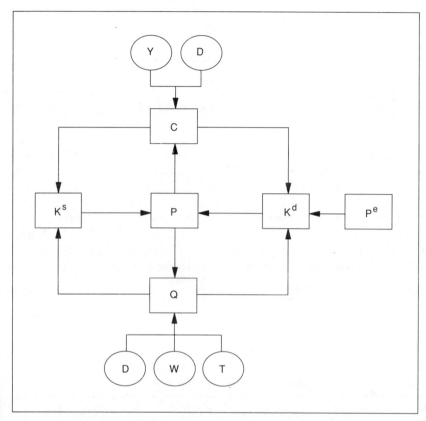

FIG. 16.1 Schematic diagram of price formation in commodity markets with stocks

C	consumption	P^e	expected market price
D	deflator	Q	production
K^d	demand for stocks	T	secular trend
K^s	supply of stocks	W	major disturbance
P	market price	Y	income

Section 16.3 describes the results for stock demand in storable commodity markets; Section 16.4 presents the results for the market price equation of each commodity; and the final section provides a sensitivity analysis of the market models.

16.1 Consumption

The empirical results for the consumption equation indicate that the general dynamic specification provides a good representation of the data-generating process in commodity markets (see Annex C). A test of parameter constancy, based on the Chow test, showed the coefficients to be stable at the 5 per cent level of significance in all the estimated relationships.

The main parameters of the consumption function are presented in Table 16.1. The income elasticities of eight of the commodities that were estimated in their structural form were, on average, equal to 0.83.[1] However, individual elasticities vary between 0.6 and 1.0. As a result, it would be misleading to make generalizations about elasticities on the basis of an average. None the less, a pattern can be discerned among the commodities. Copper, soybeans, cotton, and maize have income elasticities that are equal to unity. In contrast, sugar, coffee, and cocoa have income-inelastic consumption functions.[2] These findings conform to the characterization of the markets for these commodities in Chapter 12. The description of the market for sugar in that chapter also suggested that, because of the large amount of protection given to producers, which alters the price to producers and consumers and, consequently, the behaviour of these agents, separate consideration should be given to consumption and production in the USA, which is

Table 16.1 Consumption functions

	Income elasticity[a]	Price elasticity	Income growth elasticity[b]
Beef	0.56	−0.34	−3.10
Maize	1.00	−0.27	2.87
Sugar	*0.89*[c]	*−0.07*[c]	*−1.28*[c]
USA	1.00	−0.12	0.44
EEC	0.11	−	−0.09
Rest-of-world	1.00	−0.07	−1.63
Coffee	0.62	−0.17	0.35
Cocoa	0.62	−0.19	0.20
Soybeans	1.00	−0.28	−1.31
Cotton	1.00	−0.75	−2.87
Copper	1.00	−0.08	7.28

[a] Measured with respect to a change in aggregate, rather than per capita, real GDP.
[b] The income growth elasticity is the percentage change in consumption brought about by a 1% change in the rate of growth of economic activity, g_4, in equation (8.30).
[c] Trade-weighted average.

Note: − Not significant at the 5% level.
Source: Annex C.

one of the largest markets for sugar, and in the EEC, which has become the largest sugar-producing area.

My own estimated income elasticities are generally similar to those calculated by others.[3] In coffee, Akiyama and Duncan (1982) and Hwa (1985) calculated the long-term income elasticity to be 0.45, compared with 0.62 in the present study. The elasticity estimate for cocoa is very similar to that of 0.6 calculated by Hwa, though it is significantly higher than that of 0.25 estimated by Weymar (1968). The unitary income elasticities for cotton and copper are close to the estimate of 1.25 by Monke and Guisinger (1985) and to that of 1.2 by Adams and Behrman (1976) for cotton and of 0.9 by Hwa (1985) for copper.[4]

In addition to the effects that changes in the *level* of economic activity have on consumption, changes in the *rate of growth* of economic activity have generally strong dynamic effects on the demand for the commodities. These effects, which are measured by the income growth elasticity, produce either a positive or a negative impact on demand. When changes in the long-term rate of growth of income, denoted g_4, bring about a less-than-proportional change in the rate of growth of consumption, g_1, the demand for the commodity will decrease in response to an acceleration in the rate of economic growth. In contrast, when changes in g_4 cause a more-than-proportional change in g_1, the demand for the commodity will increase as a result of an acceleration in the rate of economic growth. As Table 16.1 shows, positive changes in the rate of growth of economic activity produce upward shifts in the demand for copper, maize, and (to a lesser extent) coffee and cocoa. The same changes bring about downward shifts in the demand for beef, cotton, soybeans, and sugar.

As expected, the price elasticity of demand for all the commodities is less than unity. The average elasticity for the commodities is equal to −0.27, although the individual elasticities range from a low of −0.07 for sugar to a high of −0.75 for cotton. However, despite this wide range, most elasticities are between roughly −0.15 and −0.3. The short-run price and income elasticities of demand for the commodities are presented in Table 16.2. The difference between the price and income elasticities in the short run and those in the long run is important since the effects of some policies, such as commodity price stabilization, depend on short-run elasticities while those of others, such as protectionism and subsidies in foreign markets, depend on long-run elasticities.

Other studies have also calculated long-term price-inelastic demand functions for these commodities. The very low price elasticity for sugar is consistent with that of −0.02 estimated by Hwa (1985), although it is somewhat lower than the −0.12 estimated by Adams and Behrman (1976).[5] The present calculation for copper is similar to that of −0.07 by Hwa, but significantly lower than the estimate of −0.4 by Fisher *et al*.

Table 16.2 Short-term price and income elasticities of consumption

	Price elasticity	Income elasticity[a]
Beef	−0.11	0.18[b]
Maize	−0.11	1.60
Sugar	*−0.04*[c]	*0.66*[c]
USA	−0.07[b]	0.76
EEC	—	0.92
Rest-of-world	−0.04	0.62[b]
Coffee	−0.25	1.34
Cocoa	−0.14	1.10
Soybeans	−0.22	0.76[b]
Cotton	−0.26[b]	0.35[b]
Copper	−0.02[b]	2.79

[a] Measured with respect to a change in aggregate, rather than per capita, real GDP.
[b] One-period lag.
[c] Trade-weighted average.

Note: — Not significant at the 5% level.
Source: Annex C.

(1972); the estimate for cotton lies between the estimate of −3.4 by Monke and Guisinger (1985) and that of −0.29 by Adams and Behrman (1976); similarly, the present result for soybeans lies between the estimate of −0.35 by Houck and Mann (1968) and that of −0.13 by Augusto and Pollak (1981). Somewhat lower elasticities were obtained in the present study for coffee and cocoa. The elasticity calculations for coffee by Adams and Behrman (1976), by Akiyama and Duncan (1982), and by Hwa (1985) range from −0.24 to −0.48, compared with −0.17 in the present study. In cocoa, the estimates by Adams and Behrman (1976), by Hwa (1985), and by Weymar (1968) lie between −0.33 and −0.41, compared with −0.19 in the present study.

The demand for most commodities begins to adjust in the same period as that in which income and price changes occur. Exceptionally, there is a one-year delay in the adjustment of the demand for beef, soybeans, and cotton to a change in income, and there is the same delay in the response of the consumption of copper, cotton, and sugar in the USA to a change in price. Once the adjustment begins, the speed with which consumption moves from one steady-state equilibrium solution to another depends on the coefficient of the error-correcting term in the demand relationship. The coefficient is close to unity in absolute terms in half of the commodities (sugar, coffee, soybeans, and cocoa), which reflects the relatively quick response of consumers to changes in income and prices; in other demand relationships (those for beef, maize, cotton, and copper) the smaller absolute values of the error-correcting coefficients indicate a relatively slow adjustment from one steady-state equilibrium solution to another. The adjustment of consumption to a new level of income follows an oscillating path in the case of maize, coffee, cocoa, copper, and sugar in the US market, and has a dampened smooth response in that of beef, soybeans, and cotton. In markets for sugar other than that of the USA, sugar too has a dampened smooth response.

Finally, as mentioned earlier in the discussion of the empirical results for international trade relationships, major disturbances tend to be less frequent in demand than in supply. Those that did occur in consumption in the sample period were concentrated around the mid-1970s and early 1980s, a period when sharp price movements in commodity markets caused unusually large, albeit temporary, shifts in consumption patterns.

16.2 Production

The two fundamental issues in the supply of commodities—price elasticities and lag structures—have already been extensively addressed in Chapter 15. This section is therefore limited to the presentation of the estimated price elasticities and lag structures of the eight commodity markets whose system of equations was estimated in its structural form.

16.2.1 Price Elasticities

The first two columns of Table 16.3 present the impact and total price elasticities of supply for the commodities. The impact elasticity is the first-period response of supply to a change in the market price. It is similar to the short-run elasticity in that it measures the first response of supply to a price change; it is different in that the first response to a price change does not take place for several years in some commodities. The number of periods that elapse before the first response occurs is indicated in parentheses beside the impact elasticity. For example, it takes four years for Brazil's coffee production to respond to a price change. The total price elasticity of supply is the same as the long-run price elasticity.

Production is inelastic with respect to prices in all the commodities. The average elasticity of the estimated production functions is 0.23, and the range is 0.05 (for sugar) to 0.6 (for copper). Two commodities (cotton and soybeans) have elasticities of around 0.35; the other four

Table 16.3 Production functions

	Price elasticity		Average lag		No. of periods for:	
	Impact[a]	Total[b]	Mean	Median	75% of response	90% of response
Beef	−0.08 (t−1)	0.13	10.4	7.2	10.0	13.6
Maize	0.09 (t−2)	0.10	2.1	1.5	1.8	2.0
Sugar	*0.04*[c]	*0.05*[c]				
USA	0.09 (t−2)	0.18	2.6	1.9	2.8	3.9
EEC	0.11 (t−1)	0.12	1.1	0.5	0.8	1.0
Rest-of-world	0.03 (t−3)	0.03	3.0	2.5	2.8	2.9
Coffee	*0.14*[c]	*0.16*[c]				
Brazil	0.28 (t−4)	0.29	4.0	3.5	3.8	3.9
Rest-of-world	0.08 (t−3)	0.10	3.2	2.6	2.9	3.6
Cocoa	0.09 (t−8)	0.10	8.1	7.5	7.8	8.0
Soybeans	0.25 (t−1)	0.33	1.3	0.7	1.0	1.8
Cotton	0.13 (t−3)	0.38	4.8	3.7	5.2	7.4
Copper	0.13 (t)	0.61	17.4	13.8	27.0	40.0

[a] The impact elasticity measures the first-period response of production to a change in price. The notation in parentheses indicates the period in which production first responds to a change in price.
[b] The total, or long-run, elasticity measures the cumulative response of production to a change in price.
[c] Trade-weighted average.

Source: Derived from Annex C.

(maize, coffee, cocoa, and beef) have elasticities of around 0.1 to 0.15. The average impact elasticity of the commodities is 0.1, and all but one of the commodities have a dampened smooth response to a price change. The exception is beef, which has a dampened cyclical response. The exceptional lag structure of this commodity is described in detail below.

These results are generally consistent with estimates by others.[6,7] Nevertheless, the range of elasticity estimates in other studies tends to be fairly wide, and therefore generalizations can be misleading. For instance, the long-run price elasticity of sugar production estimated in the present study is similar to that estimated by Hwa (1985), but is significantly below the 0.2 estimate by Adams and Behrman (1976).[8] My calculated elasticities for cocoa and cotton are also lower than those of Adams and Behrman, who estimated the elasticity for cocoa at 0.34 and that for cotton at 0.66. In contrast, in coffee my own elasticity estimate is fairly close to that of Adams and Behrman, but substantially lower than the estimate of 0.9 by Hwa and that of 0.7 by Akiyama and Duncan (1982). My calculated elasticity for copper is substantially higher than the estimate of 0.1 by Hwa, but lower than that of 2.3 calculated by Fisher et al. (1972). In soybeans, my own elasticity estimate is fairly similar to that of 0.45 calculated by Houck and Mann (1968), whereas Augusto and Pollak (1981) obtained a much larger estimate of 2.8. Accordingly, the present elasticity estimates are within the range of those calculated in other studies, although the range of those estimates tends to be wide.

A trend variable was included in the production relationship in order to capture various forms of productivity improvements and institutional policies oriented towards influencing producer decisions in each of the commodities. The variable was found to be statistically significant in all commodities except cotton and copper, and the signs of the coefficients were all positive. The coefficients of soybeans and maize were particularly high. Soybean production began to be affected by technological changes when access to traditional Asian suppliers was cut off during the Second World War. The USA expanded its production and initiated major improvements in the method of processing the product. Later, Brazil took the lead in expanding production, and one of its government's major agricultural policy objectives has been to increase the country's share of the market. In the case of maize, the trend coefficient is almost entirely a reflection of major technological innovations. Photosynthesis enhancement has improved the growth rate of crops by speeding up the natural process through which plants absorb nitrogen for protein synthesis. Other innovations, according to Simpson and Farris (1982), have included improved crop hybrids, new pest control methods, and better fertilizer and water management systems. Finally, a quadratic form of the trend variable was included in the relationship for EEC sugar production in order to capture a dampening effect on increased productivity changes that institutional policies have had on producer decisions in the EEC sugar industry.

Supply disturbances related to weather and disease in the case of agricultural products and labour disputes in that of minerals occur with frequent but unpredictable regularity. Variables that account for major random disturbances have been included in the estimated equations and

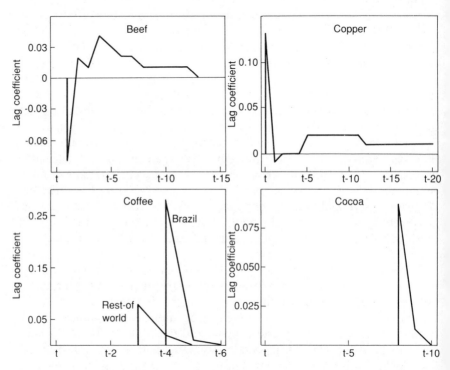

FIG. 16.2 Response profiles of production to one-time price change

	Lag											
Lag coefficient	t	$t-1$	$t-2$	$t-3$	$t-4$	$t-5$	$t-6$	$t-7$	$t-8$	$t-9$	$t-10$	$t-11$
Beef		−0.08	0.02	0.01	0.04	0.03	0.02	0.02	0.01	0.01	0.01	0.00
Coffee												
Brazil				0.28	0.01	0.00						
Rest-of-world				0.08	0.02	0.00						
Copper	0.13	−0.01	0.00	0.00	0.00	0.02	0.02	0.02	0.02	0.02	0.02	0.01
Cocoa							0.09	0.01	0.00			

are reported in Annex C. Each of these variables uses a binary series in which a value of one has been assigned to the year in which a disturbance in output occurred and of zero to all other years.

16.2.2 Lag Distributions

There is usually a short-run response of production to price variations based on variable input changes within existing capacity, and there is a long-run response based on capacity changes. These long delays in

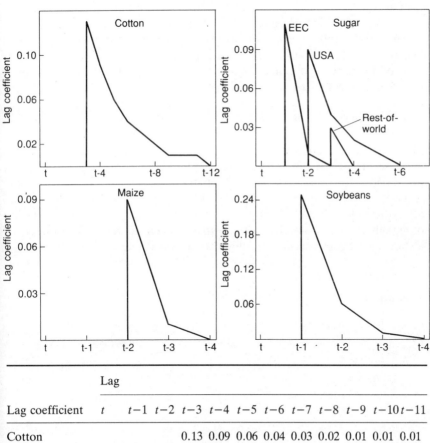

Lag coefficient	t	t−1	t−2	t−3	t−4	t−5	t−6	t−7	t−8	t−9	t−10	t−11
Cotton			0.13	0.09	0.06	0.04	0.03	0.02	0.01	0.01	0.01	
Maize		0.09	0.01	0.01								
Sugar												
USA			0.09	0.04	0.02	0.01	0.00					
EEC		0.11	0.01	0.00								
Rest-of-world				0.03	0.00							
Soybeans		0.25	0.06	0.01	0.00							

capacity adjustment give rise to observed long-term cyclical swings in commodity prices. However, the length and shape of these cycles depend on the lag structures that characterize individual commodity responses. Important differences in the estimated lag structures of the commodities reflect differences in the production cycle of each commodity. Figure 16.2 shows the lag structures for production in each of the commodities. These structures are briefly described below.

Beef Production has a negative short-run response to price changes and a positive long-run response. These estimates are consistent with the 'cattle cycle' explanation of the beef market. Although the length of the cycle can vary among countries, the phases are similar, according to Jarvis (1974) and others.[9] Figure 16.3 illustrates a typical cycle. The beginning of the cycle occurs when a price fall causes losses in revenue to producers, who are then forced to reduce inventories to lower their costs and meet their expenses. Phase I in the cycle is therefore characterized by a reduction in inventories concurrent with the price fall. The reduction in inventories is accompanied by an increase in slaughter, i.e. in production. Increased production induces a further fall in price, which stimulates a reduction in inventories and a further increase in production (Phase II). Hence the short-run response of production to price changes is negative, unlike the positive supply response of most other commodities. Phase III begins two to three years after the initial downturn in prices. It is characterized by a deceleration of slaughter rates once the cattle herd has been depleted. The effects of a shortage of production then begin to appear in the market.

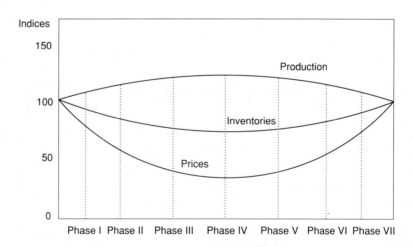

FIG. 16.3 The cattle cycle

An upswing in prices marks the beginning of Phase IV. At first prices rise moderately, but then they accelerate (Phase V) as producers hold back heifers in order to breed them to increase inventories. As slaughter cattle prices improve, the price expectations of producers become more optimistic. This optimism stimulates an increase in inventories arising from a reduction in slaughter, which causes the price rise further to accelerate (Phase VI). The increase in inventories requires three to five years because of the time it takes for cattle to reach marketable weight (Phase VII).[10] During the upswing in prices, therefore, the production response is negatively related to price changes. Because more and more cattle are held back from the market, slaughter eventually becomes unavoidable. The slaughter rate rises, supply fills the market, and prices eventually begin to fall, which marks the beginning of a new cycle.

Maize The planting cycle for maize is short, and harvesting occurs year-round throughout the world. Production decisions are based on prices in the previous year because the marketing year of maize is a split year. For example, in the USA, where the harvest occurs in August–November, the decision about the quantity of maize to be planted the following year is based on the market prices in effect either during the harvest or in the period prior to the harvest. As a result, production tends to respond to changes in prices with a lag of either one or two years. In the estimated relationship, the effect of prices lagged two years was found to be significant. As expected, the full impact of the price change in the estimated relationship occurs very rapidly.

Sugar The adjustment of cane production to price changes is slow because cane occupies the soil from five to twenty years, depending on the ratooning practices followed (Grissa 1976: 79–82). The large fixed cost of sugar production also slows down the response to market price changes, and when prices outstrip mill capacity, lengthen the time for output to come on stream. The estimated response of producers to price changes was found to be consistent with a slower response than would be suggested by only the planting cycle. The response of producers, other than those in the USA and the EEC area, begins to occur three years after a price change. Thereafter, the effect of the price change on production declines exponentially. Producers of sugar beet in the USA and the EEC area have a faster response to price changes than those in other producing countries, which are mainly the developing countries; a response was found to occur after one year in the EEC area and after two years in the USA. Adams and Behrman (1976: 38–40) also found the supply response in the industrialized countries to be faster than in the developing countries. Binary variables were included to account for the change in US sugar policy in 1974, the introduction of incentives for domestic farmers to expand production in the year preceding the change, and the entry of the UK into the EEC.

Coffee Unlike the relatively short planting and harvesting cycle of many agricultural commodities, that of coffee is long-term and is often interrupted by climatic factors. The cycle arises from the interim between the time of planting and the first harvest. For the robusta-type the average lag for the first harvest is two years; for the arabicas it is four years. However, mature yields do not occur until a few years later, the average lag between planting and the first mature yield being four years for robusta trees and seven years for arabica trees (Edwards and Parikh 1976). Once a coffee tree begins to produce mature yields, it continues to produce beans regularly for the remaining 12–30 years of its normal life (Starbird 1981). As expected, the output of Brazil, which produces arabica coffee, was found to respond to prices with a longer lag—between four and five years—than that of the rest of the world, where the response of production to a change in price was found to be between three and four years. Binary variables were introduced to capture years in which particularly severe frost damage affected output. Finally, although production responds with a fairly long lag to a change in the market price, adjustment is very rapid once it begins.

Cocoa Traditional varieties of cocoa trees produce cocoa beans approximately five years after planting and the hybrid variety has a lead time of about three years. After the initial crop, yields increase sharply until the eighth or tenth year, when disease begins to affect some of the weaker trees. Yields continue to increase noticeably until the twentieth year, when they level off and begin a gradual decline. Some trees may live up to 30 or 50 years, although maintenance and harvesting costs often outweigh the revenue they produce.

Most of the major cocoa-producing countries administer the price received by domestic producers. It would therefore seem more appropriate to use producer prices than a representative world market price in estimating the production function. However, among the five major producing countries, only Brazil and Ghana had significant responses to changes in constant local currency cocoa prices; the other three producing countries did not have a statistically significant response to variations in constant local currency prices. The final equation for world production showed that production responds to a change in cocoa market prices after a lag of eight years. As Fig. 16.2 shows, full adjustment to a price change takes ten years and the adjustment to the new equilibrium relationship is characterized by a dampened smooth response.

Soybeans Production decisions are based on the previous year's prices and the price guaranteed to producers by governments. A one-year lagged price influences production because the marketing year of soybeans is a split year. For example, in the USA, where the beans are harvested in August, the decision on the quantity of soybeans to be planted in the following year is based on the market price in effect during the harvest, or during the period immediately preceding the harvest. In major producing countries, such as the USA and Brazil, production decisions are also based on government guaranteed minimum prices. According to Griffith and Meilke (1982), the floor, or minimum, price in Brazil is maintained by intervention purchases; non-recourse loans are the price support mechanism in the USA.

Cotton Cotton is an annual crop in most areas of the world, although it is produced as a perennial crop in certain tropical areas such as Brazil. However, the estimate for the production relationship indicates that there is a three-year lag before production adjusts to changes in prices. Moreover, as Fig. 16.2 shows, the full adjustment to the change in price takes several years to complete. Chinese government policies in the first part of the 1980s created anomalies in the behaviour of global cotton production patterns, for which binary variables were introduced. These anomalies were due to the Chinese government's incentive programmes that strongly encouraged domestic cotton production.

Copper Mine capacity expansion is fixed in the short run. The response of mine capacity to changes in copper prices can be separated into the capacity obtained from existing facilities and the capacity to be derived from entry into production of new mines. In general, the expansion of output from existing facilities occurs in the same year, or within one year, of the price change; the entry into production of new mines takes six to eight years. The estimated production relationship for copper includes the short-term output response arising from changes in the use of existing capacity, and a five-year lagged response that reflects the time required to expand existing capacity and to bring new capacity into production.

16.3 Demand for Stocks

The estimates of the stock demand functions for eight of the ten commodities covered in this study are given in Table 16.4. Stocks have a proportional response to changes in the level of production or consumption in the long run. However, the ratio of stocks to production or consumption is shown to vary with expected prices. In some cases the

Table 16.4 Stock demand functions

	Elasticity w.r.t. output or consumption		Elasticity w.r.t. expected prices[a]	Production or consumption growth elasticity	K/Q ratio at g[b]
	Short-run	Long-run			
Beef	0.8	1.0	2.0 (E_t)	−4.3	n.a.[c]
Maize	1.4	1.0	—	1.6	0.16 ($g = 5.0\%$)
Sugar	0.6	1.0	0.1 (E_{t-1})	−1.2	0.59 ($g = 3.4\%$)
Coffee	0.5	1.0	1.1 (E_{t-2})	−3.0	0.76 ($g = 3.3\%$)
Cocoa	2.0	1.0	1.2 (E_{t-2})	10.8	0.35 ($g = 1.6\%$)
Soybeans	1.4	1.0	2.7 (E_{t-2})	1.3	0.16 ($g = 5.9\%$)
Cotton	0.8	1.0	0.3 (E_{t-2})	−0.5	0.38 ($g = 2.8\%$)
Copper[d]	0.4[e]	1.0	3.1 (E_{t-2})	−2.8	0.16 ($g = 2.9\%$)

[a] The expectations operator E, shown next to the elasticity, indicates whether expectations are based on the information available at period t, $t-1$, or $t-2$.
[b] The ratio of stocks K to production Q or consumption C is given at the level at which the production or consumption growth rate g equals the average rate over the period in which the equation was estimated.
[c] Beef stock data are in index form since data on actual levels of stocks are unavailable; consequently, the ratio of the stock index to the quantity of production or consumption would be meaningless.
[d] The current interest rate, based on the 3 month US Treasury bill rate, was also included in the relationship.
[e] One-period lag.
Source: Annex C.

estimates of the relationship indicate that expectations are formed with information available at period t; in others they are formed with information available at period $t-1$ or $t-2$. In all but one relationship, the demand for stocks was found to be very responsive to expected prices, the mean average elasticity being 1.3 and the range, between 0.1 for sugar and 3.1 for copper. Only in maize were expected prices not found to be significant in explaining the demand for stocks.

The autoregressive integrated moving average (ARIMA) estimates of market prices are based on the procedures for identification, estimation, and diagnostic checks suggested by Box and Jenkins (1976). In general, the price series have ARIMA processes of a low order. All series except one are integrated of order one, in so far as they need to be first differenced to become stationary. The sugar price series, which is the exception, is integrated of order zero; i.e. sugar prices tend to fluctuate around their long-term average. The constant terms in the ARIMA estimates of market prices represent the amounts by which the market prices increase in each period, and therefore define the trends. Most commodity prices have autoregressive processes of order one or two, i.e. $AR(1)$ or $AR(2)$; as expected, the autoregressive processes of coffee and cocoa prices, which are characterized by very long cycles, have autoregressive processes of a much higher order, i.e. $AR(11)$ for coffee and $AR(12)$ for cocoa. The moving-average components indicate that some processes that generate price movements have long memories of innovations. Sugar and cotton prices are both affected by innovations in the process that occurred six periods earlier. Nevertheless, other commodity prices are influenced only by current innovations and those of one period earlier.

Interest rates were included in the relationship for copper. The interest rate motivates part of the transactional demand for stocks and part of the speculative demand. As a transactions motive, interest rates affect the cost of stockholding; as a speculative motive, interest rates induce investors to move from one financial and commodity market to another, and have done so particularly since the mid-1970s when they moved from commodity markets to the Eurodollar market in response to interest rate variations. The elasticity of demand for copper stocks with respect to interest rates was found to be -2.1.

In steady-state growth, the demand for stocks of these commodities varies not only with the level, but also with the rate of growth, of production or consumption. In particular, the production or consumption growth elasticity ranges between -4.3 and 10.8. Thus, for example, if the growth of cocoa production were to rise from an average annual rate of 1.6 per cent (as it did in 1960–87) to 3.0 per cent, the long-run equilibrium ratio of stocks to consumption would increase from 0.35 to 0.41. The results shown in Table 16.4 indicate that, in the case of cocoa, maize, and soybeans, the demand for stocks increases when changes in consumption or production lead to a more-than-proportional change in stock demand ($\alpha_{71} > 1$ in (8.38)), whereas in the case of copper, beef, coffee, sugar, and cotton the demand for stocks decreases when consumption or production causes a less-than-proportional change in stock demand ($\alpha_{71} < 1$ in (8.38)).

16.4 Market Prices

The properties of the dynamic system of equations for commodity trade can be derived from the reduced form or, more generally, the final form of the model. The reduced form represents the solution to the structural system of equations in which the current values of the endogenous variables are related to both the lagged endogenous variables and the current and lagged exogenous variables in the system. The final form of the model relates the endogenous variables only to the exogenous variables.

The final form can be obtained from matrices of lag polynomials for the endogenous and the exogenous variables of the structural form of the model. The representation of the structural form of the model in terms of vectors and matrices requires the relationships in the system to be linear in parameters. When stocks are not maintained for a commodity, as in the case of bananas and iron ore,[11] the equilibrium condition is given by the equality between production and consumption. In this case, the autoregressive final form of the system of equations is (Wallis 1977; Harvey 1981: 339–43)

$$|A(L)|w_t = [\text{adj}\, A(L)]B(L)y_t + [\text{adj}\, A(L)]u_t, \qquad (16.7)$$

where $A(L)$ is a matrix of lag polynomials for the endogenous variables, $[\text{adj}\, A(L)]$ is the adjoint matrix of $A(L)$, and $B(L)$ is a matrix of lag polynomials for the exogenous variables. The resulting systematic dynamics of the final form of the system, obtained from the consumption and production functions in (8.29) and (9.47), are

$$q_t = \frac{(2 + \beta_{51})\beta_{34} - (2 + \beta_{32})\alpha_{53} + \gamma_{54} - \beta_{35}}{(\beta_{34} - \alpha_{53})} q_{t-1}$$
$$+ \frac{(\beta_{35} - \beta_{34})(1 + \beta_{51}) - (\gamma_{54} - \alpha_{53})(1 + \beta_{32})}{(\beta_{34} - \alpha_{53})} q_{t-2}$$
$$+ \omega_1 y_t + \omega_2 y_{t-1} + \omega_3 y_{t-2} + \omega_4 d_t + \omega_5 d_{t-1} + \omega_6 d_{t-2} + \omega_0 \quad (16.8)$$

and

$$p_t = \frac{(2 + \beta_{51})\beta_{34} - (2 + \beta_{32})\alpha_{53} + \gamma_{54} - \beta_{35}}{(\beta_{34} - \alpha_{53})} p_{t-1}$$
$$+ \frac{(\beta_{35} - \beta_{34})(1 + \beta_{51}) - (\gamma_{54} - \alpha_{53})(1 + \beta_{32})}{(\beta_{34} - \alpha_{53})} p_{t-2}$$
$$+ \kappa_1 y_t + \kappa_2 y_{t-1} + \kappa_3 y_{t-2} + \kappa_4 d_t + \kappa_5 d_{t-1} + \kappa_6 d_{t-2} + \kappa_0, \quad (16.9)$$

when both consumption and production have contemporaneous responses to price changes. Both quantity q and price p have the same autoregressive coefficients, a common characteristic of any final form of a model.

When commodities are storable and stocks are maintained, the relationships in the system of equations (5.33), (8.29), (8.38), and (9.47) are nonlinear in parameters. The reason is that (5.33) contains additive terms while the other three equations contain multiplicative terms. In this case the structural form of the model cannot be represented in terms of vectors and matrices, and (16.7) cannot be invoked in order to find the reduced form. Instead, the dynamic properties of the model are examined from the reduced form.

There are three possible solutions for the reduced form, each of which has important consequences for the dynamic properties of a commodity market model. The first possibility occurs when production and consumption are both related to the current price of a commodity. Its solution is a polynomial in price equal to $aP^{\gamma_1} + bP^{\phi-1} = K_t$, where a and b are constant terms composed of relatively complicated expressions of predetermined variables and their coefficients. The polynomial in price has powers that are real numbers which can be approximated through numerical optimization methods.

The other two possible solutions occur when either the consumption or the production relationship has a contemporaneous response to the price of the commodity and the other relationship has a lagged response to price. In commodity markets, delays typically arise in the response of production to market price changes. The empirical findings in Section 16.2 have confirmed the expectation that production in most commodities responds to price variations with a delay. These delays occur because of the time that elapses between the change in the price of a commodity and the change in the level of production. Lags in production were found in seven of the eight storable commodities covered in this study. Only copper production was found to have a same-period response to price changes, whereas its consumption was found to adjust to price changes with a lag.

When production has a lagged response to market price changes and consumption has a contemporaneous response, the reduced form for the market price relationship is

$$P_t = [(Q_t - \Delta K_t)/g'(C_t)]^{1/\beta_{34}}, \tag{16.10}$$

where the expression is raised to a power that is the inverse of the short-term price elasticity of demand for the commodity, the numerator of the expression in brackets is the total supply of the commodity, and $g'(C_t)$ is the inverse demand function, whose coefficients are defined as in the consumption equation (8.29). In particular,

$$g'(C_t) = e^{\alpha_{30}} Y_t^{\alpha_{31}} Y_{t-1}^{\beta_{33}-\beta_{32}-\alpha_{31}} D_t^{-\beta_{34}} (P/D)_{t-1}^{\beta_{35}-\beta_{34}} C_{t-1}^{1+\beta_{32}}.$$

Equation (16.10) describes how price changes can take place in a market. First, it should be noted that the expected sign of the short-term price elasticity of demand, β_{34}, in the exponent of the term is negative, and therefore the term inside the bracket is inverted. (The numerator becomes the denominator and vice versa.) An increase in the difference between total supply and desired stocks will induce a fall in the market price, whereas any decrease in the difference will cause the market price to increase. It can also be observed from the equation that the market price will move in the same direction as a change in either income or the general price level.

When consumption responds with a lag to market price changes and production has a contemporaneous response, the reduced form for the market price relationship is

$$P_t = [(C_t + \Delta K_t)/h'(Q)]^{1/\alpha_{53}}, \tag{16.11}$$

where α_{53} is the short-term price elasticity of supply, the numerator of the expression in brackets is the total availability of the commodity, and $h'(Q)$ in the denominator is the inverse supply function. In particular,

$$h'(Q) = e^{\alpha_{50}+\beta_{55}T+\beta_{56}W} Q_{t-1}^{1+\beta_{51}} Q_{t-2}^{\alpha_{52}} D_t^{-\alpha_{53}} (P/D)_{t-m-1}^{\alpha_{54}+m} (P/D)_{t-n}^{\alpha_{53}+n}.$$

The sensitivity of commodity prices to changes in market conditions can be appreciated from the empirical results presented in the first two sections of this chapter. As noted above, the expression in brackets of the price equation for storable commodities is raised to the power of the inverse of either the short-term price elasticity of demand or the short-term price elasticity of supply. For example, the power of the expression for copper is -33; i.e. $1/-0.03$, where -0.03 is the short-term price elasticity of demand for copper. Thus, a significant change in one of the variables in the bracketed term has a strong effect on the market price.

Prices in major protected markets have been estimated by means of a combination of binary variables reflecting policy instruments and market prices. For instance, the US price for sugar has been explained by an equation that links it to the world price and also accounts for the domestic support system. The US sugar price follows the world price closely when the world price is above the domestic floor price, but it departs from the world price when that price falls below the floor price in the USA. Particularly large divergences between the two prices occurred in 1965 and 1982–4, and these two divergences were captured by binary variables. In both cases the coefficients had the expected positive sign.

16.4.1 Time Path Convergence

The dynamic properties of the commodity market models determine the response characteristics of the system. In particular, they determine whether the system is stable when subjected to a single alteration in one of the predetermined variables, which path is followed in the return to steady-state growth, and the degree of response to the external perturbation.

In general terms, a system of equations describing the data-generating process of a commodity market will be stable if the roots of $A(L)$ in (16.7) lie outside the unit circle. Suppose production and consumption both depend on the current market price of a commodity. Let g represent the coefficients of the endogenous variables with a one-period lag, and let h represent the coefficients of the endogenous variables with a two-period lag. Then, as noted by Griliches (1967: 27), the roots of $A(L)$ in (16.7) will lie outside the unit circle if the following conditions are satisfied: $0 < g < 2$; $(1 - g - h) > 0$; and $g^2 \geq -4h$.

More often, the dynamic stability conditions for commodity markets are determined by the lagged response of production to price changes. When production has a non-contemporaneous response to price changes and consumption is related to current price changes, the cobweb theorem states that, for the system to be stable, the absolute value of the price elasticity of demand ϵ^p must be greater than the price elasticity of supply γ_1, i.e. $|\gamma_1/\epsilon^p| < 1$. For situations in which production is related to current price changes and consumption is related to past price changes, the opposite dynamic stability condition holds, i.e. $|\gamma_1/\epsilon^p| > 1$.

The dynamic stability conditions for models that are nonlinear in parameters are more complicated than for models that are linear in form and whose dynamics are of a relatively low order. The stability conditions for these models, which are those that represent the underlying data-generating process for commodities with stocks, have been evaluated by means of numerical methods. The dynamic stability conditions are that the jth interim supply and demand elasticities satisfy the cobweb theorem. When production has a non-contemporaneous response to price changes and consumption is related to current price changes, the condition is

$$|\gamma_{1j}/\epsilon_j^p| < 1, \quad j = 0, 1, 2, \ldots \quad (16.12a)$$

Alternatively, when consumption is related to past price changes and production has a contemporaneous response to price changes, the condition is

$$|\gamma_{1j}/\epsilon_j^p| > 1, \quad j = 0, 1, 2, \ldots \quad (16.12b)$$

When the conditions for stability are met, a steady-state equilibrium solution is possible in the commodity market models.

16.4.2 Time Path Oscillations

The dynamic nature of each commodity market has been examined in terms of its response to changes in both demand and supply. Changes in demand arising from changes in economic activity in the principal consuming countries would be expected to produce cycles in those markets with long lags in their production response to price changes, whereas rather smooth rates of expansion would be anticipated when production and consumption adjust fairly rapidly to changes in market conditions. Interactions among variables in the system of equations are likely to produce responses in supply and demand different from those that arise from direct estimates of single equations in the system. Secondary, or indirect, responses arise from the feedback effects on market prices of other relationships in the system. The response pattern of the markets for commodities covered in this study will be examined in Chapter 18 in the context of the effect that changes in economic activity in the industrialized countries have on international trade and prices of primary commodities. (For an application of these models to an analysis of the effects of the 1981 recession on the world markets for the primary commodities covered in this study, see Lord 1990.)

Changes in supply in commodity markets often arise from major disturbances resulting from natural disasters and labour or political disruptions. A recent incident lends itself to an analysis of the dynamic properties of the system of equations that characterize an adjustment caused by a major disturbance. Towards the end of 1985, Brazil's coffee crop suffered from a drought that lowered 1986 production from an anticipated level of 1.62 million metric tons to less than 1.0 million tons. The simulation has been performed by adding a dummy variable in the production function of Brazil so as to lower production by 0.62 million tons in 1986.

The effect of the disturbance on market prices of coffee is depicted in Fig. 16.4. Lowered production induces the price of coffee to move sharply above the level to which it would have risen otherwise, which causes the quantity of coffee demanded to fall. Producers respond to the higher prices after a delay associated with planting and harvesting. The increase in the amount of coffee supplies dampens the original price increase, and lower prices bring about an increase in the quantity of coffee demanded. The effect of a disturbance such as that which occurs as a result of a drought in Brazil is transitory; eventually, prices would converge to their steady-state path were no other disturbances to occur in the market.

16.5 Conclusions

This chapter has derived the dynamic specification of commodity market relationships for adjustment processes towards long-run equilibrium solutions that are consonant with the equilibrium specification of the commodity trade theory formulated in Part II. In particular, the

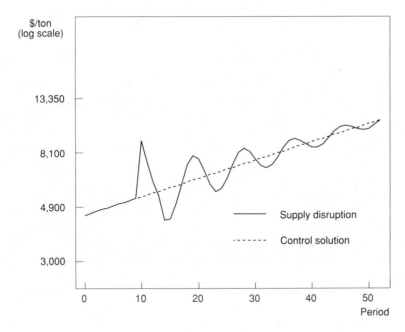

FIG. 16.4 Estimated response of coffee prices to a major supply disturbance

relationship between commodity market theory and observations of market behaviour is that observed behaviour eventually converges to steady-state equilibrium growth if all changes cease in the variables used to characterize the underlying processes of commodity markets. However, the dynamics in the system introduce an additional effect that shows that dependent market variables can respond not only to changes in the level of the explanatory variables, but also to changes in the rates of growth of those variables that generate a steady-state response.

The dynamic processes of adjustment have been described by stochastic difference equations. The supply relationship is usually characterized by long and complicated lag structures. Since the nature of the response to price changes is central to the dynamic specification of the relationship, the stochastic difference equation framework provides a convenient means by which to move from a general to a particular lag structure. For the consumption and stock demand relationships, the ECM offers a particularly appropriate means of characterizing their data-generating processes within this class of equations. Finally, in the common modelling framework of commodity markets, a rational expectations process has been built into the dynamic specification.

Differences between the commodity markets covered in this study appear in the model structures, which define the parameters that have been included in the models, and in the model specifications, which determine the values that have been assigned to the parameters. The structures of the models have been shown to vary with the degree of market fragmentation and the degree of substitution among commodities. The model specification has been shown to vary in the magnitude of the price and income elasticities and in the lag structures of economic agents, which are particularly important to these commodity markets since differences between lag structures in supply and demand functions give rise to observed price cycles in the markets.

Differences in the responses of production and consumption to price changes give rise to stability conditions that depend on the cobweb model. The stability conditions cannot be easily derived from the final form of the system of equations since the models are nonlinear in parameters. Consequently, numerical methods have been used to ensure that the models satisfy the dynamic stability condition, that condition usually being that the absolute value of the price elasticity of demand be greater than that of supply for all interim multipliers generated by a unit change in any of the exogenous variables in the model.

Notes

1 The income elasticities reported in this chapter are based on total, rather than per capita, real GDP.
2 Sugar consumption in the USA decelerated in the sample period as a result of the rapid substitution of high-fructose corn syrup for sugar in the beverage industry. In contrast, world beef consumption accelerated in the sample period. These occurrences were captured by the inclusion of a quadratic trend variable in each of the two markets.
3 Comparisons of elasticity estimates with different sample periods are valid when the estimated parameters in the relationships are stable. As noted above, the tests for stability of coefficients did not reject the null hypothesis of parameter constancy in any of the estimated relationships.
4 The reported elasticity estimates of Adams and Behrman (1976) for cotton are trade-weighted averages of estimates for individual regions.
5 The reported elasticity estimates of Adams and Behrman (1976) for sugar, as well as those by the same authors for cotton, and of Fisher *et al.* (1972) for copper, are trade-weighted averages of estimates for individual countries or regions.
6 For these commodities numerous estimates of their production functions are available (see the surveys by Labys and Hunkeler 1974, and by Askari and Cummings 1977). However, my own results have been compared only with estimates of other studies in which supply and demand have been calculated in order to determine the market price of a commodity, since, as will be

discussed later in the chapter, the joint consideration of these parameters is central to the dynamic stability of equilibrium in commodity markets.

7   As with estimates of the consumption function, comparisons of elasticity estimates with different sample periods are valid when the estimated parameters in the relationships are stable. As noted above, the tests for stability of coefficients did not reject the null hypothesis of parameter constancy in any of the estimated relationships.

8   The reported elasticity estimates for sugar and cotton, both by Adams and Behrman (1976), and those for copper by Fisher *et al*. (1972), are trade-weighted averages of estimates for individual countries or regions.

9   See DeGraff (1960), McCoy (1979), and Simpson and Farris (1982).

10 For example, when a rancher decides to increase production by holding back cows, there is a 9-month gestation period, followed by a period of 15–24 months before a heifer enters the breeding phase. After that heifer is bred, there is another period of gestation. This newly born calf must go through a growth period of 15–36 months before it is marketable (Simpson and Farris 1982). In areas where cattle breeding is intensive, this lag may total 4 years; however, the lag is 7 years in Argentina and Australia, and 7–8 years in the EEC area; in North America it may require up to 10 years from the time the decision is made to produce additional beef to the time the beef is produced (IBRD 1981).

11 Annual stocks of bananas do not exist because of the perishability of the product. As a result, production is equal to consumption plus waste. In modelling the banana market, I have assumed that world production is equal to consumption, which presupposes that waste is negligible and insignificant in the determination of market prices. The planting cycle of bananas is relatively short; on average, 11–15 months are necessary to produce a bunch of fruit.

Iron ore is used almost exclusively for the production of pig iron, which in turn is used to produce steel. Steel is produced and stored in the producing countries. The steel industry uses either available supplies or on-line production.

17
Summary

This part of the study has dealt with the specification and estimation of trade models for Latin America's major commodity exports. It has sought quantitatively to characterize trade in those commodities in light of economic theories underlying statistically estimated relationships from time-series data. The empirical results indicate that the model specification provides a good representation of the data-generating process in international trade of Latin America's major commodity exports. Some of the more important statistical findings are as follows.

1 The average income elasticity of demand estimated for imports in the region's principal geographic markets is small relative to that of total imports by the industrialized countries; it is 1.2 in the long run, or about three-quarters of the elasticity estimated for the total import demand by the industrialized countries.

2 The average price elasticity of import demand is -0.5 in the long run. This average lies in the low range of elasticity estimates of total merchandise imports of the industrialized countries computed in other studies, and is equal to about one-fourth of the price elasticity of demand for manufactured imports of those countries.

3 The average price elasticity of export demand for the region's major commodity exports is -0.5 in the short run and -1.4 in the long run. By way of comparison, the average estimated price elasticity of demand for exports of the industrialized countries computed in other studies is around -1.2 in the short run and -2.0 in the long run.

4 For nearly all commodities, export supply is generally inelastic with respect to price in the short run and elastic in the long run. The average long-run elasticity of all the region's major commodities is 1.4, compared with 3.2 estimated by others for exports of all types of goods by the industrialized countries. As was to be anticipated, perceptible responses occur in the export supply of sugar, bananas, cotton, soybeans, iron ore, and copper within at most three years of a price change, after which the response decays exponentially. In contrast, there are long delays in the responses of beef, cocoa, and coffee supplies to price changes.

5 There are considerable variations in the sensitivity of export supply and demand to price changes in commodities, both among exporters of the same commodity and, particularly for export demand, in the

geographic markets of the same exporter. These differences suggest that caution should be exercised in the above-mentioned generalizations about price and income elasticities in the international trade of Latin America's major commodity exports.

6 Export demand varies not only with the level of imports and relative prices, but also with the rate of growth of imports. Variations in the growth rate of imports in foreign markets induce substantial changes in export market shares of countries. In four of the commodities, an increase in the rate of growth of imports leads to an increase in the average market share of the countries; in another four, it induces a decrease in their market shares; and in the remaining two, it produces no change.

7 The import demand of Latin America's principal export markets is, in most cases, positively influenced by changes in the rate of economic growth in those markets. As was to be anticipated, an increase in the rate of growth of domestic economic activity will cause the level of imports to increase, whereas a decrease in the growth of economic activity will cause the import level to decline.

These statistical results suggest several reasons for Latin America's historically slow rate of growth in primary commodity exports. On the one hand, the fact that the income elasticity of import demand for primary commodities has been about three-quarters of that for total goods explains why the growth rate of world trade in primary commodities has been 0.7 times the growth rate of trade in manufactured goods. On the other, the fact that Latin America's commodity export growth has been different from that of the world average is explained by the changes in the region's competitive position in world commodity trade and, to a lesser extent, by the positive influence on the region's exports of any acceleration in foreign import demand.

From the results of this part of the study, it can be concluded that the generalizations found in the development trade literature about the price elasticities of supply and demand in commodity trade are empirically unfounded when such results are based on analyses of aggregated data; that, in nearly all Latin America's commodity exports, relative prices play an important role in determining export demand; that the slower long-term expansion of Latin America's primary commodity exports compared with the expansion of manufactured exports of the industrialized countries has, for the most part, been attributable to their smaller income elasticities of import demand in the geographic markets; and that a contributing factor in this slower overall expansion of nearly half of the region's major commodity exports has been the negative influence on the demand for exports associated with any acceleration in the rate of import growth in the region's geographic markets, although this negative influence has sometimes been offset by the generally positive effect on the region's export demand associated with improvements in the growth rate of economic activity in the geographic markets.

These conclusions suggest that the Latin American countries, as well as other developing regions, have considerable scope for increasing their exports by more than the growth in overall world trade in primary commodities, and thus for achieving growth rates more in line with those of exports of other types of traded goods. In the last part of this study, we shall consider how econometric models of commodity trade can be brought to bear on the empirical analysis of economic policies.

Part V

Modelling Trade Repercussions and Trade Policies

Introduction

In this part we examine the impact of changes in the international economic environment on commodity trade, and analyse the effects of alternative economic policies on trade between the developing and the industrialized countries. Two types of policies are considered: those that influence the long-term growth of exports, and those that have a one-time impact on exports. As demonstrated earlier, the long-term growth of trade is influenced by policies that change the rate of growth of economic activity in the principal geographic markets. In contrast, changes in prices, either of particular exporting countries relative to the overall price in their export markets or of importing countries acting individually or in unison, can influence the growth of trade for a limited time.

The first chapter in this part traces the impact on the exports of the developing countries of economic policies that produce income changes in the industrialized countries. The system of equations used to represent international commodity trade takes into account possible feedback effects between import demand changes and commodity market price formation. These feedback effects have important consequences for the relationship between economic growth in the industrialized countries and export growth in the developing countries.

The next two chapters examine the trade policies of the developing and the industrialized countries. The first analyses optimal export policies of the developing countries, and the second, trade liberalization policies of the industrialized countries. Once again, the response characteristics of relationships in the system to variable or parameter changes will be measured. In this case, however, we are concerned with the system as a whole. Like the analysis of the characteristics of the response to the predetermined economic activity variables in the system, analyses of international commodity policies measure the response characteristics of the system when endogenous variables are explicitly altered by predetermined values. My own earlier findings, set forth in Part IV, suggested that the dynamics underlying the data-generating processes of commodity trade make the effects of these policies contingent not only on their final impact on trade volumes and market prices, but also on the pace at which they take place. In this part, empirical evidence will be provided of the dynamic effects underlying the adjustment process of trade, as well as of the final effects produced by trade policy measures in the developing and industrialized countries.

18

The International Transmission of Income Changes

The link between the performance of primary commodity exports and economic activity in the industrialized countries has been one of the central themes of the development trade literature. The overall results for the estimated import demand functions for the commodities covered by this study support global generalizations to the effect that the response of primary commodity exports from the developing countries to income changes in the industrialized countries is lower than that of exports of manufactured goods. However, these results refer to instances in which income changes do not significantly affect market prices and where, consequently, trade itself is unaffected by possible feedback effects between price changes and production and consumption decisions in the developing and industrialized countries. The results presented in this chapter underscore the importance of feedback effects on trade.

When income changes are sufficiently widespread to affect market prices, new channels emerge through which changes in the economic activity in industrialized countries influence the commodity exports of the developing countries. In the industrialized countries an expansion of economic activity, for example, directly increases import demand; it may also reduce the quantity of imports demanded when the expansion in economic activity raises world commodity prices and the cost of the imported goods. For the commodity exporting countries, an increase in the economic activity in the industrialized countries that brings about higher market prices influences exports in several ways: (1) it increases the demand for exports as a result of higher import levels in the industrialized countries; (2) it expands the quantity of exports supplied as a result of changes in market prices; and (3) it may influence the quantity of exports demanded because of variations in relative export prices. The transmission of income changes in foreign markets is therefore complex.

The magnitude of these different effects on the level of trade can be readily calculated through multiplier analysis. The results indicate how macroeconomic policies producing income changes in the industrialized countries influence the primary commodity exports of the developing countries. Multiplier analysis also provides us with an opportunity to evaluate the dynamic properties of the system of equations for commodity trade in relation to the process of adjustment of the system from one steady-state growth path to another when changes in economic activity in the foreign markets take place. The stability conditions required for the system of equations to converge to an equilibrium solution are described in the Appendix to this chapter.

18.1 Transmission Mechanisms

The effect of changes in economic activity on commodity trade has been measured by the difference between two solutions obtained from dynamic simulations of the commodity market models. The difference between the two simulations in their predetermined variables occurs in the values assumed by the economic activity variable. The first set of values for the economic activity variable generates the control solution. In the second simulation, the original values of the economic activity variable in the control solution are increased by an amount that is kept constant in all subsequent periods. Comparison of the two solution paths provides information about the contemporaneous response, or impact multiplier, and the total response, or cumulative multiplier. Analyses of this type measure real and nominal value differences between base and alternative simulations and are often used to evaluate the response characteristics of econometric models. (For a discussion of conventional multiplier analysis, see Goldberger 1964: 373–6; Klein 1974: 240–8; and Theil 1971: 465–8.) When calculated, rather than actual, values are used for lagged endogenous variables in the system, dynamics are introduced. These show the time path of the trade variables generated by changes in economic activity.

Multiplier analysis that compares changes in the dollar magnitudes of commodity exports in one country with those in incomes in other countries is less meaningful than multiplier analysis that compares changes in macroeconomic variables within a country. In multiplier analysis in international trade, therefore, it is more advisable to use elasticities, which are dimensionless measures. The short-run elasticity is the same-period effect resulting from a change in economic activity. The response of the production of many commodities to a change in price induced by a sustained change in economic activity tends to be slow, so the adjustment from the initial to the new solution is not fully realized within the same period. Convergence to the new steady-state growth path occurs only after several periods. None the less, most of the response tends to occur in the first few periods after the change in economic activity, after which the new steady-state solution is approached asymptotically. The long-term elasticity measures the total effect of a change in economic activity. It is calculated from the response of the market variable to a sustained new level of economic activity, the sustained change being constant in its unit (US dollar) amount.

18.1.1 Non-Contemporaneous Changes in Income by Importers

A change in the level of economic activity in a small importing country will have the same effect on import demand as that calculated for the income elasticity of import demand reported in Chapter 13. The reason is that the change in economic activity affects only import demand, and has no impact on the price of the commodity. In the exporting country, the change in foreign market imports brought about by a new level of economic activity will alter the level of export demand. The effect measures the *foreign income elasticity of exports*, which is defined as the percentage change in exports brought about by a 1 per cent change in economic activity in the foreign market. If the commodity market were to approximate perfect competition, a change in import demand of a foreign market would induce a proportional change in the export demand of all foreign suppliers, since the price elasticity of demand would be infinity. Imperfect competition in commodity trade introduces the possibility that foreign income changes will also bring about changes in relative export prices.

The channels through which foreign market income changes affect a country's commodity exports can be readily discerned in the export demand function. If we let the subscript for the foreign market be implicit in order to simplify notation, the export demand function in monopolistic competition may be specified as

$$X_i^d = f\{M(Y), R_i[M(Y)]\}, \tag{18.1}$$

where foreign market imports M are themselves a function of domestic income Y, and the country's relative export price R_i depends on foreign market imports. Differentiation of (18.1) with respect to foreign market income yields (a) a direct effect on the commodity producing country's export demand, and (b) an indirect effect on the country's export demand through induced relative export price variations:

$$\frac{dX_i^d}{dY} = \underbrace{\frac{\partial X_i^d}{\partial M}\frac{dM}{dY}}_{\text{direct demand effect}} + \underbrace{\frac{\partial X_i^d}{\partial R_i}\frac{\partial R_i}{\partial M}\frac{dM}{dY}}_{\text{indirect demand effect}}. \tag{18.2}$$

Transmission of Income Changes

This expression is the *foreign income multiplier*. The foreign income elasticity of demand for exports is obtained by multiplying both sides of the expression by Y/X_i^d.

Whereas the direct effect on export demand is positively related to changes in foreign market incomes, the indirect effect associated with the change in the exporter's relative price is negative since the export price will rise relative to the unchanged import price in the geographic market (see equation (5.30)). As the competitiveness of the country decreases, exports will increase by a smaller amount than they would have increased under perfect competition. Thus, the magnitude of the foreign income elasticity of demand for exports is smaller than it would be in a perfectly competitive market.

18.1.2 Concurrent Income Changes by Importers

The effects are more complex when changes in the economic activity of several importing countries take place at the same time. The overall change in economic activity will produce a change in the commodity's market price, which in turn will alter the quantity demanded of the commodity in the importing countries. In accordance with the previous notation, the result of a change in economic activity in all importers is

$$\frac{dM}{dY} = \frac{\partial M}{\partial Y} + \frac{\partial M}{\partial P}\frac{dP}{dY}, \tag{18.3}$$

where P denotes the world market price of a commodity.

In the exporting country, the change in market price has three additional indirect effects on the foreign income multiplier, or the foreign income elasticity of exports. Two of them affect export demand. The first is the change in export demand brought about by the effect of the new price level on foreign import demand. The second is the impact of import demand changes on relative export prices. Both of these effects are negatively related to the change in foreign income. Finally, export supply is influenced by the new world market price. This new price level has a direct effect on exports, which is positively related to the change in foreign market incomes.

The channels through which concurrent changes in all foreign markets affect a country's exports can be found from the solution of the total derivative of the equilibrium identity between the country's export demand, X_i^d, and its export supply, X_i^s, with respect to foreign income, Y:

$$\underbrace{\frac{\partial X_i^d}{\partial M}\frac{\partial M}{\partial Y}}_{\text{direct demand effect}} + \underbrace{\frac{\partial X_i^d}{\partial R_i}\frac{\partial R_i}{\partial M}\frac{dM}{dY} + \frac{\partial X_i^d}{\partial R_i}\frac{\partial R_i}{\partial M}\frac{\partial M}{\partial P}\frac{dP}{dY} + \frac{\partial X_i^d}{\partial M}\frac{\partial M}{\partial P}\frac{dP}{dY}}_{\text{indirect demand effects}}$$

$$\underbrace{-\frac{\partial X_i^s}{\partial P}\frac{dP}{dY}}_{\text{direct supply effect}} = 0, \tag{18.4}$$

where, as before, M denotes foreign market imports, P is the commodity's market price, and R_i is the country's relative export price. The foreign income multiplier is obtained from the simultaneous solution of the export demand and export supply functions with respect to foreign market income (see Annex D for the derivation):

$$\frac{dX_i}{dY} = \frac{(\partial X_i^s/\partial P)(\partial X_i^d/\partial Y)}{(\partial X_i^s/\partial P) - (\partial X_i^d/\partial P)}, \tag{18.5}$$

the expected sign of which is positive.

18.2 Market Price Responses to Income Changes

The dynamic properties of the trade models have been assessed in relation to the time path and the degree of response of prices and traded volumes of each of the commodities to changes in economic activity in the principal geographic markets. The effect of a change in the level of world economic activity on commodity market prices has been measured by the difference between two solutions obtained from dynamic simulations of the trade models. The difference between the two simulations in their predetermined variables occurs in the values assigned to economic activity. The first set of values for the economic activity variable generates the control solution. In the second simulation, the original value of the economic activity variable in the control solution is increased by 1 per cent in one year. Comparison of the two solution paths then provides information about the price response to income changes.

The effects of a one-time 1 per cent increase in economic activity of the industrialized countries on the market prices of the commodities covered in this study are shown in Fig. 18.1. It is evident that the one-period change effects on the markets occur over several years, and that, for most commodities, restoration of market prices to their long-term equilibrium solutions takes a long time.

The time path followed by the market price variable of each commodity is closely associated with the speed with which production responds to price variations. Commodities in which production and consumption both respond within a short period of time to changes in macroeconomic conditions follow a dampened smooth path back to the steady-state equilibrium. Commodities in which production responds to price changes with a significantly greater delay than consumption are distinguished at the beginning of the period by cycles that tend to dampen over time. Thus the systems are stable, but in most of them there is a dampened cyclical, rather than a smooth, response to a change in economic activity.

As Fig. 18.1 shows, the short-term response tends to be greater than the long-term response. The reason for this difference is shown in Fig. 18.2. Initially, equilibrium is at quantity Q_0 and price P_0. Higher income causes demand for a commodity to increase from D to D'. The price of the commodity rises to P_1. At this new price the quantity that producers are willing to supply expands. Additional output lowers price and eventually a new equilibrium is reached at Q_2 and P_2. The new equilibrium price P_2 is below the price, P_1, that was reached when the demand for the commodity initially increased. However, the degree of response to a shift in demand in the short run and in the long run would be the same if supply were highly inelastic with respect to price. When supply is perfectly inelastic, the price rise brought about by a shift in demand will cause the initial rise in price to remain unchanged.

The simulation results show that particularly large price changes occur in sugar and maize, both in the short run and in the long run; copper and cocoa also have large short-term responses, but their long-term responses are around the mean average of 3.3 per cent of the commodities. The responses of cotton, beef, soybeans, and coffee are around 1–2 per cent in the long run. The pattern and magnitude of these market price responses to income changes in the industrialized countries have important consequences for trade in these commodities.

18.3 Trade Responses to Income Changes

Tables 18.1 and 18.2 present the effects estimated by the model of increased economic activity in the principal geographic markets for the commodities. The first of these tables shows how the income changes affect the commodity imports of the industrialized countries; the second

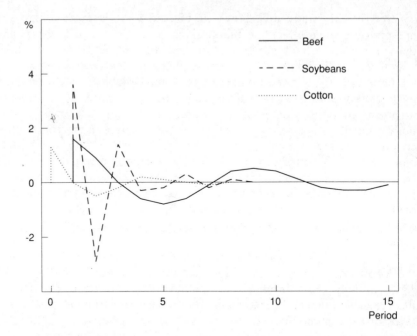

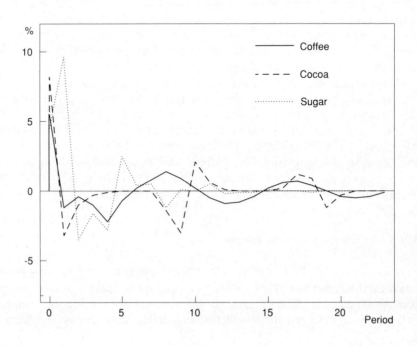

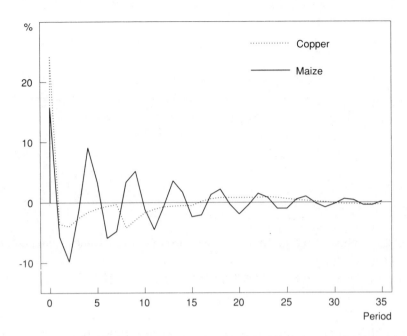

FIG. 18.1 Response path of market prices to one-time 1 per cent increase in world economic activity

	Multiplier			Multiplier	
	Impact	Total		Impact	Total
Cocoa	8.2	2.7	Maize	15.8	5.8
Coffee	5.5	1.9	Soybeans	3.6[a]	1.8
Copper	24.2	3.6	Sugar	9.6[a]	8.7
Cotton	1.3	0.9	Beef	1.6[a]	1.2

[a] One-period lag.

shows how the income changes in the industrialized countries affect the exports of the Latin American countries. The simulations were performed with a one-time 1 per cent increase in the income of each of the importing countries, and the amount by which income increased in the first period was maintained in subsequent periods.

The results presented in Table 18.1 show that in most countries the expansion in imports after the income increase is considerably smaller when there are concurrent changes of incomes in all the importing

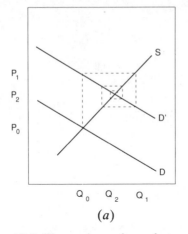 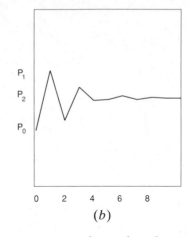

FIG. 18.2 Illustration of market price response to demand and supply changes
(a) Sustained increase in income (b) Time path of market price

Table 18.1 Total import response to one-time 1 per cent increase in income
(Percentage change)

	Non-contemporaneous income change by each importer	Concurrent income changes by all importers	
	Volume effect	Volume effect	Value effect
Beef	*1.49*	*0.60*	*1.99*
USA	0.19	−0.98	0.22
Japan	3.46	1.76	2.96
France	4.60	4.60	5.80
W. Germany	1.08	0.77	1.97
Greece	1.89	1.24	2.44
Maize	*1.00*	*1.00*	*6.80*
Italy	1.00	1.00	6.80
Bananas	*1.00*	*−0.36*	*1.69*
USA	1.00	1.00	3.05
Japan	1.00	−1.67	0.38
W. Germany	1.00	−4.43	−2.38
Italy	1.00	0.01	2.06

Table 18.1 (*cont.*)

	Non-contemporaneous income change by each importer	Concurrent income changes by all importers	
	Volume effect	Volume effect	Value effect
Sugar	*0.84*	*−2.54*	*6.17*
USA	0.58	−1.58	7.13
UK	1.00	−4.69	4.02
Coffee	*0.74*	*0.45*	*2.40*
USA	0.18	−0.08	1.87
Japan	2.27	1.86	3.81
Belgium	1.13	0.67	2.62
France	0.70	0.63	2.58
W. Germany	1.00	0.73	2.68
Italy	1.20	0.85	2.80
Netherlands	1.22	0.90	2.85
Finland	0.74	0.48	2.43
Sweden	0.05	−0.40	1.55
Switzerland	0.48	0.08	2.03
Cocoa	*0.75*	*0.06*	*2.72*
USA	1.02	−0.07	2.59
Japan	0.34	−0.56	2.10
W. Germany	0.51	0.38	3.04
Italy	0.23	−0.41	2.25
Netherlands	1.00	0.05	2.71
Spain	0.41	0.22	2.88
Soviet Union	0.59	0.24	2.90
Soybeans	*2.34*	*1.70*	*3.45*
W. Germany	1.97	1.97	3.72
Netherlands	3.26	2.16	3.91
Spain	1.92	1.04	2.79
Switzerland	1.16	−0.53	1.22
Cotton	*0.39*	*−0.64*	*0.27*
USA	1.00	1.00	1.91
Japan	0.02	−0.52	0.39
Belgium	1.19	−2.66	−1.75
France	1.00	−0.39	0.52
W. Germany	1.00	−1.17	−0.26
Italy	0.28	−0.12	0.79
Netherlands	1.00	1.00	1.91
UK	1.00	1.00	1.91

Table 18.1 (cont.)

	Non-contemporaneous income change by each importer	Concurrent income changes by all importers	
	Volume effect	Volume effect	Value effect
Iron ore	*1.28*	*0.61*	*1.93*
USA	1.09	−0.72	0.61
Japan	1.42	0.67	1.99
France	2.82	2.82	4.14
W. Germany	0.91	0.91	2.23
Italy	1.29	1.29	2.61
UK	0.36	0.36	1.68
Copper	*1.07*	*0.62*	*5.05*
USA	1.22	0.16	4.59
Japan	1.69	1.69	6.12
Belgium	1.00	0.58	5.01
W. Germany	0.83	0.83	5.26

Note: Commodity totals are trade-weighted averages.

Table 18.2 Total export response to one-time 1 per cent increase in foreign income
(Percentage change)

	Non-contemporaneous income change by each importer	Concurrent income changes by all importers	
	Volume effect[a]	Volume effect[a]	Value effect
Beef	*0.55*	*0.39*	*1.59*
Argentina	1.48	1.72	2.92
Costa Rica	0.14	5.27	6.47
Honduras	0.20	−1.04	0.16
Nicaragua	0.20	−1.00	0.20
Uruguay	1.26	0.87	2.07
Maize	*1.00*	*1.00*	*6.82*
Argentina	1.00	1.00	6.82

Table 18.2 (cont.)

	Non-contemporaneous income change by each importer	Concurrent income changes by all importers	
	Volume effect[a]	Volume effect[a]	Value effect
Bananas	*0.57*	*0.36*	*2.41*
Costa Rica	0.47	1.13	3.18
Ecuador	0.43	3.63	5.68
Honduras	0.72	0.95	3.00
Panama	0.50	−1.48	0.57
Sugar	*0.66*	*−2.15*	*6.56*
Barbados	1.00	−4.69	4.02
Brazil	0.60	−1.60	7.11
Costa Rica	0.56	−1.61	7.10
Dom. Republic	0.58	−1.58	7.13
El Salvador	0.56	−1.62	7.09
Guatemala	0.60	−1.60	7.11
Guyana	0.57	−1.60	7.11
Jamaica	0.86	−3.72	4.99
Panama	0.56	−1.62	7.09
Peru	0.58	−1.58	7.13
Coffee	*0.19*	*0.87*	*2.82*
Brazil	0.06	0.70	2.65
Colombia	0.14	1.23	3.18
Costa Rica	0.27	1.10	3.05
Dom. Republic	0.18	−0.08	1.87
Ecuador	0.36	0.87	2.82
El Salvador	0.67	0.67	2.62
Guatemala	0.40	0.61	2.56
Haiti	0.91	1.14	3.09
Honduras	0.05	1.05	3.00
Mexico	0.34	−0.05	1.90
Cocoa	*0.74*	*0.06*	*2.72*
Brazil	0.54	0.23	2.89
Dom. Republic	1.00	−0.10	2.56
Ecuador	0.75	0.01	2.67
Soybeans	*1.81*	*2.30*	*4.05*
Brazil	1.39	2.48	4.23
Paraguay	2.68	1.95	3.70

Table 18.2 (*cont.*)

	Non-contemporaneous income change by each importer	Concurrent income changes by all importers	
	Volume effect[a]	Volume effect[a]	Value effect
Cotton	*0.58*	*−0.75*	*0.16*
Brazil	0.26	−0.69	0.22
El Salvador	0.00	0.02	0.93
Guatemala	0.04	0.07	0.98
Mexico	0.06	0.54	1.45
Nicaragua	0.19	−0.55	0.36
Paraguay	1.08	−1.25	−0.34
Peru	0.54	−0.84	0.07
Iron ore	*0.78*	*0.94*	*2.26*
Brazil	0.79	1.03	2.35
Peru	0.69	−0.12	1.20
Copper	*0.60*	*0.32*	*4.75*
Chile	0.48	0.45	4.88
Peru	1.06	−0.08	4.35

[a] Foreign income elasticity of demand for exports.

Note: Commodity totals are trade-weighted averages.

countries than when there is a non-contemporaneous change in income in each of the importing countries. On average, a 1 per cent increase in income leads to a 1.1 per cent increase in imports when income changes do not occur at the same time in the importing countries, whereas it leads to only a 0.2 per cent increase in imports when income changes occur at the same time in all the importing countries.[1] However, the smaller increase in imports because of concurrent changes in income in the industrialized countries is more than offset by the rise in import prices, which increases the value of imports above what it would be when there are non-contemporaneous changes in incomes. The rise in prices that simultaneous income changes generate leads to an average increase in the value of imports of the commodities equal to 3.2 per cent. In contrast, prices remain unchanged when income changes are non-contemporaneous, and the increase in the value of imports reflects only the 1.1 per cent greater volume of imports.

These generalizations do not always apply to individual countries. In several countries, price rises lead to a more-than-proportional decline in import demand. The dampening effect of price rises on import demand is particularly apparent in Belgium's imports of cotton and West Germany's imports of cotton and bananas. In contrast, when import price changes do not influence import demand—as in Italy's imports of maize—the only effect on imports is that which is brought about directly by the increased income, and the world market price rise is fully transmitted to the country in the form of an increase in the value of imports.

These differences between countries importing the same commodity suggest that an exporter's assessment of its markets should not be based on income elasticity estimates alone, but should also take into account the dynamic feedback effects in each market arising from both income and price changes. For instance, an income change in the Netherlands produces a substantially greater expansion in demand for coffee than it does in Belgium when concurrent income changes in the industrialized countries affect the world price of that commodity. This difference arises from the feedback effects of higher prices on import decisions, despite the fact that the income elasticity of import demand of Belgium is greater than that of the Netherlands. A similar situation occurs in US imports of copper, which respond strongly to both income and price changes. Although the high income elasticity would suggest that the USA is a growing market for copper, the high price elasticity indicates that growth in the demand for this commodity would be sluggish when concurrent increases in income take place in the industrialized countries. For this reason, it would be more expedient for the developing countries to assess their export markets in relation to solutions to systems of equations designed to represent trade in commodity markets, rather than in relation to estimates of individual behavioural relationships alone.

Table 18.2 presents calculations of the foreign income elasticities of exports, as well as calculations of the changes in the value of exports resulting from foreign income changes. When concurrent increases in the economic activity of the industrialized countries raise the price of a commodity, the effects on the volume of exports differ from those brought about by non-contemporaneous changes in income in those countries that leave the market price unchanged. The magnitude of the difference in the change in exports when the market price changes, and when it remains unaffected by income changes, depends on the price elasticities of both imports and exports. In coffee, soybeans, bananas, and iron ore, Latin America's total export expansion is more than proportional to the total import growth in its geographic markets, since exports have been concentrated in those markets that have above-average increases in import demand resulting from increased economic activity. In contrast, Latin America's total exports of beef and copper increase by a less-than-proportional amount of the total increase in imports of its geographic markets since these goods have been exported to markets that have below-average increases in import demand as a consequence of increased economic activity. These regional-level generalizations do not, of course, reflect the country-specific export responses of the commodities.

There are three types of responses of exporting countries to a concurrent increase in the income of the industrialized countries. The first occurs when both import and export demand is unresponsive to price movements, i.e. when price changes do not influence import demand and when relative-price changes do not influence export demand. In this case the change in export demand is proportional to the change in the import demand of the geographic market; the changes in both imports and exports are of the same amount as when changes in economic activity in the export markets are non-contemporaneous. This type of response is exemplified by the maize exports of Argentina; a 1 per cent change in income of Argentina's export markets leads to a proportional change in both imports and exports, regardless of whether the income change takes place at the same time as income changes in all the other importers of maize.

The second occurs when import demand is responsive to price changes but export demand is not influenced by relative-price changes. In this case, the change in the import demand of the principal geographic markets is lower when income changes occur at the same time (which produces feedback effects between market prices and import prices) than when income changes are non-contemporaneous. However, as in the first type of response, the change in export demand is proportional to the change in import demand. This situation is illustrated by the sugar exports of all the Latin American countries to both the USA and the UK. Both importing countries respond to import price changes. Since sugar market prices have a strong response to income changes, the quantity demanded of sugar imports decreases in response to the higher import prices after the increase in the income of the industrialized countries. Since both of these importing countries have established preferential markets for sugar, the Latin American countries are price-takers in these markets, and have perfectly price-elastic demand functions. Thus, the change in the exports of the Latin American countries to these two markets is proportional to the change in imports in each of these markets.

The third type of response occurs when relative export price changes influence export demand. Whether a country's market share increases, decreases, or remains unaltered depends on whether its export price changes by less than, more than, or the same amount as the overall import price of the commodity in its foreign market. In the Latin

American countries changes in relative export prices brought about by changes in imports and foreign market prices augment the magnitude of the foreign income elasticity of export demand. Where there are concurrent changes in income that increase the commodity's market price, and where relative-price changes influence the demand for exports, the calculations show that there is a more-than-proportional increase in exports of those countries in which relative export price changes influence export demand. For example, the percentage increase in Brazilian exports of soybeans to Spain is higher than the percentage increase in Spain's soybean imports after the income and price rises.

The time path and the degree of response of exports to changes in economic activity are summarized in Fig. 18.3. The response of commodity trade to a sustained change in foreign economic activity tends to be slow, so the adjustment from the initial to the new solution is not fully realized within the same period, and convergence to the new steady-state growth path occurs only after several periods. In beef the dampened cyclical movements reflect the second-order difference equation for the export supply of Argentina and Costa Rica, which captured the cattle cycle inherent in the exports of those countries. The delayed response of export supply to changes in prices is particularly evident in the long, dampened cyclical response of both Chilean and Peruvian exports of copper. In all the commodities, most of the response tends to occur in the first few periods after the change in economic activity, after which the new steady-state solution is approached asymptotically.

These results provide a number of insights into the behaviour of the models used to represent commodity trade. First, the models themselves have been shown to be stable when subjected to exogenous shocks. Second, trade prices and volumes generally follow a dampened cyclical path in their return to a steady-state equilibrium. Finally, the response of trade to isolated events in one geographic market can be very different from its response to events that give rise to feedback effects from income and price changes. These different responses are often critical to the outcomes of the types of trade policies considered in the next two chapters.

Appendix: Dynamic Stability

As described in Chapter 16, the final form of the model, which relates the endogenous trade variables to only the exogenous variables, can be obtained from matrices of lag polynomials for the endogenous and exogenous variables of the structural form of the model. The representation of the structural form of the model by vectors and matrices requires the relationships in the system to be linear in parameters. Nonlinearity

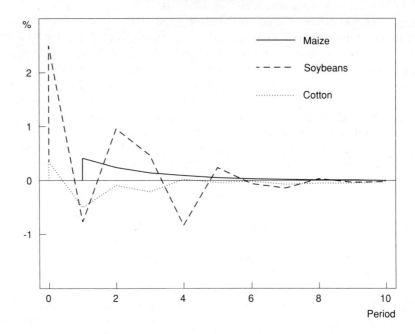

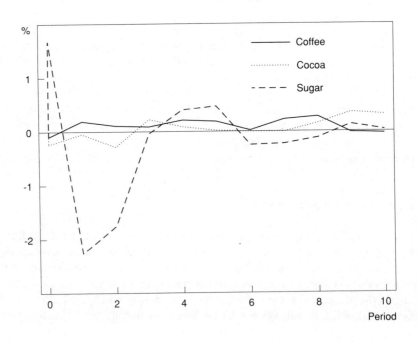

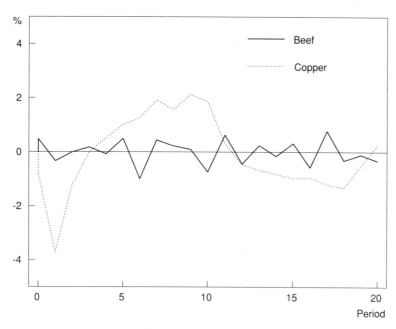

FIG. 18.3 Response path of exports to one-time 1 per cent increase in world economic activity

	Multiplier			Multiplier	
	Impact	Total		Impact	Total
Cocoa	−0.2	0.1	Maize	0.4[a]	1.0
Coffee	−0.1	0.9	Soybeans	2.5	2.3
Copper	−0.8	0.3	Sugar	1.7	−2.2
Cotton	0.4	−0.8	Beef	0.5	0.4

[a] One-period lag.

is introduced into the model for exports when there is a logit transformation of the demand for exports in equation (8.16).

Properties of the Linear System of Equations

Without the logit transformation of the demand relationship in (8.16), the final form for the export price and quantity relationships can be readily obtained. The procedure used is the same as that employed to find the final form of the system of equations of markets for non-storable commodities in Chapter 16. The autoregressive final form of the system of export supply and demand equations is

$$|A(L)|w_t = [\text{adj } A(L)]B(L)y_t + [\text{adj } A(L)]u_t \tag{18.6}$$

where, as before, $A(L)$ is a matrix of lag polynomials for the endogenous variables, $[\text{adj } A(L)]$ is the adjoint matrix of $A(L)$, and $B(L)$ is a matrix of lag polynomials for the exogenous variables. Equation (18.6) yields

$$x_{it} = \frac{(2 + \beta_{41})\alpha_{23} - (2 + \beta_{22})\alpha_{43} + \gamma_{44} - \beta_{24}}{(\alpha_{23} - \alpha_{43})} x_{i,t-1}$$
$$+ \frac{(\beta_{24} - \alpha_{23})(1 + \beta_{41}) - (\gamma_{44} - \alpha_{43})(1 + \beta_{22})}{(\alpha_{23} - \alpha_{43})} x_{i,t-2}$$
$$+ \gamma_1 m_t + \gamma_2 m_{t-1} + \gamma_3 m_{t-2} + \gamma_4 p_t + \gamma_5 p_{t-1} + \gamma_6 p_{t-2} + \gamma_7 d_t$$
$$+ \gamma_8 d_{t-1} + \gamma_9 d_{t-2} + \gamma_0$$

and

$$p_{it} = \frac{(2 + \beta_{41})\alpha_{23} - (2 + \beta_{22})\alpha_{43} + \gamma_{44} - \beta_{24}}{(\alpha_{23} - \alpha_{43})} p_{i,t-1}$$
$$+ \frac{(\beta_{24} - \alpha_{23})(1 + \beta_{41}) - (\gamma_{44} - \alpha_{43})(1 + \beta_{22})}{(\alpha_{23} - \alpha_{43})} p_{i,t-2}$$
$$+ \theta_1 m_t + \theta_2 m_{t-1} + \theta_3 m_{t-2} + \theta_4 p_t + \theta_5 p_{t-1} + \theta_6 p_{t-2} + \theta_7 d_t$$
$$+ \theta_8 d_{t-1} + \theta_9 d_{t-2} + \theta_0 \tag{18.7}$$

when both export supply and demand have contemporaneous responses to price changes. Both export quantity X_i and export price P_i have the same autoregressive coefficients, which is a common characteristic of the reduced form of a model.

In general terms, a system of equations describing the data-generating process of commodity exports will be stable if the roots of $A(L)$ in (18.6) lie outside the unit circle. Suppose export supply and demand both depend on the current export price of the commodity. Let λ_1 represent the coefficients of the endogenous variables with a one-period lag and λ_2 the coefficients of the endogenous variables with a two-period lag. Then, as noted by Griliches (1967: 27), the roots of $A(L)$ in (18.6) will lie outside the unit circle if the following conditions are satisfied: $0 < \lambda_1 < 2$; $(1 - \lambda_1 - \lambda_2) > 0$; and $\lambda_1^2 \geq -4\lambda_2$.

More often, the dynamic stability conditions for commodity exports are determined by the lagged response of exportable production to price changes. When export supply has a non-contemporaneous response to price changes and export demand is related to current price changes, the cobweb theorem states that, if the system is to be stable, the absolute value of the price elasticity of export demand, ϵ_x^p, must be greater than the price elasticity of export supply, γ_1, i.e. $|\gamma_1/\epsilon_x^p| < 1$. For situations in which export supply is related to current price changes and export demand is related to past price changes, the opposite dynamic stability condition holds, i.e. $|\gamma_1/\epsilon_x^p| > 1$.

Properties of Nonlinear Systems of Equations

The logit transformation of the export demand relationship causes the relationships in the system of equations for exports to be nonlinear in parameters. The reason is that the transformed export demand equation (8.17) contains additive terms while the export supply equation (9.41) contains multiplicative terms. Thus, the structural form of the model cannot be represented by vectors and matrices, and (18.6) cannot be used in order to find the reduced form. Instead, the dynamic properties of the model need to be derived from the reduced form, where the current values of the endogenous trade variables are related to both the lagged endogenous variables and the current and lagged exogenous variables in the system.

For the reduced form there are three possible solutions, each of which has important implications for the dynamic properties of the models. The first solution occurs when export supply and demand are both related to the current export price of the commodity. The solution is then a polynomial in price equal to $\lambda_3 P^{\xi_1} + \lambda_4 P^{\xi_2} = M_t$, where λ_3 and λ_4 are constant terms composed of relatively complicated expressions of predetermined variables and their coefficients. The polynomial in price has powers that are real numbers capable of being approximated through numerical optimization methods.

The other two possible solutions occur when either the export supply or demand relationship has a contemporaneous response to the price of the commodity, and the other relationship has a lagged response to price. In commodity trade, delays typically occur in the response of exportable production to changes in the price of exports, and the empirical findings in Chapter 15 confirmed these expectations. These delays occur because of the time that elapses between the change in the price of a commodity and the onset of the changes in the level of export supplies.

When export supply has a lagged response to market price changes and export demand has a contemporaneous response to relative-price changes, the reduced-form equation of the export price relationship to a particular market is

$$P_{it} = P_t[(X^s_{it}/M_t - 1)/f'(X^d_i)]^{1/\alpha_{23}} \qquad (18.8)$$

where the expression in brackets is raised to a power that is the inverse of the short-term price elasticity of demand for the commodity exported by a country i, and $f'(X^d_i)$ is the partial inverse of the export demand function.[2] The export quantity is given by the export supply relationship (equation (9.41)).

On the other hand, when export supply responds to contemporaneous prices and export demand is unresponsive to relative-price changes in the short run, the reduced-form equation of the export price relationship is

$$P_{it} = [M_t/(1 + Z^{-1}_{it})g'(X^s_i)]^{1/\alpha_{43}}, \qquad (18.9)$$

where the expression in brackets is raised to a power that is the inverse of the short-term price elasticity of supply, $g'(X^s_{it})$ is the inverse supply function,[3] and, as before, $Z_i = (X^d_i/M)/[1 - (X^d_i/M)]$. The reduced-form equation of the export quantity relationship is

$$X_{it} = M_t/(1 + Z^{-1}_{it}). \qquad (18.10)$$

The dynamic stability conditions for models that are nonlinear in parameters are more complicated than for models that are linear in form and have dynamics of a relatively low order. In practice, the stability conditions for the system of export equations that includes a logit transformation of the export demand relationship have been evaluated by numerical methods. The conditions required for dynamic stability are that the jth interim export supply and demand elasticities satisfy the cobweb theorem. When export supply has a non-contemporaneous response to price changes and export demand is related to current price changes, the condition is

$$|\gamma_{1j}/\epsilon^p_{xj}| < 1, \qquad j = 0, 1, 2, \ldots \qquad (18.11)$$

Alternatively, when consumption is related to past price changes and production is related to current price changes, the condition is

$$|\gamma_{1j}/\epsilon^p_{xj}| > 1, \qquad j = 0, 1, 2, \ldots \qquad (18.12)$$

When the conditions for stability are met, a steady-state solution is possible for the commodity exports of a country.

Notes

1   The results for the aggregate of each of the commodities reported in Table 18.1 are trade-weighted averages, whereas the aggregate income elasticities of import demand for each of the commodities reported in Table 13.2 above are unweighted averages of elasticity estimates for individual countries. The

reason for this distinction is that in Chapter 13 each of the importing countries was equally important as a potential export market, while in this chapter we are concerned with the simulations of the system of equations as a whole and the resulting changes that occur in overall trade volumes.

2. In particular,

$$f'(X_i^d) = \{e^{\alpha_{20}} M_t^{\beta_{21}} M_{t-1}^{-\beta_{21}} Z_{t-1}^{1+\beta_{22}} P_t^{-\alpha_{23}} P_{t-1}^{\alpha_{23}-\beta_{24}} P_{i,t-1}^{\beta_{24}-\alpha_{23}}\},$$

the coefficients of which are defined in (8.17).

3. In particular,

$$g'(X_i^s) = \{e^{\alpha_{40}+\beta_{45}T} X_{i,t-1}^{1+\beta_{41}} X_{i,t-2}^{\alpha_{42}} D_t^{-\alpha_{43}} (P_i/D)_{t-n}^{\gamma_{44}-\alpha_{43}}\},$$

the coefficients of which are defined in (9.41).

19
Optimal Export Policies

In a competitive market, factor prices are equalized across sectors, so there are no gains from the redistribution of resources from one sector to another. In international trade, industry-oriented policy measures such as export subsidies not only distort the allocation of resources, but also produce a welfare loss in the countries implementing the subsidies, and these losses are transferred to foreign markets in the form of welfare gains of equivalent magnitudes.

When imperfect competition is inherent in a market, for instance as a result of product differentiation, the removal of trade intervention measures designed to bring about export changes, though it increases competition, does not necessarily bring about gains similar to those that occur in perfect competition. If there are non-zero profits in a market with imperfect competition, the removal of intervention measures by a country can, in fact, shift excess returns from that country to its competitors. It is for this reason that countries may find it advisable to apply certain types of policy measures to their export industries when there is imperfect competition in international trade.

This chapter examines optimal export policies for developing countries that are based on the model of imperfect competition used to characterize international commodity trade in Parts II and III. Thus we consider, in general, the export policies of developing countries under conditions of monopolistic competition, and, in particular, the large-group case of monopolistic competition where there are many exporting countries in a commodity market that can none the less exert some influence on the price of their exports. A market structure of this type is appropriate for the analysis of trade in most commodities, since there is usually a large number of countries that export a given commodity and the actions of those countries have a negligible effect on the world market.

In practice, however, not all the features of monopolistic competition are always present. For example, with product differentiation, the action of an exporting country may elicit a response from its close competitors, despite its negligible overall market share. The competing exporters that are most likely to respond are those whose exports have characteristics similar to those of the country initiating a change in its export policy, since importers buying from these countries will substitute those suppliers that approximate their most preferred export type. With respect to the characteristics space of a good described in Chapter 3, the exporters that would respond to a price change by a country are its neighbours in the characteristics space.

Free entry is another feature of monopolistic competition that may be absent. Free entry in the market ensures that pure profits are zero in the long run, although above-normal returns to factors of production are not ruled out in the short run. In practice, the possibility of long-run excess returns exists because of the so-called integer constraint (see Helpman and Krugman 1985: 146–51). In industrial economics, this constraint is associated with the indivisibility of firms in an industry. In international trade, where goods are normally distinguished by country of origin, the integer constraint may be greater since the number of countries that can enter a market is usually more limited than the number of firms. Consequently, in markets where imperfect competition and product differentiation exist, the export price may exceed the average cost for the exporter.

These different assumptions concerning competitors' responses and excess returns in a market with imperfect competition and product differentiation underlie much of the recent literature on 'strategic' trade policies. Strategic trade policies refer to the actions taken by a country's government to influence exports in a market where there is imperfect competition. It has become clear that changes in the characterization of imperfect competition often have important implications for policy outcomes. Consequently, the analysis of policy formation should be consistent with the models used to represent commodity trade.

Ultimately, the usefulness of strategic export policies must be proved by empirical evidence. The recent studies by Dixit (1988) on the automobile industries in the USA and Japan and by Baldwin and Krugman (1988) on random access memories provide valuable insights into the nature of trade policies in imperfectly competitive markets. Using conjectural variations models, these studies offer a useful framework for evaluating complicated policy responses to imperfect competition. However, these static models fail to provide analyses of the consequences for industries of transitional effects and of delays in achieving the final outcomes of policy initiatives. In this chapter we use econometric models to examine the processes of dynamic adjustment to trade policy measures.

Preliminary evidence from the recent literature suggests that the potential gains to large countries from strategic policy measures may be too small to warrant the efforts associated with the identification of strategic sectors. Krugman (1987: 219) has provided an illustration of the insignificance for an economy such as that of the USA of a strategic trade policy designed to capture the economic rents of an industry. Nevertheless, for small, open economies that rely on a few primary commodity exports for the bulk of their export earnings, the potential gains from export policy measures may be considerable.

The plan of this chapter is as follows. Section 19.1 examines alternative policy instruments and describes the way in which they can be represented in our econometric models of commodity trade. Section 19.2 analyses alternative export policies, particularly in the light of the recent literature on 'strategic' trade policies, and Section 19.3 provides a quantitative analysis of optimal trade policies for the sample of commodities in this study. The final section summarizes the findings.

19.1 Policy Instruments

The intervention measure usually adopted by an exporting country is either a subsidy—an export subsidy (e.g. export credit or tax rebate), an output subsidy (e.g. a subsidy on inputs to exportable production), or a research and development (R and D) subsidy—or an export tax. Sector policies are also used to promote exports through development of infrastructure (e.g. construction of roads and ports, development of marketing facilities, or the establishment of specialized financial institutions). These policies affect the supply of exports by influencing either the unit cost or the price of exports. Jung and Lee (1986) have shown that the effects on export supply of changes in the unit cost and in the price of exports are the same; they both shift the export supply curve, which thereby changes relative export prices and alters the quantity of exports demanded by foreign buyers. Consequently, it is not necessary to distinguish these channels when the effects of export policies are analysed.

A government export subsidy reduces the marginal cost of the exported commodity and, as a result, increases a country's exports. The resulting shift in the export supply curve, in turn, lowers the country's export price, and consequently increases the quantity of exports demanded. This effect can be shown in the first-order condition for foreign exchange earnings in which a country i exports the amount at which its marginal cost, $E'(X_i)$, is equal to its marginal revenue, $F'(X_i)$. A subsidy, denoted S, lowers the per-unit cost of the commodity export and reduces the marginal cost of exports to $E'(X_i) - S$. The introduction of this subsidy into the first-order condition of equation (4.17) yields the following export supply schedule (see Annex D for the derivation):

$$X_i^s = k_4' \left[\frac{(1 + 1/\epsilon_x^p)P_i + S}{D_i} \right]^\gamma t^{\phi_1} w^{\phi_2}, \qquad (19.13a)$$

where, as before, ϵ_x^p is the price elasticity of export demand, γ is the price elasticity of export supply, and $k'_4 = k_4/(1 + 1/\epsilon_x^p)^\gamma$, k_4 being defined as in equation (4.18). The subsidy rate, denoted s, is calculated as the total value of the subsidy, SX_i, divided by the total value of exports, P_iX_i, after the subsidy. Hence $s = (SX_i)/(P_iX_i) = S/P_i$.

An export tax, denoted T, has the opposite effect. The tax increases the per-unit cost of the commodity export by T, so that the marginal cost of exports is $E'(X_i) + T$. The resulting export supply schedule is

$$X_i^s = k'_4 \left[\frac{(1 + 1/\epsilon_x^p)P_i - T}{D_i} \right]^\gamma t^{\phi_1} w^{\phi_2}. \qquad (19.13b)$$

Like the export subsidy rate, the export tax rate, denoted t, is calculated as the total value of the export tax, TX_i, divided by the total value of exports, P_iX_i, after the subsidy. Hence, $t = (TX_i)/(P_iX_i) = T/P_i$.

There is a widespread perception that export promotion policies have significantly influenced the growth of Latin American commodity exports. For example, Morsink-Villalobos and Simpson (1980) reported that export subsidies have been a major stimulus to Costa Rica's banana industry. Export taxes are commonly imposed in Latin American countries and have had a strong impact on several export industries. Schiff and Valdes (1986) measured the levels of taxation on particular agricultural exports in five countries of the region. They found that, for the commodities covered by the present study, Argentina has imposed taxes on beef and maize production, as well as on wheat exports; Brazil has taxed soybean production; Colombia has taxed coffee production; and the Dominican Republic has taxed coffee production. These policies, according to Schiff and Valdes, have reduced the incentives of exporters.

Macroeconomic policies have also had a bearing on the competition and growth of traditional exports. The effects of policies of this type have recently been examined with CGE models, and according to Valdes (1986) their effects have been unfavourable to the expansion of some traditional agricultural exports. The reason is that import substitution policies have lowered the competitive position of traditional agricultural exports, since policies aimed at protecting semi-manufacturing and manufacturing industries have raised the cost of imported inputs for agriculture. The resulting higher cost of imported products has required reductions in the real exchange rates of these countries. These policies, according to Valdes, have also eventually led to increased costs of exportables as a result of indirect taxation designed to finance the more expensive imported protected goods. Consequently, the supply of traditional agricultural goods has decreased and relative prices have risen. Valdes found that the effects of Latin American import substitution policies on traditional agricultural exports have been particularly marked in Colombia and Peru.[1]

Thus, macroeconomic and sector- or industry-specific policies appear to provide considerable scope for influencing exports. In many cases, governments can bring about favourable changes merely by rescinding policies that have adversely affected exports. Whether export policies can, in fact, provide developing countries with concrete opportunities for influencing their export growth is examined in the following section.

19.2 Export Policies

In commodity trade, particularly that in the specific commodities covered in this study, trade policies need to be analysed against the background of the 'large-group' case, where there are many small exporting countries. As with the 'small-group' case, where there are a few large exporting countries, the types of strategic trade policy decisions that can be taken depend on the nature of the competition, and in particular on whether there is Cournot (quantity) competition or Bertrand (price) competition.

In Bertrand competition, Eaton and Grossman (1986) have shown that a government seeking to maximize export revenue and domestic income should tax, rather than subsidize, the country's export industry. The strategic steps for the exporting country would be to select an appropriately higher price, and raise prices through policies designed to shift the country's export supply curve back along the demand curve. As Deardorff and Stern (1987: 50) point out, the gains to the exporting country would be reflected in an improvement in its terms of trade. More importantly, however, are the revenue gains that can accrue to a country that has a price-inelastic demand function, since a price rise would lead to an expansion in export earnings. An alternative strategy would be for a small country to become a free-rider of policy measures taken by large exporting countries that had an inelastic export demand function. In this case, the demand curve of the small country would simply shift up to a new price level. The classic example of this situation, as Dixit (1987) points out, is the non-OPEC countries, for example Mexico, which became free-riders of the price hikes by OPEC members.

In Cournot competition, export quantity changes by a small country would be unnoticed by competitors that did not have similar export characteristics and were therefore not close rivals in particular geographic markets. When the exporter initiates strategic action and the competing exporters with similar characteristics initiate a response, two outcomes are feasible. In the first, if the conjectural variation by competing exporters consists of reductions in their output, the overall export earnings of the original country would increase. In the second, if close competitors also pursue strategic policy initiatives designed to increase exports, a 'prisoner's dilemma' arises. In this instance, the country and its competitors find that they are better off by actively promoting their export sectors; each loses, but the outcome is better for the country that would not have taken an initiative had the other been the only one to promote its exports. (For an illustration, see Richardson 1986: 270–4.) In this type of competition among small countries, it pays a country to know which are its closest competitors and what their most likely responses would be to possible policy initiatives on its part.

Strategic policy alternatives for small countries have recently been considered by Dixit (1987) against the background of policies oriented at increasing exports likely to bring about a response by other exporters that had similar characteristics and were close rivals in a market. This type of analysis had been undertaken with the use of game theory in which the non-cooperative solution for export earnings (pay-offs) of a country and its competitors in a particular geographic market is compared with that of the Nash bargaining (cooperative) solution. A country has an incentive to enter into a cooperative arrangement with its competitors when it can increase its earnings by doing so. However, unless its bargaining strength is commensurate with its gain, the country may have to compensate its competitors to ensure their participation.

Brander and Spencer (1985) have provided an alternative analytical framework in which an export subsidy is offered by a government. This framework lends theoretical support for the use of this policy instrument since it enables a country to increase both its market share and its earnings, provided its price elasticity of export demand is greater than unity. It can also be adapted to policy analyses of international trade under monopolistic competition. In their analysis, Brander and Spencer assume that the government uses a subsidy for strategic purposes and that the export industry takes the subsidy as given.[2] This assumption is essential to the outcome since otherwise the domestic industry would have no incentive to increase its output (exportable production being set at the level where the industry's marginal cost equals its marginal revenue). The subsidy lowers the marginal cost of the industry, which consequently increases the amount of exports it supplies to the market. The expansion in exports will increase the export earnings of the country provided its price elasticity of export demand is greater than unity. The promotion of exports will cause the country's national income to expand, not only because of the shift in production of a good from competing exporters to the domestic economy, but also because of the domestic multiplier effect that accompanies an increase in the real exports of an economy. Thus, when a subsidy is granted, there is not just a transfer from the government to the exporter; the country as a whole stands to gain from it.

We can illustrate the Brander–Spencer results for unilateral strategic export promotion in commodity trade with reference to a particular geographic market, j, and two foreign supply countries, denoted i and k respectively, which provide the unit-equivalent characteristics and the degree of diversification most preferred by the importing country. Thus, $X_{ij} + X_{kj} = M_j$. Unilateral action is taken by the government of the foreign supplying country i, while no action is taken by the government of the foreign supplying country k. Hence equation (19.13a) represents the export supply function for country i with a subsidy, and the corresponding export supply function for country k without a subsidy is given by (9.42). The export demand functions of both countries are described by (8.19).

By setting the export supply function of a country with a subsidy in (19.13a) equal to its export demand function in (8.19), we can find the effect of a change in the subsidy rate s on the volume and price of the export. The impact of a 1 per cent change in the export subsidy rate on the export volume of the country is

$$\frac{\partial X}{\partial s}\frac{s}{X} = -\frac{s\epsilon_x^p}{(1 - \epsilon_x^p/\gamma)(1 + 1/\epsilon_x^p + s)}, \qquad (19.14a)$$

and the accompanying effect on the export price is

$$\frac{\partial P}{\partial s}\frac{s}{P} = -\frac{s}{(1 - \epsilon_x^p/\gamma)(1 + 1/\epsilon_x^p + s)}. \qquad (19.14b)$$

Thus, provided the price elasticity of export demand, ϵ_x^p, is greater than unity, an increase in the subsidy will generate greater exports and higher export revenues. The subsidy simply represents a transfer from the government to the exporter and does not, in and of itself, change the economic activity of the country. However, a subsidy will lower the exporter's marginal cost and increase the quantity demanded of country i's exports, and thus will provide an overall revenue gain for the country.[3] The increase in the quantity demanded of country i's exports will, in turn, reduce the demand for country k's exports to $X_{kj} = M_j - X_{ij}$. Although this case is relatively simple, it provides insights into the options available to countries in formulating their export policies.

Despite the intuitive appeal of the Brander–Spencer model, Krugman (1987) has argued that it is based on too simple a representation of an economy. In particular, he points out that the model assumes that all output is exported, so there is no need to consider domestic consumer interests, and that each exporter has only one firm, so the effect of resource redistribution does not need to be considered. Brander and Spencer (1985: fn.6) recognized that their results could be affected by one or more of the following factors: (1) more domestic firms, as depicted in the model of Dixit (1984); (2) the existence of other industries with a greater degree of imperfect competition than the one being subsidized, as shown by Dixit and Grossman (1984); and (3) the need for there to be quantity, rather than price, competition, since Eaton and Grossman (1986) have demonstrated that an export subsidy with price competition may cause the amount of the exporter's earnings to increase by less than the subsidy.

When there is price, rather than quantity, competition, a tax, rather than a subsidy, is the appropriate policy instrument if a country has an upward-sloping supply function. The effect of a change in the tax rate t on the volume and price of exports can be found by the procedure we used to find the effect of a subsidy on exports, namely by setting the export supply function of a country with a tax in (19.13b) equal to its export demand function in (8.19) and solving for price and quantity. The impact of a 1 per cent change in the export tax rate on the export volume of a country is

$$\frac{\partial X}{\partial t}\frac{t}{X} = \frac{\epsilon_x^p t}{(1 - \epsilon_x^p/\gamma)(1 + 1/\epsilon_x^p - t)} \qquad (19.15a)$$

and the accompanying change in export price is

$$\frac{\partial P}{\partial t}\frac{t}{P} = \frac{t}{(1 - \epsilon_x^p/\gamma)(1 + 1/\epsilon_x^p - t)}, \qquad (19.15b)$$

the total revenue effect being

$$\frac{\partial PX}{\partial t}\frac{t}{PX} = \frac{(1 + \epsilon_x^p)t}{(1 - \epsilon_x^p/\gamma)(1 + 1/\epsilon_x^p - t)}. \qquad (19.15c)$$

The literature dealing with strategic trade policies in imperfect competition has shown that changes in market conditions often have important implications for policy outcomes. In the next section, we examine the optimal level of export promotion policies for commodity exporting countries when trade is characterized by monopolistic competition and non-retaliation by close rivals in the market.

19.3 Calculating Optimal Export Policies and Measuring their Impact

Econometric models of commodity trade can be used to assess the results of trade policies when they have been estimated in their structural form. The alternative methods used for policy evaluation with econometric models are the instruments targets approach and the social welfare function approach. In the social welfare function approach, the optimal policies are those that maximize the value of well-defined objective functions. This approach to policy evaluation requires specific presuppositions about the objective functions, however. In the instruments targets approach, optimal export policies are derived from the set of estimated econometric models for commodity trade once the export target (goal) is given.[4] In commodity trade under monopolistic competition, export policies are usually designed to take into account the quantity competition by countries that have price-elastic export demand functions. Thus, an export subsidy is an appropriate instrument (policy variable) for obtaining the export target.[5] For countries with price-inelastic export demand functions, export policies would instead be designed for a target level, or growth rate, of export earnings, and an export tax would be used as the policy instrument as long as the countries had upward-sloping supply functions.

In the case of quantity competition, small commodity exporting countries can usually expand their exports to a market without fear of protectionist measures being imposed by the importing countries.[6] As a consequence, strategic trade policies can be formulated as non-cooperative games involving a large number of countries exporting differentiated products. Each country establishes its export quantity in accordance with Brander and Spencer's approach, in which a unilateral export subsidy is offered by the government to the export industry. In the absence of well-defined objective functions for the countries covered by this study, we use the instruments targets approach to formulate optimal export policies with the estimated trade models. A government's target can be either a desired level of exports or a desired export growth rate, denoted g_x^*.

If the desired target is the level, or the growth rate, of exports, the structural form of the commodity trade models can be solved for the optimal export subsidy. Calculation of an optimal export subsidy for each exporter in the estimated trade models has been based on a common set of target growth rates for exports during a period of five years. The target growth rate for all Latin American countries has been established as that which would generate a 3 per cent higher rate of growth than that of imports in each of the geographic markets. Overall, this higher rate would mean that the region's long-term historical growth rate of 4 per cent a year in these commodities would rise to 7 per cent, which is equivalent to the long-term historical growth rate of the industrialized countries. The results of simulations with $g_x^* = 0.07$ provide estimates of optimal subsidy levels in each country, as well as of the changes that would occur in their relative export prices.

These simulations assume that the export policies of the Latin American countries are formulated independently of one another. This assumption limits the analysis to unilateral government incentives to subsidize exports. It implies either that the unilateral incentives of governments to subsidize exports occur in different time periods, or that the close competitors of the Latin American countries are outside the region. Certainly, the achievement by one Latin American country of higher export growth rates need not be at the expense of either countries within the region or other developing countries since, as Kravis (1970a) has noted, the industrialized countries compete in many of the same commodities. In either case, the assumption allows us to eliminate the consideration that a government's export subsidy level is dependent on the subsidy level of the other exporting countries in the Latin American region.

The results of the simulations are presented in Table 19.1 and are summarized in Fig. 19.1.[7] Overall, the 3 per cent higher average annual export growth rate needed for the Latin American countries if g_x^* is to equal 0.07 during a five-year period would be accompanied by a 1.8 per cent average annual fall in export prices. In addition to the increase in earnings resulting from these changes, the subsidy would expand the combined export market share of these countries from 18 per cent at the beginning of the period (according to data for 1987) to 20 per cent after the full impact of the five-year subsidy increases had made itself felt.

Since several periods are usually required for exports to adjust to a change in an export subsidy, the dynamics underlying the adjustment from one export growth rate to a target export growth rate need to be considered. Moreover, since subsidies take several years to exercise their full impact on export demand, the effect would continue past the time when the export subsidy had ceased. Overall, an 11 per cent increase in annual subsidies leads to a 16 per cent increase in export quantity and a 9 per cent fall in export prices during a five-year period. In the first period, the subsidy increases exports by only 0.8 per cent. After five periods, the subsidy causes the quantity of exports to expand by 2.5 per cent. This expansion incorporates both the same-period effect of the subsidy and the lagged effects of the subsidy granted in the earlier years.

These aggregate results, however, conceal the considerable variations in the amount of a subsidy required to achieve the target growth rate of exports. The more imperfectly competitive the export market, the greater the price change from the export subsidy. Thus, a 3 per cent higher growth rate is accompanied by a greater reduction in export prices in those countries that have less price-elastic export demand functions; the maize exporters of Argentina, the banana exporters of Ecuador, and the copper exporters of Peru must reduce prices by more than 2 per cent a year to achieve the target growth in exports. In contrast, it takes a smaller reduction in export prices of those countries that have more price-elastic export demand functions; a 3 per cent annual expansion in export volumes is accompanied by a less than 1 per cent fall in export prices by the coffee exporters of Costa Rica, El

Salvador, and Honduras, the beef exporters of Argentina and Costa Rica, and the cotton exporters of El Salvador and Peru.

For those countries that have inelastic export demand functions and upward-sloping export supply functions, the desired target can be

Table 19.1 Export subsidy required to achieve 3 per cent higher export growth rate[a]
(Percentages)

Commodity Exporter	Change in subsidy rate[b] (%)	Change in export price (%)	Initial market share[c]	Market share after 5 yrs
Beef				
Argentina	10.5	−0.8	14.2	16.5
Costa Rica	5.1	−0.6	3.2	3.7
Maize				
Argentina	1.7	−2.1	20.6	23.9
Bananas				
Ecuador	0.2	−2.4	17.8	20.5
Coffee				
Brazil	10.7	−1.7	16.9	19.6
Colombia	8.8	−1.1	13.3	15.4
Costa Rica	21.4	−0.7	2.8	3.2
Ecuador	2.1	−2.1	2.0	2.3
El Salvador	8.7	−0.9	6.9	8.0
Guatemala	8.5	−1.3	3.8	4.4
Haiti	4.4	−1.7	2.1	2.4
Honduras	12.6	−0.5	1.8	2.1
Nicaragua	9.7	−1.0	2.2	2.6
Cotton				
El Salvador	10.0	−0.9	3.2	3.7
Guatemala	3.2	−1.7	1.4	1.6
Peru	9.2	−0.6	3.5	4.1
Copper				
Chile	7.6	−1.6	17.5	20.3
Peru	1.6	−2.2	7.7	8.9

[a] For comparative purposes, it is assumed that there is a uniform initial subsidy rate equal to 20%.
[b] The subsidy rate equals the total amount of the subsidy divided by the country's earnings from the commodity export.
[c] Trade-weighted average of market shares of each principal geographic market of the exporting country.

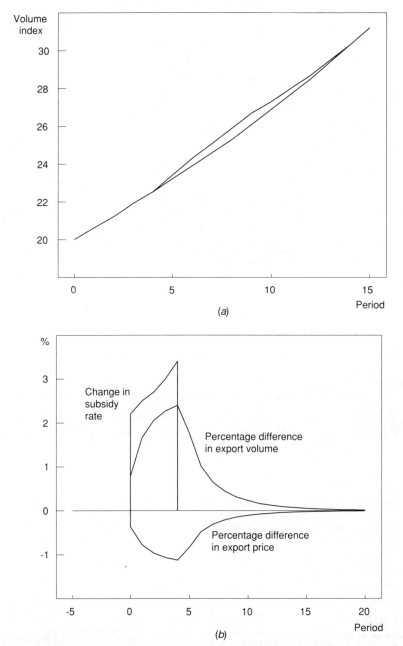

FIG. 19.1 Illustration of 5-year export subsidy needed to achieve 7 per cent target growth rate of exports
(a) Dynamic adjustment process of exports
(b) Export subsidy and percentage difference in export volumes and prices

established in export earnings. The structural form of the commodity trade models can also be solved for the optimal export tax rate t^*, and the effect of a tax on export price and quantity can be calculated from estimates of (19.13b). In this case, Dixit (1987) has pointed out that, since competition is of the Bertrand type, the small country is in a better strategic position than that previously examined; by remaining a small exporter because of the higher price of its exports, the country does not subject itself to the possibility of retaliation by its competitors, which might feel threatened lest the small country expand its market share at their expense in Cournot competition.

Calculations of the optimal export tax for each exporter in the estimated trade models have been based on a common set of target growth rates for export earnings. The target growth rate for those Latin American countries with price-inelastic demand functions was established 3 per cent higher than the steady-state growth of their export earnings without policy intervention. The results of simulations with this set of target growth rates provide estimates of the annual rates of change needed for export prices and, consequently, the optimal change in tax levels needed if the target growth rates are to be reached. The

Table 19.2 Export tax needed to achieve 3 per cent higher growth rate of export earnings[a]
(Percentage change)

Commodity Exporter	Change in tax rate[b]	Change in export price	Change in export volume
Beef			
Uruguay	11.9	6.3	−3.1
Bananas			
Honduras	7.1	5.3	−2.2
Cocoa			
Brazil	15.4	5.7	−2.5
Ecuador	11.5	4.5	−1.4
Soybeans			
Paraguay	4.7	4.0	−1.0
Cotton			
Mexico	15.3	9.7	−6.1
Iron ore			
Brazil	16.0	9.4	−5.8

[a] For comparative purposes, it is assumed that there is a uniform initial tax rate equal to 20%.
[b] The tax rate equals the total amount of the tax divided by the country's earnings from the commodity export.

results are presented in Table 19.2. Overall, a 3 per cent higher average annual growth rate of export earnings for the Latin American countries would be accompanied by a 5.8 per cent average annual rise in their export prices and a 2.6 per cent average annual fall in their export volumes.

Policies that either expand exports through increased export subsidies or expand earnings through increased export taxes could not be continued indefinitely, since relative export prices must be stationary. Instead, during a medium-term period they could be used to generate more foreign exchange earnings for servicing the external debt and sustaining the economic growth of the developing countries. Moreover, such policies could offset the temporary downturn in exports of manufactured goods from the developing countries that usually accompanies a world-wide recession, since their market outlets—other developing countries—tend to place severe restrictions on imports of goods of these types when their own exports decline.

19.4 Conclusions

This chapter has examined optimal trade policies for developing countries in a model of imperfect competition in international commodity trade. It has been assumed that the policies of the commodity exporting countries are formulated independently of one another. The analysis suggests that, in most cases, the appropriate policy instrument is an export subsidy; in the few cases in which countries have an export demand that is inelastic with respect to price and an upward-sloping export supply function, a tax would be the appropriate instrument to use in order to obtain higher export revenues. The use of these instruments is limited in so far as they provide a one-time effect on exports; they do not influence the long-run, or steady-state, growth of commodity exports.

Notes

1 The adverse effects of import substitution policies and the favourable opportunities opened up by export promotion policies have also been examined in the light of the developing countries' industrialization policies at the OECD by Little *et al.* (1970), at the World Bank by Balassa (1982), and at the National Bureau of Economic Research (NBER) by Bhagwati (1978) and Krueger (1978). These studies have found that import substitution policies in developing countries create a bias against their export growth.
2 In an earlier work, Brander and Spencer (1983) analysed strategic actions taken by the domestic industry in the traded good. Both the domestic

industry and the foreign competing industry were assumed to be aware of possible benefits arising from the use of strategic policies and were able to take the strategic actions themselves. That approach gave rise to a different analytical approach from the one followed here.

3 For example, suppose a country exports 100,000 units of a commodity at \$100 a unit. If $\epsilon_x^p = -2$ and $\gamma = 1.5$, an increase in an existing subsidy of \$5 to \$7 a unit will increase exports to over 103,000 units as the price falls to \$98 a unit. Since the subsidy represents only a transfer from the government to the export industry, the country has achieved a \$151,000 increase in its foreign exchange earnings as a result of the subsidy.

4 When the target choice or the welfare function to be maximized is unknown, an alternative approach is to calculate policy multipliers, i.e. the impact of changes in policy variables on endogenous variables in the model. Lord (1989*b*, 1989*c*) provides estimates of policy multipliers for the models of commodity trade developed in this study.

5 The number of instruments must be at least equal to the number of targets. When the number of instruments exceeds the number of targets, the excess instruments are assigned a priori values at the time the model is simulated.

6 In the commodities covered by this study, Chilean copper exports to the USA in 1984 are an exception. The domestic copper industry in the USA unsuccessfully attempted to convince the government that it should apply protectionist measures against Chilean copper imports since that country had expanded its market share by increasing export volumes and lowering export prices.

7 The results for sugar are not shown in the table since a reduction in its relative export price would have a minimal impact on demand for all exporting countries. As noted earlier, most shipments are directed to the USA, where the price is administered. Consequently, the price elasticity of demand for exports of Latin American countries in the US sugar market is infinite with respect to price.

20
Trade Liberalization

Intervention measures aimed at expanding exports were shown in the previous chapter to lead to gains for one country at the expense of others when international commodity trade is imperfectly competitive. These results do not generally hold in either perfect or imperfect competition for policies that restrict imports. Tariffs are often used to shift profits from foreign suppliers to the domestic economy, and Brander and Spencer (1984) have attempted to demonstrate that they can be used to increase a country's domestic welfare in imperfect competition. However, these results have been criticized by Dixit and Grossman (1984) on the grounds that the promotion of one industry in a country can undermine other industries in the same country that also operate in markets with imperfect competition, the result being that the losses can outweigh the gains from intervention. Subsequent studies, such as that of Flam and Helpman (1987), recognized that general economic structures that include more than one non-competitive industry can weaken the case for trade intervention.

The goal that a tariff aims to achieve can also be attained by imposing a subsidy on domestic output or a tax on consumption, which do not give rise to the excess producer or consumer distortions that usually accompany a tariff. Deardorff and Stern (1987: 38–45) have argued that a tariff is inefficient because it always causes producer and consumer distortions in an economy, and that a better policy is one that brings about changes in the targeted sector, through either a subsidy on production or a tax on consumption. The potential effectiveness of such an approach was demonstrated in the previous chapter, where a subsidy was used to expand a country's foreign exchange earnings from a specific commodity export in Cournot competition, and where an export tax was applied to a specific commodity when competition was of the Bertrand type.

These considerations suggest a different approach to the analysis of trade policies. In contrast to the previous chapter's focus on the effects of the imposition of subsidies or taxes on commodity exports, this chapter examines the impact of the removal of trade barriers on commodity trade and on market prices. When trade barriers are high, trade liberalization has a significant impact on foreign demand for primary commodities in their principal geographic markets, as evidenced by the price elasticities of import demand presented in Chapter 13. On the whole, the average price elasticity of import demand of goods of these types lies within the lower range of other estimates for total merchandise imports of the industrialized countries. However, the range of elasticities varies greatly, for both primary commodities and manufactured goods, within each type of good and among geographic markets. With respect to geographic markets, my own estimates for primary commodities showed that, for a given reduction in tariffs, the generally stronger response of imports to price changes would have a greater impact on the trade of the USA than on that of Japan, and that it would have a greater impact on the trade of Japan than on that of the EEC.

In addition, the USA, the EEC, and Japan together absorb a large part of the total world output of these commodities. Were trade to be concurrently liberalized in these three important market areas, feedback effects between domestic production and consumption decisions and world market prices could give rise to complex dynamic effects. The path to a new steady-state solution would depend on the dynamics underlying the production and consumption responses to price changes. The often long lags between production adjustments to price changes would give rise to a cyclical price path when feedback effects occur between market prices and production and consumption decisions in the group of countries pursuing trade liberalization policies. These responses in commodity market prices, as well as their accompanying effects on trade volumes, have important implications for the evaluation of the consequences of trade liberalization.

Consideration of the dynamics underlying the adjustment processes of commodity markets after trade liberalization is what distinguishes the present analysis from previous attempts to quantify the effects of the removal of barriers to primary commodity trade. Earlier studies have sought to measure the impact of reduced barriers to trade on the agricultural exports of the developing countries that compete with the domestic production of the same goods in the industrialized countries. (For a critical survey of earlier studies, see Gardner 1985; Valdes 1987; and Quizon et al. 1988.) The present analysis examines many of these same agricultural commodities; it also treats tropical products, which are not produced in the industrialized countries, as well as mineral exports in order to evaluate the effects of trade liberalization on different commodity categories.

The major distinction between the present analysis and earlier studies of trade liberalization lies in the modelling framework. Most other studies that have measured the effects of trade liberalization have relied on CGE models or fairly simple comparative-static models. (For a comprehensive review and critique of the application of CGE models to the analysis of trade liberalization, see Shoven and Whalley 1984.)

These models suffer from a number of shortcomings. In the first place, the coefficients in the equations of the models are not statistically estimated. Instead, they are selected from specialized studies, often from literature surveys of elasticities (such as the compendium by Stern *et al.* 1976) or, when actual estimates are unavailable, from best-guess estimates. Once the elasticity values of the model have been chosen, these parameters must be adjusted to the system of equations (so the model can replicate the actual data in the base period) and must be used as an equilibrium solution of the model (the so-called 'calibration' procedure). Given the generally low price elasticities in commodity markets and the sensitivity of the price path of a commodity to these elasticities, the selection of the elasticities is crucial, and the results of the models become highly dependent on the selected parameters.

Second, by their very nature, CGE models fail to consider intertemporal issues in trade. We have seen (in Chapter 16) that long lags in adjustment processes by economic agents often give rise to cyclical behaviour in commodity markets. The long-run comparative-static approach used in CGE models, as well as in simpler models of the same type that have recently been used to analyse the commodity-specific effects of trade liberalization, fail to consider the dynamics of market price and trade adjustment processes. The results may therefore be misleading. Dynamic effects that include large cyclical movements in the adjustment process of the market can significantly influence the welfare effects of the trade liberalization policies of countries and their trading partners. As the simulation results in this chapter demonstrate, the initial effects of trade liberalization policies are often very different from their long-run effects. For instance, the initial effect of trade liberalization is to lower market prices, but the long-run effect is to raise them. Not only is this distinction between the initial and final outcome important, but the price path of the commodity after trade liberalization can have an important effect on the potential gains and losses from such a policy.

These shortcomings of CGE models, as well as of simple comparative-static models, used to analyse trade liberalization policies are overcome by the use of an econometric modelling framework. Econometric models are based on statistically estimated parameters that are consistent with the model used to represent data-generating processes, and the dynamic effects are deduced from numerical solutions of the equations that make up the model. Nevertheless, CGE models provide an analytical capacity that usually lies outside the partial equilibrium framework of these dynamic econometric models, in so far as they are capable of analysing the welfare effects of trade liberalization in relation to the impact that the elimination of trade barriers and domestic price support systems would have on the economies of the major trading areas. (See Trela *et al.* 1987, for the case of multilateral trade liberalization effects in the grain trade.) Unfortunately, dynamic econometric models do not lend themselves to being integrated into CGE models. Thus, the decision about the adoption of a modelling framework must be based on the objective of the analysis.

In this study, our objective is quantitatively to assess the potential effects of trade liberalization on trade and market prices for particular commodities. In accordance with the data sample selected in Part III, the commodities examined are those that represent the most important non-fuel exports of Latin America. In the industrialized countries, protection of those commodities that are domestically produced has taken the form of tariffs and NTBs to trade, with the result that a 'price wedge' has been created between the domestic producer price and the import or export price of the goods. This price wedge has been used in this study, as in others, as a measure of the *ad valorem* tariff-equivalent of a country, since the NTB component of trade is notoriously difficult to quantify. This chapter investigates the effects of the removal of these price wedges on commodity market prices and trade.

The chapter consists of four parts. The first section explicitly introduces tariff and non-tariff measures in importing countries into the commodity trade model developed in Parts II and III in order to examine protectionist-related policy issues. Section 20.2 examines the types and magnitude of trade restrictions that are imposed on primary commodities by the industrialized countries. Sections 20.3 and 20.4 evaluate the impact of lower intervention levels on trade and market prices for the commodities. The first of these two analytical sections assesses the impact of unilateral trade liberalization and the second, that of multilateral trade liberalization. Although different results are to be expected from multilateral and unilateral trade liberalization, the empirical findings show how important the difference can be for some traded goods. The final section presents a summary of the findings and the conclusions.

20.1 Modelling Trade Liberalization

Explicit recognition of imperfect competition in international commodity trade is essential for modelling trade liberalization. In the pure Heckscher–Ohlin model for homogeneous products, complete trade liberalization would lead to extreme specialization of production, and trade flows would be based on transportation costs. In contrast, heterogeneity of goods by country of origin permits the use of price elasticities for measuring the effects of trade liberalization by trading partners. (For a discussion of the market structure underlying much of the analysis of trade liberalization, see Whalley 1985.)

The theoretical approach used in this chapter is similar to those developed by Flam and Helpman (1987) and Venables (1987) for examining the effects of protectionist measures in international trade. Like the model developed in Parts II and III, their models demonstrate that product heterogeneity by country gives rise to a given amount of foreign spending on the differentiated good and to downward-sloping foreign, as well as domestic, demand functions for that good.

Expenditure-switch policies in the form of tariffs and NTBs to trade create a 'price wedge' between the domestic price to the consumer and the world market price of the commodity. These measures effectively impose a tax on the consumer. The effective tax rate, denoted t, raises the price of the commodity to $(1+t)P$ in the geographic market. Hence the demand function for the traded commodity in the long-run dynamic equilibrium relationship implicit in equation (8.13) is

$$M^d = k_1^* Y^{\epsilon_m^y} \left[\frac{(1+t)P}{D} \right]^{\epsilon_m^p}. \tag{20.16}$$

The incidence on the consumer of the effective price rise of the commodity by $(1+t)P - P$ would be to reduce the quantity of the commodity demanded.

The relative prices of domestic and foreign suppliers to the market remain unaltered: that of the domestic producer is $(1+t)P_i/(1+t)P$, and that of each foreign supplying country is $(1+t)P_k/(1+t)P$. Thus, the demand function for each foreign supplier is

$$X_k^d = k_2 M \left[\frac{(1+t)P_k}{(1+t)P} \right]^{\epsilon_x^p}. \tag{20.17}$$

A change in the quantity demanded of the commodity because of protectionist measures would cause a proportional change in the demand for the commodity supplied from foreign sources.

Equation (20.16) shows that the effect of trade liberalization will depend on the price elasticity of import demand, ϵ_m^p, and the tariff-equivalent rate, t, of protection in the importing country. In addition, feedback effects can occur between previously supported domestic prices and the world market price of a commodity when trade liberalization takes place either in a large importing country or in several countries at the same time, and these effects will influence production and consumption decisions in both exporting and importing countries. The consequences of trade liberalization with and without feedback effects between previously supported domestic prices and the world market price are analysed after the next section, which deals with protectionism affecting the demand for commodity imports of the industrialized countries.

Table 20.1 Protection in selected commodities by the USA, the EEC, and Japan, 1984–1986

Market Product	Tariff	NTBs to trade	Ad valorem tariff-equivalent (%)
USA			
Sugar	0.625 cents a pound import fee for raw sugar for Canada, Australia, Gabon, and Brazil 2 cents a pound import fee	(1) Price support system (2) Global quota established annually; country-specific quotas	319.2
Beef		(1) Global quota triggered at determined export levels; country-specific quotas (2) Voluntary export restrictions imposed when imports exceed trigger level (3) Prohibition of imports from S. America because of foot-and-mouth disease (4) Health and sanitary regulations (5) Standards for labelling of meat products	16.6
Maize	5–25 cents a bushel	Price and income support systems	33.5
Cotton	(none)	(1) Direct payments to producers (2) Global and country-specific quotas	0.0
Copper	1% ad valorem	(none)	1.0
EEC			
Sugar	Variable import duty	(1) Production quotas for EEC members (2) Price support system	174.8

Commodity	Trade barrier	Other measures	Tariff equivalent (%)
Beef	20% *ad valorem* and variable import levy	(1) Country-specific quotas allocated when deemed necessary (2) Health and sanitary regulations (3) Standards for labelling of meat products	75.3
Soybeans	(none)	Price support system	93.7
Coffee	5% *ad valorem*	(none)	5.0
Cocoa	3% *ad valorem*	(none)	3.0
Bananas	20% *ad valorem*	Bilateral trade agreements	20.0
Copper	6% *ad valorem*	(none)	6.0
Japan			
Sugar	Variable import duty; import fee	Production support and stabilization system	483.3
Beef	25% *ad valorem* for quota imports; 18.8% for non-quota imports	(1) Global import quota established bi-annually (2) Health and sanitary regulations (3) Standards for labelling of meat products	187.0
Bananas	30% *ad valorem*	(none)	30.0
Copper	8.5% *ad valorem* on unrefined copper; concentrates enter duty-free	(none)	8.5

Note: No barriers to trade are currently in effect for iron ore.

Sources: *Ad valorem* tariff-equivalents are calculated as the difference between the domestic producer price and either the border (export or import) price or the world market price of the good. Data for 1979–81 are from OECD, *National Policies and Agricultural Trade*, 1987; country studies for the USA, the EEC, and Japan. Data for 1982–6 are from USDA Foreign Agricultural Service, 'Preliminary Estimates of Producer and Consumer Subsidy Equivalents (PSE's and CSE's)', mimeo.; data for tropical goods are from the International Coffee Organization (ICO); International Cocoa Organization (ICCO), *Obstacles à l'accroissement de la consommation et du commerce du cacao*, ICC/28/7, 28 February 1985; and GATT, *Tropical Products: Background Material for Negotiations*, MTN.GNG/NG6/W/6/Rev.1, 18 January 1988. See Annex A for details.

20.2 Protectionism in the Industrialized Countries

The USA, the EEC, and Japan place severe restrictions on trade in most of the major commodity exports of the Latin American region. The objective of these measures is to protect domestic producers from foreign competition and to maintain the level of their earnings. Of the commodities covered in this study, only coffee, cocoa, and bananas are tropical products and therefore are not produced in these market areas; none the less, they compete with goods that are domestically produced, and so their entry into some of the markets is restricted.

Table 20.1 summarizes the tariff and NTBs to trade affecting the major commodity exports of Latin America. In accordance with the conventional procedure for incorporating NTBs to trade, I have calculated the *ad valorem* tariff-equivalents, which are measured as the difference between the domestic producer price and either the border (import or export) price or the representative world price for the particular commodity market (as defined in Annex A). In the table, the average for each commodity is equal to the consumption-weighted share of the USA, the EEC, and the Japanese *ad valorem* tariff-equivalents multiplied by their respective shares of world consumption. Thus, the averages are only for the three main market areas and preclude trade barriers in the rest of the world. NTBs to trade include quantitative controls on imports and all non-tariff policies that affect trade flows. (For a review of measures applied to the commodities covered in this study, see Lord and Boye 1987: 17-30; and OECD 1987*a*: 17-25.) Restrictions to trade can arise as a result of administrative procedures for quality standards and health regulations, even though there is no declared intention to limit trade. As a result, the levels of protection on domestically produced commodities such as sugar, beef, and soybeans tend to be higher than those on tropical products such as coffee, cocoa, and bananas.

In the *US market*, producers are assisted by both price and income support programmes that include non-recourse loans, stock programmes, deficiency payments, and target prices. The 1985 Food Security Act made comprehensive changes in domestic agricultural policies that considerably influenced the world markets for several commodities. The objective of the Act is to make US exports more competitive by reducing support prices to producers of grains, oilseeds (including soybeans), cotton, and other products, while compensating them for these reductions through income support payments.

In the *EEC market*, agricultural production and trade are regulated by the Common Agricultural Policy (CAP). Producers are protected from world market fluctuations through guaranteed prices and receive preferential treatment in agricultural trade. Pricing policies and protective barriers against third countries have been supported by rules of competition, Community-wide financing of agriculture, and removal of intra-Community barriers to trade in agricultural products. These policies have resulted in surpluses in commodities such as sugar, grains, dairy products, and meat.

The present EEC support system is designed to work to the advantage of both participant exporters and importers. It is based on a target price for each commodity, which is a theoretical market price established by representatives of all the member countries and is usually well above world market prices. As long as the EEC market price for a commodity is below the target price, EEC producers have an advantage over importers and will be the preferred suppliers to the market. On the other hand, if internal supply decreases to the point where the market price rises above the target price, importers fill the market to close the gap. If world market prices rise above the threshold price (a minimum import price), export levies can be imposed on EEC producers so as to discourage output and thus raise internal prices. Variable levies are imposed on most agricultural commodities. They are calculated as the difference between the minimum import price and the lowest cost, insurance, freight (c.i.f) offer price. Import quotas are also used to protect the EEC market, as are health and sanitation restrictions for certain commodities.

In 1975 the Lomé Convention established preferential trade arrangements with 46 ACP countries. It was renegotiated in 1980, when ACP membership increased to 66 countries, and again in 1984, when the Lomé III Convention was adopted for the period 1985-90. That Convention specifies that goods imported into the EEC from ACP countries enter free of customs duties. While imports of the ACP countries enter the Community duty-free, ACP imports from EEC countries are given most-favoured nation (MFN) treatment. The supply arrangements for products such as bananas and sugar, which are described below, were improved in the Lomé III Convention.

In the *Japanese market*, protection is given to both agricultural commodities and some of the minerals covered by this study. The agricultural policy, which fosters self-sufficiency and guarantees farm incomes, was established in 1961 by the Agricultural Basic Law. One of the principal objectives has been to achieve and maintain income parity between farm and urban households. To that end, income support and price stabilization policies are based on administered prices that help keep Japanese prices above levels prevailing in international markets. A system of public and private trading agencies has been established to ensure that domestic producers of commodities competing with foreign goods obtain adequate returns. It purchases commodities from producers at a price that is established annually, and resells the products to a wholesaler at a price that is also established annually. Charges are also levied on imports in order to raise the domestic prices of foreign products and thus improve the competitive position of Japanese farmers in the domestic market.

The last column of Table 20.1 shows the tariff-equivalent levels of the combined tariffs and NTBs to trade in the USA, the EEC, and Japan in 1984–6. The combination of these measures gives the price wedge between the domestic price in the geographic market and the foreign supply price to that market. The *ad valorem* tariff-equivalent level of all these measures has been calculated as the percentage difference between domestic and foreign supply prices (for details, see Annex A). These rates are based on recent data compiled by the US Department of Agriculture (USDA 1987) and the OECD (1987*a*, 1987*b*, 1987*c*) for the purpose of calculating producer and consumer subsidy-equivalents in the industrialized countries.

In recent years, price wedges for many commodities have increased, since domestic support prices have remained stable whereas world market prices have fallen. This trend was particularly noticeable in sugar (see Fig. 20.1). When the market price for this good was at a record high in 1980, the US and EEC domestic producer prices nearly matched the world market price (although Japanese producer prices were set at two-thirds higher than the world market price). When world prices

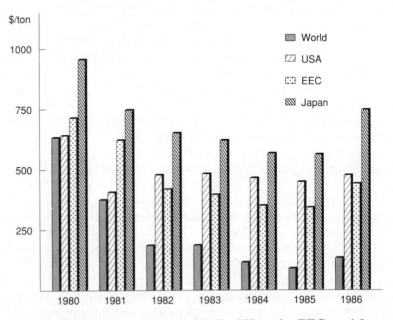

FIG. 20.1 Sugar prices of the world, the USA, the EEC, and Japan

Table 20.2 Ad valorem tariff-equivalent levels in selected commodities in the USA, the EEC, and Japan, 1979–1986 (Percentages)

	1979	1980	1981	1982	1983	1984	1985	1986
Beef	*55.4*	*77.6*	*53.2*	*36.3*	*43.7*	*55.9*	*27.7*	*51.9*
USA	39.0	77.8	23.2	15.5	17.5	13.8	14.7	21.3
EEC	72.6	75.4	91.6	60.0	74.9	109.6	37.2	79.2
Japan	139.6	105.8	105.2	115.2	122.3	129.1	153.9	277.9
Maize	*5.6*	*1.2*	*5.3*	*2.6*	*40.4*	*12.4*	*50.6*	*51.4*
USA	0.0	0.1	2.5	0.0	45.2	13.2	50.0	37.2
EEC	52.4	11.1	29.6	54.0	92.0	76.5	122.9	335.5
Sugar	*119.1*	*4.9*	*55.3*	*140.5*	*122.0*	*283.1*	*206.8*	*185.9*
USA	9.3	7.3	76.1	185.5	223.7	471.3	238.6	247.7
EEC	161.3	1.0	39.5	121.0	68.0	197.6	173.2	153.6
Japan	40.5	66.2	229.0	238.1	360.3	555.3	564.3	330.4
Soybeans	*15.5*	*26.0*	*6.7*	*7.5*	*25.9*	*0.2*	*9.9*	*9.3*
USA	15.5	26.0	6.7	7.5	25.9	0.2	9.3	6.7
EEC	0.0	0.0	0.0	9.5	32.4	5.4	109.2	166.6

Note: The average shown for each commodity is equal to the production-weighted shares of the USA, the EEC, and, where applicable, Japanese *ad valorem* tariff-equivalents multiplied by their respective shares of world production.

Sources: Ad valorem tariff-equivalents are calculated as the difference between the domestic producer price and either the border (export or import) price or the world market price of the good. Data for 1979–81 are from OECD, *National Policies and Agricultural Trade*, 1987, country studies for the USA, the EEC, and Japan. Data for 1982–6 are from USDA Foreign Agricultural Service, 'Preliminary Estimates of Producer and Consumer Subsidy Equivalents (PSE's and CSE's)', mimeo.

plummeted from $632 to $115 a metric ton between 1980 and 1984, domestic price supports in each of these three areas caused wide disparities between domestic producer prices and the world market price. In fact, the EEC emerged as the world's largest producer during this period.

Increased levels of protectionism in these industrialized countries affected most temperate zone commodities. As Table 20.2 shows, the implicit *ad valorem* tariff level rose sharply in goods of these types between the early and the middle part of the 1980s. Consequently, the impact of trade liberalization depends, among other factors, on the period used to measure the degree of protectionism in international trade.

20.3 Unilateral Trade Liberalization

The effect of trade liberalization in a small geographic market is simply determined by the amount of the reduction in the price wedge and by the price elasticity of import demand. The elimination of the price wedge of a particular geographic market j would lower the domestic price from $(1 + t)P_j$ to P_j, where, as before, t is the tariff-equivalent rate of protection and P_j is the import price of the commodity. This reduction has been shown to lead to an increase in the quantity demanded of that good and to a reduction in the quantity of the good supplied by domestic producers.

20.3.1 Empirical Findings

The effects of the one-time change in price in the period of trade liberalization on the import demand for the agricultural and non-fuel mineral exports of Latin America are presented in Table 20.3. Import demand takes several periods to adjust to the price change. For example, it takes four periods for beef imports of the USA to adjust to 95 per cent of their new steady-state solution, and it takes 12 periods for sugar imports of the UK to adjust to 90 per cent of their new steady-state solution.

As expected, the largest increase in imports would generally occur in those exports of Latin America that are also produced in the industrialized countries, namely sugar, beef, and soybeans. The exception is the banana imports of EEC member countries, particularly West Germany; imports of this good would increase significantly as a result of the high level of protection given to domestic producers and the relatively high price elasticity of import demand for the products in that market. None the less, imports of sugar would expand far more than those of any

Table 20.3 Effect of trade liberalization on US, EEC, and Japanese imports (Percentage changes)

Market area Commodity Importer	Initial effect		Total effect			
			Unilateral liberalization		Multilateral liberalization	
	Import price	Import volume	Import price	Import volume	Import price	Import volume
USA						
Sugar	−76.1	34.3	−76.1	44.1	−40.6	36.4
Beef	−14.2	4.9	−14.2	17.4	−7.7	11.3
Copper	−1.0	0.6	−1.0	1.5	4.6	−6.7
EEC						
Sugar	−63.6	13.6	−63.6	74.2	−28.1	54.0
UK	−63.6	13.6	−63.6	74.2	−28.1	54.0
Beef	−43.0	15.8	−43.0	12.3	−36.4	11.2
France	−43.0	0.0	−43.0	0.0	−36.4	0.0
W. Germany	−43.0	9.1	−43.0	15.9	−36.4	14.7
Greece	−43.0	109.7	−43.0	76.2	−36.4	71.7
Soybeans	−48.4	44.3	−48.4	36.1	−46.4	35.3
W. Germany	−48.4	0.0	−48.4	0.0	−46.4	0.0
Netherlands	−48.4	94.1	−48.4	74.5	−46.4	73.4
Spain	−48.4	45.2	−48.4	43.4	−46.4	42.4
Switzerland	−48.4	64.0	−48.4	87.2	−46.4	84.8

Table 20.3 (cont.)

Market area Commodity Importer	Initial effect		Total effect			
			Unilateral liberalization		Multilateral liberalization	
	Import price	Import volume	Import price	Import volume	Import price	Import volume
Coffee						
Belgium	−4.8	0.6	−4.8	0.7	−4.3	0.7
France	−4.8	1.7	−4.8	1.3	−4.3	0.8
W. Germany	−4.8	0.5	−4.8	0.2	−4.3	0.1
Italy	−4.8	0.5	−4.8	0.7	−4.3	0.6
Netherlands	−4.8	0.7	−4.8	1.0	−4.3	0.8
Cocoa	−4.8	1.8	−4.8	0.7	−4.3	0.7
W. Germany	−2.9	0.5	−2.9	0.8	−2.6	0.1
Italy	−2.9	0.1	−2.9	0.2	−2.6	0.1
Netherlands	−2.9	0.9	−2.9	0.9	−2.6	0.7
Spain	−2.9	0.8	−2.9	1.3	−2.6	0.8
Bananas	−2.9	0.4	−2.9	0.3	−2.6	0.3
W. Germany	−16.7	10.4	−16.7	37.8	−16.6	38.0
Italy	−16.7	16.2	−16.7	51.6	−16.6	51.9
Copper	−16.7	7.6	−16.7	9.2	−16.6	9.2
Belgium	−5.7	0.3	−5.7	0.8	−0.1	0.0
W. Germany	−5.7	0.7	−5.7	1.8	−0.1	0.1
	−5.7	0.0	−5.7	0.0	−0.1	0.0

Japan					
Sugar	−79.3	11.5	−79.3	246.5	−43.8	202.8
Beef	−65.2	72.2	−65.2	184.7	−58.6	180.5
Bananas	−23.1	21.4	−23.1	36.6	−23.0	36.7
TOTAL[a]					
Sugar	−74.8	21.5	−74.8	127.0	−39.3	103.1
Beef	−31.0	13.2	−31.0	30.8	−24.0	26.9
Soybeans	−13.9	12.8	−13.9	10.4	−7.3	9.1
Bananas	−6.3	4.8	−6.3	12.4	−6.1	12.4
Coffee	−2.2	0.3	−2.2	0.3	−2.6	0.4
Cocoa	−1.3	0.2	−1.3	0.3	−1.2	0.3
Copper	−4.1	0.3	−4.1	0.8	1.7	−1.0

[a] Total of the USA, the EEC, and Japan.

Note: Percentage changes in EEC product totals are trade-weighted averages of the specified member-countries, which are those that represent the major markets for Latin American exports of the product.

other good in all three market areas. The level of protection of domestic sugar is greater than that of any other product. Moreover, Japan has a relatively high price elasticity of import demand for sugar.

The effects of trade liberalization on the demand for exports of the Latin American countries would be proportional to the change in the total imports of each commodity in the geographic market. When relative prices remain unchanged, the change in export demand is proportional to the change in import demand in each geographic market of the countries in the region (see equation (20.17)). Since relative prices would be unaffected by trade liberalization, the market shares of the Latin American countries and other exporters in the USA, Japan, and the EEC would remain unaltered.

20.3.2 Comparison of Findings with Those of Other Studies

The empirical findings presented here differ from those of other similar analyses for one or more of the following reasons: differences in the price elasticity of import demand for the commodities used for the calculations; differences in the effective levels of protection of the commodities in the periods in which the calculations were made; and differences in the type of model used to analyse the consequences of trade liberalization. Moreover, as Valdes (1987) points out, the product definitions used in the studies often vary.

These differences are brought out by a comparison of the findings presented in this chapter and those reported by Kirmani *et al.* (1984) for trade liberalization in sugar and meats by the industrialized countries. In the first place, Kirmani *et al.*'s price elasticities of demand for sugar, taken from Cline *et al.* (1978), differ significantly from those found in this study: for the USA they used an elasticity of -0.82, compared with -0.3 estimated in this study; for the EEC they used -1.06 versus -0.07 estimated for the UK in this study; and for Japan they applied -0.56, in contrast to -1.6 calculated in this study. There are even greater differences between the price elasticities used by Kirmani *et al.* for meat in the EEC (-1.09) and those for beef estimated in this study for individual EEC countries (-0.1 trade-weighted average), although the price elasticity for Japan (-1.13) is closer to that estimated for that country in this study (-1.6). Moreover, the product definition for beef used in this study (SITC 011.1) is narrower than the product definition for meat used in the aforementioned study (SITC 01).

The other major source of discordance between the findings of this study and those of Kirmani *et al.* is the effective levels of protectionism for the commodities. The reason for the difference is that the spread between world market prices and domestic producer prices in 1979–80, the period in which the other study calculated the *ad valorem* tariff-equivalents of protectionist measures in the industrialized countries, was considerably narrower than in 1984–6, the year used to calculate the *ad valorem* tariff-equivalent levels of intervention in the industrialized countries in this study. For instance, the spread was four times greater for sugar in the USA than in the world market. In 1979–80, the US price averaged 23 cents a pound compared with 19 cents a pound in the world market; in 1984–6, the US price averaged 21 cents a pound compared with 5 cents a pound in the world market. In contrast, the spread in beef was considerably smaller in the USA, and about the same in the EEC and Japan in 1979–80 and 1984–6. (See Table 20.2 for details on the spread in these and other commodities.) Consequently, the levels of protectionism used in this study and in that of Kirmani *et al.* were dissimilar in sugar and, to a lesser extent, in beef.

These differences serve to demonstrate why estimates of the effects of trade liberalization can vary significantly among empirical studies. They also show how dissimilarities in the findings can be resolved. It would be possible to obtain consistent estimates of the effect on import demand of trade liberalization in the industrialized countries once inconsistencies in the price elasticities of import demand in the various studies were reconciled. Primary data are now available for calculating *ad valorem* tariff-equivalent levels of intervention in agricultural trade, and these will ensure consistency in the studies.

The findings of this analysis demonstrate the extent of the time lag between the initial reduction in import prices after trade liberalization and the time required for imports to adjust fully to the new price level in the market area. Comparative-static analyses provided by other studies based on CGE models, or simple comparative-static models, are unable to take into account the delay in the adjustment of imports, which is particularly important when the effects of trade liberalization are considered. These delays affect the cost of liberalization and expectations about the consequences of liberalization. The adjustment process will be shown in the following section to have even greater consequences for the effects of multilateral trade liberalization.

20.4 Multilateral Trade Liberalization

When several geographic markets simultaneously reduce the amount of protection given to domestic producers, the impact of trade liberalization depends not only on import price elasticities, but also on the transmission of domestic production and consumption adjustments to the world market for the commodity, and on the feedback effects between market prices and production and consumption decisions in the group of countries pursuing trade liberalization policies. These feedback effects are examined in this section in the light of the most recent multilateral trade negotiations (MTNs). The results of the simulations of multilateral trade liberalization are then compared with those of unilateral trade liberalization, and with the results of other studies of multilateral trade liberalization in some of these commodities.

The Uruguay Round of MTNs being held under the auspices of the General Agreement on Tariffs and Trade (GATT) began in 1986. Unlike previous rounds, the Uruguay Round put agriculture and NTBs to trade high on the agenda. The first six rounds of MTNs had dealt almost exclusively with tariffs, and the seventh, the Tokyo Round, included NTBs to trade. In their product coverage, the previous negotiations had dealt mainly with barriers to trade in manufactured goods, since the most active participants were the industrialized countries, and consequently trade issues centred on disputes between them. The Uruguay Round was initiated by the industrialized countries, but this time agriculture emerged as a major trade problem for those countries; trade disputes over access to agricultural export markets had grown and the financial burden of the domestic support programmes described earlier in this chapter had increased. In their comprehensive study of NTBs to trade, Nogues *et al.* (1986) found that the incidence of NTBs on agricultural imports in the industrialized countries was more than twice as high as on manufactured imports in the industrialized countries during the early part of the 1980s.

20.4.1 The Effect on World Market Prices

The impact of trade liberalization on a commodity market and the dynamics underlying the adjustment process are illustrated in Fig. 20.2. Part (*a*) of the figure illustrates the difference between the initial and the final solutions in a comparative static framework; Part (*b*) demonstrates the price path from one steady-state solution to another in a dynamic framework.

In free trade, the amount of excess demand (imports) in the world market is equal to excess supply (exports). Figure 20.2(*a*) represents excess demand, *ED*, as originating from three market areas—the USA, the EEC, and Japan—and excess supply, *ES*, as originating from the rest of the world. In the three market areas, domestic suppliers produce q_0^s and consumers demand q_0^d; the difference, $q_0^d - q_0^s$, is imported. In the rest of the world, excess supply is equal to exports, or $q_0^{s'} - q_0^{d'}$.

Trade intervention in the form of a tariff *t* raises the domestic price in the USA, the EEC, and Japan to $P_2 = (1 + t)P_0$. At the higher price, domestic production increases to q_1^s and consumption decreases to q_1^d, which reduces excess demand to Q_1 in the world market. The lower

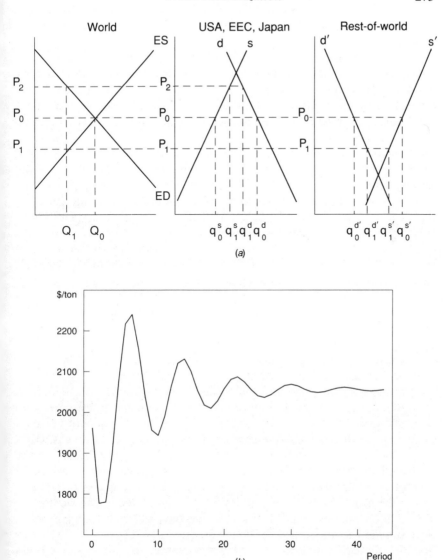

FIG. 20.2 (a) Illustration of sustained multilateral decrease in *ad valorem* tariff
(b) Beef prices after multilateral trade liberalization

excess demand drives world market prices down to P_1, so that, in the rest of the world, the quantity supplied is reduced to $q_1^{s'}$, and the quantity demanded is increased to $q_1^{d'}$. At price P_1, rest-of-world exports, $q_1^{s'} - q_1^{d'}$, are equal to net imports, $q_1^d - q_1^s$, by the USA, the

EEC, and Japan, both of these amounts being equal to Q_1 in the world market.

As discussed in Section 20.2 above, in the USA, the EEC, and Japan trade barriers to the commodities covered by this study are usually designed to maintain the level of earnings of domestic producers through such intervention measures as guaranteed prices. Under these conditions, the results in Fig. 20.2(a) illustrate the difference between the initial and final impacts of trade intervention measures. The initial difference between the price in the three market areas and the world market price is $P_2 - P_0$, but the final difference is larger since the wedge between the domestic price in the protected markets and the world market price has grown to $P_2 - P_1$. Both the domestic prices in the protected markets and the price in the world market are different from the prices that existed with free trade.

The process of adjustment in the world market to the removal of trade barriers in the USA, the EEC, and Japan is shown in Fig. 20.2(b). In the beef market, the adjustment from one equilibrium level with tariffs in several protected markets to another one without tariffs follows the typical Walrasian adjustment process when production responds with a lag to a change in price and consumption has a concurrent response to a change in price.

The price path in Fig. 20.2(b) reflects the actual empirical findings for the market price of beef derived from simulations of the estimated relationships for the system of equations of this market. The elimination of trade barriers to beef imports in the three market areas reduces domestic prices to the world market price. Initially, the quantity of beef demanded in the previously protected markets increases, whereas world production remains virtually unchanged. (Some slaughter occurs in the protected markets as a result of lower prices, which drives the world market price down in the first two periods.) The expansion in demand raises the price from its initial level of $P_1 = \$1,962$ a metric ton to $P_2 = \$2,240$ a metric ton. After six periods have elapsed, the quantity of beef supplied to the world market increases from Q_1 to Q_2. With this quantity the price falls to $P_3 = \$1,990$ a metric ton. At this price less beef will be produced, and after five periods the amount of beef supplied to the market will decrease to Q_3. These adjustments will continue until a new steady-state solution of \$2,055 a metric ton is eventually reached as a result of the elimination of trade barriers.

The free trade price is higher than the price that prevailed in the unprotected markets before the removal of trade barriers. However, the price path follows a dampened cyclical response, and a considerable period of time elapses before the new steady-state solution is reached. (In fact, 27 periods elapsed before price movements remained within ±15 per cent of the free trade price in the simulations.)

These results have a bearing on the literature dealing with the effects of trade liberalization on the stabilization of commodity markets. Comparative-static analyses have substantiated expectations that the opening of protected markets widens the area in which price adjustments take place in the world market, which reduces price fluctuations. Bale and Lutz (1979) have provided a formal theoretical treatment, and Tyers and Anderson (1986: 41) have offered empirical support. However, my own findings indicate that trade liberalization could increase the instability of commodity markets during the transition period, and that this period could be very long. Thus, consideration of dynamic transition effects demonstrates that long-run comparative-static analyses offer misleading results.

The results of simulations of the effect of the complete removal of trade barriers in the three main markets on the world market prices of the commodities covered by this study are shown in Table 20.4. In each case, the initial effect is to lower the average price of the commodity, whereas the long-term effect is to raise the market price. A long period of time is required for the markets to adjust to lower average prices for coffee, cocoa, beef, maize, and copper. Considerably shorter time periods are needed for the markets to adjust to the lower prices of sugar, bananas, and soybeans. The average number of periods required for the adjustments to remain within ±15 per cent of the new steady-state price is 20. In most cases, these price adjustments show oscillations similar to those of beef and sugar; in that of bananas, the adjustment follows a dampened smooth response.

The largest price changes resulting from multilateral trade liberalizations in the major market areas would take place in the temperate zone goods and in the one mineral commodity. For temperate zone goods, large price changes result from the high trade barriers imposed by the industrialized countries to protect their domestic producers, compared with the usually lower trade barriers that are imposed on tropical goods, even though these goods compete with other domestically produced goods. Among individual commodities, the largest change, by far, would occur in the market price of sugar. The principal reason is that the average *ad valorem* tariff-equivalent levels in the major market areas are very high when market prices are depressed. In 1984–6 the weighted average *ad valorem* tariff-equivalent in these markets was equal to 225 per cent, whereas those of beef and maize were near 40 per cent, and that of soybeans was less than 7 per cent (see Tables 20.1 and 20.2). Within each market area, domestic sugar prices in the USA, the EEC, and Japan averaged 319, 175, and 483 per cent, respectively, above the prevailing world market price during this period. With these *ad valorem* tariff-equivalents, the simulated impact on the market of complete trade liberalization in the major market areas would cause the price of sugar

Table 20.4 Simulated world price and quantity effects of trade liberalization in the USA, the EEC, and Japan (Percentage changes)

	Price effect		No. of periods to reach ±15% stability[a]	Quantity effect	
	Impact	Total		Impact	Total
Temperate Zone Goods					
Beef	−15.6	4.8	27	0.8[b]	0.6
Maize	−16.1	25.9	28	−1.8[c]	2.6
Sugar	−32.5	31.5	7	1.4[c]	1.6
Soybeans	−3.5	2.0	2	0.8[b]	0.7
Tropical Zone Goods					
Coffee	−1.4	0.4	38	−0.1[d]	0.1
Cocoa	−0.9	0.1	34	−0.1[e]	0.1
Bananas	−1.7	1.8	7	0.1	0.2
Minerals					
Copper	−1.5	5.7	17	0.2	3.5

[a] Number of periods that elapse before price oscillations stay within ±15% of new steady-state path.
[b] One-period lag.
[c] Two-period lag.
[d] Three-period lag.
[e] Eight-period lag.

to rise nearly 32 per cent after production and consumption had had time fully to adjust.

A sizeable rise in market prices would also occur in maize, given the large spread between world prices and domestic producer prices during periods when the market is depressed. The simulated removal of trade barriers would lead to a rise of nearly 26 per cent in the world price of this commodity. In contrast, soybean prices would increase by only 2 per cent. The reason for these very different effects is that the USA is a large producer and consumer of both commodities, and in the simulated period it provided high domestic price supports for maize, whereas for soybeans it maintained minimal domestic price supports. Although the EEC also has had high price supports for both commodities in periods of low world market prices, it has a significantly smaller share of both markets. For this reason, the policies of the USA tend to determine the impact of trade liberalization on both maize and soybeans.

For beef, there were relatively smaller domestic price supports during the simulated period in the USA although the EEC maintained fairly high domestic producer prices compared with its prevailing border price, and Japan's support prices were more than one-and-a-half times higher than the world market price. Nevertheless, the average of the *ad valorem* tariff-equivalents for the three markets remained at about the same level as in the beginning of the 1980s and was therefore lower than that for most other temperate zone goods; it tended to rise as market prices showed a sustained decline after the 1981 world-wide recession.

For tropical goods, the simulated impact of trade liberalization on market prices would, in most cases, be considerably smaller than that on temperate zone goods. Tariffs on some tropical goods are high, and although the three market areas are important in international trade, they none the less absorb a relatively small amount of total world consumption of those goods. Consequently, trade liberalization in these markets alone would have a small impact on world market prices. In contrast, the elimination of tariffs in these markets could have a significant effect on the trade of some of these commodities.

20.4.2 The Effect on Import Demand

Multilateral trade liberalization significantly alters the results of trade in temperate zone goods compared with those that would otherwise occur under unilateral trade liberalization in these markets. This result does not hold true for tropical goods, since their world market prices remain virtually unaffected by multilateral trade liberalization in the three market areas.

The rise in world market prices of temperate zone goods would cause import prices in the USA, the EEC, and Japan to fall by less than the original amount of the drop brought about by the elimination of their *ad valorem* tariff-equivalents. The smaller drop in import prices leads to less of an increase in imports. However, the change in the overall value of trade is higher because of the rise in world market prices for these commodities. The most striking results are those for sugar; with unilateral trade liberalization, the value of trade of these three markets would increase by about 50 per cent, whereas with multilateral trade liberalization it would increase by nearly 70 per cent. These results indicate that the gains to foreign suppliers to these markets would be considerably greater if trade liberalization measures were to be taken by all the major market areas than if they were to be adopted by individual geographic markets.

20.4.3 Comparison of Findings with Those of Other Studies

The two main studies on multilateral trade liberalization of primary commodities covering the 1980s are those of Tyers and Anderson (1986) and Zietz and Valdes (1986a, 1986b).[1] Both studies base their analysis on comparative-static models, although Tyers and Anderson incorporate partial adjustment equations to distinguish between short-term and long-term effects. Hence their results cannot be compared with the present findings concerning the dynamic effects of trade liberalization. Nor do those studies use trade elasticities; instead, they rely on assumptions about domestic production and consumption elasticities, and derive changes in exports and imports from differences in production and consumption changes in countries. Thus, their findings for trade cannot be assessed in the light of the underlying price elasticities of international trade. Nevertheless, my own simulations of total price changes resulting from trade liberalization can be compared with the findings of those studies.

The Tyers–Anderson and Zietz–Valdes studies provide comparative findings for sugar, beef, and maize. As with the results of unilateral trade liberalization, a major source of difference between the findings of the present study and those of others is the period in which the *ad valorem* tariff-equivalent levels were calculated. Tyers and Anderson based their calculations of *ad valorem* tariff-equivalents on world market conditions prevailing in 1979–80, when many commodity prices reached record highs; and those of Zietz and Valdes were calculated for 1980–2, which included both a period of record high prices and a period of low prices accompanying the 1981 world-wide recession.

In the present study, the period used to calculate the *ad valorem* tariff-equivalent levels is 1984–6 when nearly all world market prices of primary commodities reached a record low. In fact, according to Grilli and Yang (1988), the overall drop in commodity prices between 1980 and 1986 was larger than during the Great Depression if movements in commodity prices are measured against prices of manufactured goods. Thus, a considerable divergence occurred during this period between domestic support prices in the industrialized countries and corresponding world market prices for many commodities. For this reason, the *ad valorem* tariff-equivalents used for my simulations tend to be much higher than those in the aforementioned studies.

To provide a basis for comparing the results generated by simulations of the models developed in this study with those of Tyers and Anderson and of Zietz and Valdes, I simulated the models in the period analysed by those studies with the levels of protectionism prevailing at those times. My results were, nevertheless, considerably different. In sugar, the price changes estimated by Zietz and Valdes (1986*a*, 1986*b*) in 1979–81 ranged from 16 to 19 per cent, whereas my own simulations showed that prices would have risen by 10 per cent had trade been liberalized in that period. There are several possible reasons for this difference. First, the price elasticities estimated in this study differ from those used by Zietz and Valdes; their benchmark supply elasticity was assumed to be 0.6 for all countries (Zietz and Valdes 1986*a*: 23), whereas the elasticity estimates in the present study were, on average, equal to 0.05 for the world. Similarly, there is a large discrepancy between the price elasticity of demand of −0.4 used by Zietz and Valdes and the average of −0.06 estimated in this study.

Second, the *ad valorem* tariff-equivalents of tariff and NTBs differ considerably for the same period. The averages calculated by Zietz and Valdes were 28, 57, and 131 per cent for the USA, the EEC, and Japan, respectively, whereas those reported by the OECD (1987*a*, 1987*b*, 1987*c*) and used in the present analysis were 42, 20, and 148 per cent respectively.[2] Finally, the country coverage differs. The present approach focuses solely on the principal geographic markets of the Latin American countries, whereas Zietz and Valdes introduced *ad valorem* tariff-equivalent levels of other industrialized countries into their calculations.

Nevertheless, the findings of all the studies support the earlier suggestion that a much lower change in prices would have occurred had multilateral trade liberalization taken place at the beginning of the 1980s rather than in the middle of that decade. My simulations for 1980–2 and 1984–6 showed that the percentage change in the world sugar price with the high levels of protectionism in 1984–6 would have been three times higher than that with the much lower levels of protectionism that existed in 1979–81, when world market prices were at a record high.

The different findings for beef and maize can be attributed to reasons similar to those given for sugar. However, both the Tyers–Anderson and Zietz–Valdes studies provide incomplete information with which to examine numerically the causes of the inconsistencies. The Tyers–Anderson study provides information about the *ad valorem* tariff-equivalents, but no averages of the large number of elasticities used in their model. In contrast, the Zietz–Valdes study offers information about their average elasticities, but not on their *ad valorem* tariff-equivalents. The information provided, however, is sufficient to suggest why differences occur in the final impact of trade liberalization on these two products. My own simulations for beef in the same period used by Tyers and Anderson as their base period (1980–2) indicate that, had trade liberalization occurred, world beef prices would have risen by 4.8 per cent in the long run; in contrast, Tyers and Anderson's results showed that prices would have risen by 16 per cent. The dissimilarity can be explained in part by the fact that the *ad valorem* tariff-equivalent estimates used by Tyers and Anderson were twice as high as those reported by the OECD (1987a, 1987b, 1987c)—47 versus 23 per cent.

Zietz and Valdes's results for the 1979–81 period also yielded higher price changes of 16–18.5 per cent, compared with 6 per cent for that period in the present study. In this case the difference is due in part to the fact that the elasticities used by Zietz and Valdes were much higher than those calculated in this study. Their assumed supply elasticities range from 0.5 to 1.02, whereas the average estimated in this study is 0.13. Similarly, Zietz and Valdes used demand elasticities that ranged between -0.41 and -1.2, whereas the average elasticity estimated in this study is -0.34. Similar differences with respect to both *ad valorem* tariff-equivalents and price elasticities help to explain the dissimilarities that exist in the analysis underlying the effect of trade liberalization in the market for maize.

20.5 Conclusions

Trade liberalization usually has a greater impact on trade flows than on commodity market prices. The patterns of response to that impact demonstrate the dynamics of the price path, as well as those of trade, when the effects of the removal of barriers to trade are analysed. For instance, it has been shown that beef prices would rise by nearly 5 per cent as a result of trade liberalization; however, since it would take more than 30 years to achieve this rise, its effect would be negligible. Studies that rely on CGE models or simple comparative-static models are incapable of considering these effects because of the nature of those models.

The magnitudes of changes in import demand resulting from trade liberalization tend to be significantly higher than those of the changes in market prices. None the less, the impact on import demand of price changes after multilateral trade liberalization is significant. Multilateral trade liberalization produces a greater impact on trade flows than does unilateral trade liberalization. The greater impact is due not to world market price changes, but to the effect those market price changes have on the demand for imports in the main geographic markets for primary commodities.

The policy implications of the empirical estimates of price elasticities of import demand for the agricultural commodities covered by this study, and of the *ad valorem* tariff-equivalent levels of tariffs and NTBs to trade, are that trade liberalization could have a strong impact on the demand for imports of most of these goods in the industrialized countries. This condition emerges from the analysis even though the price elasticities of the agricultural commodities are, in most cases, quite low. In particular, the trade-weighted average price elasticity of import demand was found to be -0.5, compared with an average of around -2.0 for manufactured goods found in other studies, although the range of elasticities of both products and markets varies greatly. The reason for the potentially strong impact of trade liberalization on foreign demand is that the USA, the EEC, and Japan give a large amount of protection to agriculture, particularly domestically produced products, and have done so especially in recent years when world market prices of many commodities have been depressed. Overall, trade liberalization would have the greatest impact on the demand for sugar imports in the USA, the EEC, and Japan. There would also be particularly large effects on Japanese imports of beef, as well as on the banana imports of Japan and West Germany.

Notes

1 Earlier studies of the effects of trade liberalization during the 1970s in some of the commodities covered by this study are those of Valdes and Zietz (1980) and Koester and Schmitz (1982).
2 The *ad valorem* tariff-equivalent level for the EEC of Zietz and Valdes was calculated as a production-weighted average of the EEC member-countries.

21
Summary

This study has provided a unified approach to the specification, estimation, and simulation of international commodity trade between the developing and the industrialized countries. It began with the formulation of a general theoretical framework based on recent theories of trade in the presence of imperfectly competitive markets, in particular those in which trade takes place under conditions of large-group competition. Relaxation of the assumption of product homogeneity allowed the specification of the system of equations to incorporate the fundamental elements of monopolistic competition. This system of equations was disaggregated in such a way that the principal trading partners of each exporting country were considered separately from one another, since relative-price movements can influence a country's share of the market of each of its trading partners.

The dynamics underlying the adjustment processes in commodity trade were then specified. Several years are necessary for exporters to react fully to price changes, and for importers of raw materials and basic, pre-processed foods to adjust to income and price changes. These non-instantaneous adjustments of supply and demand are what give rise to observed cyclical responses of trade to changes in economic conditions influencing commodity markets. In the specification strategy of the adjustment process in commodity trade, I adopted the findings of recent studies on dynamic time-series models that explain observed disequilibria in the framework of steady-state solutions of behavioural relationships. The steady-state solutions of the dynamic relationships were shown to encompass the theoretical system of equations representing the key relationships in commodity trade.

Next, the validity of the system of equations was tested by applying it to trade at a highly disaggregated level for Latin America's major commodity exports. The empirical results were assessed in the light of well-established development trade theories. Estimates of individual relationships in the system of equations suggested that the main reason for the region's historically slow rate of growth in primary commodity exports had been the generally lower income elasticity of import demand for primary commodities than for other goods imported by the industrialized countries. However, estimates of disaggregated trade flows

Summary

were shown to yield much higher price elasticities than those of previous studies that relied on aggregated data. These price elasticities indicate that there is considerable potential for the expansion of commodity exports in the region.

Having treated the separate behavioural relationships in the system of equations, I turned in this last part of the study to the performance of the system as a whole. The models of the individual commodities were used to examine two types of trade policy: those that influence the long-term growth of exports, and those that have a one-time impact on exports. I first traced the effects of economic policies influencing the level of economic activity in the industrialized countries on the exports of the developing countries. The results of the estimates of individual behavioural relationships support generalizations concerning the lower response of the industrialized countries for primary commodity imports than for imports of manufactured goods when incomes change. However, the generally price-inelastic production and consumption functions of primary commodities cause large price movements in their markets when incomes change. Consequently, whereas import volumes tend to have sluggish responses to income changes, prices tend to have strong responses when there are concurrent income changes in the industrialized countries. For this reason, the results of simulations of concurrent increases in economic activity in the industrialized countries showed that, whereas the trade volume responses were generally low, the level of export earnings tended to increase substantially for the developing countries. These results underscore the importance for trade of possible feedback effects in commodity markets.

It has been emphasized throughout the empirical analysis in this study that there are significant differences both between commodities and, within each commodity, among exporters and importers. The simulations of commodity trade and world markets within the system of equations as a whole made these differences even more apparent. Differences in price movements resulting from income changes magnified the differences in the trade responses of the exporting and importing countries. Imperfect competition in international commodity trade increases the differences in the responses of exporting countries because the channels through which income changes affect exports become more complex.

Next, trade policies of the developing and industrialized countries were examined. The export policy measures considered were subsidies of various types that lowered the unit cost of exports, and an export tax. In the case of a subsidy, the formulation of export policies of developing countries in the large-group case of monopolistic competition allowed consideration of policies aimed at expanding exports without fear of retaliation by either the importing countries or the competing exporters.

Such an expansion is not ensured in the small-group case, since there might be a reaction by exporting countries that are close competitors in a particular export market. In accordance with strategic trade policy analysis, export policies were examined in a non-cooperative game among a large number of countries exporting differentiated products, with each country establishing its export quantity separately. An export subsidy was offered by the government for strategic purposes and the export industry took it as given. The subsidy enabled a country to increase both its market share and its earnings, provided it had a price-elastic export demand function. In contrast, in the case of an export tax, we examined the effect of government action designed to maximize export earnings through relative-price increases, which would reduce the country's market share. The tax would increase earnings as long as the country had a price-inelastic export demand function and an upward-sloping export supply function.

Optimal export policies for given export targets were derived from the set of estimated econometric models for commodity trade in Latin America's major exports. Export policy measures took the form of an export subsidy in quantity competition of countries having price-elastic export demand functions, whereas they took the form of an export tax in price competition of countries having price-inelastic export demand functions. The commodity trade models in their structural form were solved for the optimal export policies needed to achieve established targets. The simulations assumed that the export policies of the Latin American countries were formulated independently of one another, which limited the analysis to unilateral government incentives to subsidize or to tax exports.

This analysis suggested that, for most of the Latin American countries, the appropriate policy instrument for the commodities covered by this study was an export subsidy; for the few countries that had an export demand that was inelastic with respect to price, a tax was the appropriate instrument for achieving a target growth rate of export earnings. For an export subsidy, the results of the simulations showed that, for Latin America as a whole, relative export prices would need to fall by about two-thirds of the target export volume growth rate. The impact of export subsidies would not be fully felt for several years, and would continue after the subsidy had been discontinued. In addition to the increase in earnings resulting from these changes, the subsidy would significantly expand the market shares of the countries. The subsidy required to achieve a target growth rate of exports varied considerably among the countries of the region.

In those countries having inelastic export demand functions and upward-sloping export supply functions, the desired target was established in terms of export earnings and the commodity trade models in their structural form were solved for the optimal export tax rate. For the government the strategic choice was to select an appropriately higher price that would increase export earnings at the desired target growth rate. Calculations of the optimal export tax for each exporter in the estimated trade models were based on a common set of target growth rates for export earnings. The results showed that, on the whole, the percentage reduction in exports would be about one-half of the percentage rise in export price caused by the tax. As with the calculations of the subsidies, the estimates of the optimal export tax varied considerably from country to country.

The use of these export policy instruments is limited; since they produce a one-time effect on exports, they do not influence their long-term growth. Consequently, they serve to generate temporary increases in foreign exchange earnings, which are particularly useful for developing countries that have high debt-servicing requirements and at the same time need to sustain their economic growth. Such policies could also be used to offset the downturns in exports that usually accompany world-wide recessions. Since delays occur in the effects these policies have on exports, the policies should be so implemented that the targeted increase in exports is achieved in the desired period. The potential gains from these export policy measures may be considerably greater for developing countries, which usually have open economies and rely on a few primary commodity exports for most of their export earnings, than for large industrialized countries, which implement strategic trade policies in one of a multitude of industries.

Our analysis of trade liberalization by the industrialized countries considered unilateral and multilateral actions by the USA, the EEC, and Japan. These areas give a large amount of protection to domestically produced agricultural products, and have done so especially in recent years when the world market prices of many commodities have been depressed. Since they together absorb an important part of total world output of the commodities covered by this study, the simultaneous liberalization of trade in those areas would give rise to feedback effects between domestic production and consumption decisions and world market prices. These feedback effects were examined against the background of the Uruguay Round of MTNs. The path to a new steady-state solution was found to depend on the dynamics underlying the production and consumption responses to price changes, and to have important implications for the effects of trade liberalization.

What distinguishes the analysis of trade liberalization in this study from previous attempts to quantify the effects of the removal of barriers to primary commodity trade is consideration of the dynamics underlying the adjustment processes of commodity markets after trade liberalization. Most other studies measuring the effects of trade liberalization have relied on CGE models, or fairly simple comparative-static models. The principal shortcomings of such models are that the numerical solutions are not deduced from statistically estimated parameters in the model, that the selected elasticities are adjusted when the system of equations is calibrated, and that the resulting comparative-static analyses fail to consider intertemporal issues concerning trade. The final effects of trade liberalization shown by those studies differ from my results because of differences in the price elasticities of import demand for the commodities used for the calculations, dissimilarities in the effective levels of protection of commodities in the periods in which the calculations were made, and differences in the product definitions. The differences in the final effects can be resolved if consistent parameters and data are used in both econometric and CGE models. However, dynamic effects cannot be examined in comparative-static models, whereas my results show that these effects play a critical role in determining the efficacy of trade liberalization.

A long period of time was often found to be required for the commodity markets to adjust to the lower prices in previously protected markets after multilateral trade liberalization. Most of these price adjustments were characterized by a dampened cyclical response. The long adjustment process diminishes the present discount value of the overall gains to be derived by the developing countries. Nevertheless, large price changes would occur as a result of multilateral trade liberalization, particularly in temperate zone goods, and in some mineral commodities. The price changes resulting from multilateral trade liberalization would significantly alter the outcomes of trade in temperate zone goods from those that would otherwise occur under unilateral trade liberalization. The reason for the difference is that the rise in world market prices of temperate zone goods causes import prices in the USA, the European Community, and Japan to fall by less than the amount of the original fall brought about by the elimination of their *ad valorem* tariff-equivalents, since the world market prices of these commodities rise as a result of multilateral trade liberalization.

Multilateral trade liberalization produces a greater impact on trade flows than does unilateral trade liberalization. The greater impact is not so much due to world market price changes as it is to the effect those market price changes have on the demand for imports in the main geographic markets for primary commodities. This indicates that the gains to foreign suppliers to these markets would be considerably greater if trade liberalization measures were to be taken by all the major market areas than if they were to be adopted by geographic markets unilaterally.

The results of the final part of the study indicated that the feedback effects in commodity markets make it necessary to consider income and price effects in commodity trade by using the system of equations as a whole. Whereas the long-term growth of commodity trade depends on economic growth in the principal export markets, trade policies can serve to expand commodity trade. Export policies in the developing countries are generally more effective in expanding trade than are import deregulation policies in the industrialized countries. This finding holds for both the final effect of trade policies and the time needed for the desired changes to occur. Consideration of trade at a disaggregated level points up the need to examine economic policy effects on income, and trade policy effects on imports and exports at the country level, since these effects vary greatly in both the exporting countries and their particular export markets.

Throughout this study I have emphasized that the analysis of international commodity trade in the presence of imperfect competition underscores the close relationship between trade policies and assumptions about the behaviour of economic agents. Thus, for the market structure of international commodity trade, the analysis in this study has depended on one of several sets of assumptions. Nevertheless, it is hoped that this unified treatment of the theory, dynamics, and policy modelling of international commodity trade in the presence of imperfect competition will provide a useful framework for analyzing different features of trade in commodities and goods of other types.

Annex A
Data and Estimation Procedure

The sample of commodity trade is composed of five elements: the commodity itself, the country of origin, the geographic market (or country of destination), the economic variable, and time. The first step in drawing the sample was to choose the commodities that represent the major non-fuel exports of the Latin American region. Then the countries of origin of each commodity, that is, those countries to whose total export earnings it made a significant contribution, were selected and designated exporters of interest $i = 1, \ldots, n$. Next, their principal geographical markets $j = 1, \ldots, m - 1$ were chosen, and all others aggregated into a group, denoted m, and regarded as a single, residual market. Trade volume and value data for the period 1960–85 were gathered and unit trade prices were then calculated; production, consumption, and stock data used to determine world market prices were gathered for the period 1960–87. The structure of the data for each commodity is summarized in Fig. A1.

The system of equations that characterizes international commodity trade has been estimated in its structural form. The results provide information about the

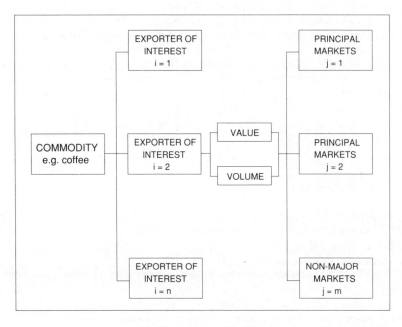

FIG. A1 Structure of trade data

parameters of the relationships that constitute the theoretical model. For the sample of commodity trade selected, it is a large-scale econometric model consisting of over 450 behavioural equations.[1]

Three limitations hampered estimation of the large number of equations: the small sample size for time-series data of annual periodicity, limited estimation techniques, and the test statistics available in the econometric software system that was used. A system that could build large-scale models capable of easily handling data transformations, and that had automated storage of estimated equations and good interface between equation estimation and model simulation, was given preference over others that had estimator-generating functions and a sequential diagnostic testing procedure without large-scale model-building capabilities. Notwithstanding these diagnostic test constraints, and within the limitations encountered in empirical econometrics, a systematic strategy was used for the identification, estimation, and testing of the commodity trade model.

A.1 Commodity and Country Coverage

The criterion for selecting the country of origin and geographic markets is their relative importance within the aggregate. The degree of importance was measured over a relatively long period of time so as to avoid transient influences on the choice. A 15-year period was considered appropriate.

The major non-fuel commodities of Latin America are considered to be those goods that in 1970–85 represented an average of at least 1 per cent of the total regional value of merchandise exports. Country coverage is delimited at the 5 per cent level. The countries of origin of each commodity are limited to those Latin American countries in which the commodity concerned accounted for an average of at least 5 per cent of their total merchandise export earnings in 1970–85. Similarly, principal geographic markets are defined as those countries of destination that absorbed an average of at least 5 per cent of a country's exports in 1970–85. The result of applying these criteria is a sample comprising 49 product exports[2] to 68 principal geographic markets. When non-principal markets are treated as a single, residual market, there are 140 bilateral trade flows in the sample (see Table A1).

A.2 The Trade Data

The United Nations defines a 'primary' or 'basic' commodity as a good 'of farm, forest, fishing or hunting, or of an extractive industry, to whose value transformation has made only a very minor contribution' (UN 1970: 34). This definition has been used to select the major primary commodity exports of Latin America. However, in the coverage of the chosen commodities, tradition sometimes overrides strict adherence to a coverage limited to the earliest stage of production of a product in which trade takes place. The commodities are described in Table A2 in accordance with the Standard International Trade Classification, Revision 2 (SITC, Rev. 2).

Table A1 Latin America: principal geographic markets[a] of regionally significant non-fuel exports

Number of major markets			
Soybeans	7	Beef	18
Iron ore	8	Sugar	19
Copper	10	Coffee	19
Bananas	14	Cocoa	20
Maize	17	Cotton	20

Number of products[b] to each market			
Developed countries			
USA	16	Switzerland	5
W. Germany	13	Finland	4
Italy	12	Portugal	3
Japan	12	Denmark	1
Netherlands	12	Ireland	1
Spain	11	New Zealand	1
UK	11	Norway	1
Belgium	8	South Africa	1
Canada	8	Turkey	1
France	5	Yugoslavia	1
		SUBTOTAL	127
Less developed countries			
Israel	2	Morocco	1
Algeria	1	Nigeria	1
Ghana	1	Taiwan	1
Hong Kong	1	Tunisia	1
Malaysia	1	Zaire	1
		SUBTOTAL	11
Centrally planned economies			
Soviet Union	6	China	2
E. Germany	3	Bulgaria	1
Poland	3	Hungary	1
		SUBTOTAL	16

[a] Principal geographic markets are defined as those country destinations that have absorbed at least 5% of a country's exports during 1970–85.
[b] A 'product' is defined as a commodity distinguished by country of origin.

Note: The distribution of countries into economic groups is based on the classification used by the International Monetary Fund.

Table A2 Description of major commodity exports of Latin America

Product	SITC Rev. 2	Description
Beef	011.1	Meat of bovine animals, fresh, chilled or frozen
Maize	044	Maize (corn), unmilled
Bananas	057.3	Bananas (including plantains), fresh or dried
Sugar	061.1; 062.2	Sugars, beet and cane, raw, solid; refined sugars and other products of refining beet and cane sugar, solid
Coffee	071.1	Coffee, whether or not roasted or freed of caffeine; coffee husks and skins; coffee substitutes containing coffee in any proportion
Cocoa	072	Cocoa, which encompasses the following types:[a]
	072.1	Cocoa beans, whole or broken, raw or roasted
	072.2	Cocoa powder, unsweetened
	072.31	Cocoa paste (in bulk or in block), whether or not defatted
	072.32	Cocoa butter (fat or oil)
Soybeans	081.31; 222.2	Oil-cake and other residues (except dregs resulting from the extraction of vegetable oils of soya beans); soya beans
Cotton	263.1	Cotton (other than linters), not carded or combed
Iron ore	281	Iron ore concentrates
Copper	287.1; 682	Copper ores and concentrates; copper matte; cement copper; copper

[a]The different cocoa types cannot be directly summed. To calculate their combined volume, they are converted into cocoa beans equivalent by the following standard conversion factors:

Cocoa powder	1.18
Cocoa paste	1.25
Cocoa butter	1.33

The first seven commodities are classified as basic foods, the last three as raw materials. Maize is used almost exclusively for animal feed throughout the world, except in Latin America, where it is an important source of food. Soybeans are processed into soybean oil, which is used mainly for edible purposes, and into soybean meal, which is widely used in the preparation of livestock feeds.

Trade data are available for a sufficiently long period to provide enough observations for estimating and testing suitable parsimonious models. The 'original' SITC was adopted by the United Nations Secretariat in 1950, and by 1960 many countries were using it in compiling their trade data. As a result, from 1962 onwards annual trade-by-commodity data at the disaggregated level needed here are available for most Latin American countries in 'Series D' of the UN Statistical Office.

However, the United Nations information system has a time delay in processing data, and some countries tend to lag behind others in reporting. It was therefore decided that, to compile export data, it would be more expedient to send questionnaires to national statistical offices throughout Latin America. This procedure has afforded access to more recent data, besides gathering data with effect from 1960 for all countries. Since the commodities of interest are the traditional exports of the countries concerned, data are more readily available for them than for non-traditional exports. The first set of questionnaires was processed in 1977, and since then the data base has been updated annually. The data obtained from national statistical offices cover the volume (in metric tons) and the value (in thousands of US dollars) of trade by commodity between the Latin American exporter and its trading partner. For each exporter there is a maximum of 48 trading partners in the sample period; if the number of trading partners exceeds this limit, minor trading partners are grouped into a single, aggregate market. In accordance with the practice of the UN Statistical Office, volume data have been compiled in metric tons, and value data in thousands of US dollars. The transaction value is the value at which goods were sold by the exporter and includes the cost of transportation and insurance to the frontier of the exporting country (free-on-board (f.o.b.) valuation).

Once the disaggregated data have been stored and edit-checks performed,[3] the criteria specified above for selecting the exporters of interest and their principal geographic markets are used to extract a subset of data. This subset constitutes the trade data base for the commodity trade models. The extraction process also serves to identify the major geographic markets. The total imports of the commodities by these markets are then compiled and stored in the trade data base. Volume and value data on imports are obtained from the UN Statistical Office. In this case, the transaction value is the value at which the goods were purchased by the importer plus the cost of transportation and insurance to the frontier of the importing country (c.i.f. valuation).

A.3 Prices

The prices of exports and imports are based on unit values of trade. Calculations based on the prices of individual commodities eliminate some of the well-known statistical inaccuracies inherent in aggregated commodity price series used as an explanatory variable for international trade. In aggregate product comparisons, the failure of the price variable adequately to reflect international transaction prices has been verified by Kravis and Lipsey (1971: 186–92; 1974). This failure arises from unequal commodity coverages, unreflected changes in the commodity composition within the groups, and dissimilar methods used by the countries for constructing indices. As a result, composite unit value indices of an exporter of interest can show variations relative to other suppliers because of differences in weighting, even when the price movements of the commodity components are identical.

However, individual product unit values are themselves subject to statistical defects when used as a proxy for price variables. In the first place, when prices are derived by dividing value by volume, an error in the measurement of the actual international transaction prices is likely. Customs data may be incorrect if export declarations are inaccurate. Although this problem affects primary products, it is much more prevalent in manufactures because of specification difficulties (Kravis and Lipsey 1971: 4). Scobie and Johnson (1975) have shown that, if the observed value and the volume data contain errors of measurement for actual transactions, the estimated elasticity of substitution will be biased towards zero.

Secondly, unit values suffer from the traditional f.o.b./c.i.f. valuation problems. Elasticity estimates are based on f.o.b. prices, which, because they exclude changes in trade resulting from transportation and distribution costs or from tariffs, do not take into account all price differences between suppliers to the ultimate consumer. In the 'standard' formulation, therefore, the observed relative prices assume fixed transfer costs. This assumption is unrealistic for the later period of the present analysis, when the rise in oil prices led to an increase in freight costs.

Import prices are derived in Chapter 3 as the trade-weighted averages of the export prices of foreign supplying countries. The value of the import price variable in the import demand schedule of equation (3.8a) is the sum of export prices of foreign supplying countries in a constant elasticity form. Unit import values would be the same as the values of the import prices in (3.8a) only in the special case where the substitution parameter, $\beta/(\beta - 1)$, approached zero and the definition of the variable included a distribution parameter that measured the market share of exporting countries in the geographic market.

The world prices used for commodities are normally those representative of trade between markets that dominate international trade. Where these markets are protected by import quotas or tariffs, a price quotation in the free market is selected instead as the indicator of movements in aggregate supply and demand. There are exceptions, however. During several periods since 1953 sugar supply in the free market has been regulated by international sugar agreements aimed at stabilizing price fluctuations. Another exception is iron ore, which is traded under long-term contracts. For example, Brazilian iron ore producers begin negotiations with European steel producers at the end of each year, then proceed to negotiate with Japanese steel producers for delivery of iron ore in the next year under long-term contracts. Table A3 describes the markets.

A.4 Exogenous Variables

The principal variable determined outside the system of equations is economic activity in the geographic markets concerned and in the world market. Specifically, the real gross domestic product (GDP) has been used as a measure of economic activity. The source of the data is the International Financial Statistics data base of the International Monetary Fund.

The other exogenous variable is the deflator for commodity prices. The deflator used is the wholesale price index (WPI). Again the source of the data is the International Financial Statistics data base of the International Monetary Fund.

Table A3 Description of world market prices for commodities

Product	Type	Description
Beef	USA	Imported frozen boneless, 85% chemical lean; before December 1975, 90% visible lean cow meat, f.o.b. port of entry.
Maize	USA	No. 2 yellow, f.o.b. Gulf ports.
Bananas	Central & South America	From 1979 onwards, first-class quality tropical pack, importer's price to jobber or processor, f.o.b. US ports; beginning January 1987, prices have been estimated based on average wholesale prices at New York City and Chicago; before 1979, Ecuadorian, c.i.f. Hamburg.
Sugar	World	ISA daily price, f.o.b. and stowed at greater Caribbean ports; before 1961, New York World Contract no. 4, f.a.s. Cuba.
Coffee	Other milds	ICO indicator price, other mild arabicas, average NY Bremen/Hamburg markets, ex-dock; before January 1984, Guatemalan prime washed, ex-dock NY for prompt shipment.
Cocoa	Ghanaian	ICCO daily price, average, New York and London, nearest 3 future trading months.
Soybeans	USA	c.i.f. Rotterdam.
Cotton	'A' index	Middling (1–3/32 in.), c.i.f. Europe; before January 1984, Mexican, middling (1–3/32 in.), c.i.f. N. Europe.
Iron ore	Brazilian	Europe, 65% c.i.f. North Sea ports; before 1975, 68%.
Copper	London Metal Exchange	Grade A cathodes, settlement price; before July 1986, high grade.

A.5 Ad Valorem Tariff-Equivalents

Data used in Chapter 20 for *ad valorem* tariff-equivalents of tariffs, NTBs to trade, and subsidies for the products that are also domestically produced in the principal export markets of Latin America have been calculated from the price difference between the domestic producer price and the c.i.f. import, or world market, price of the commodity, using information provided by staff members of the United States Department of Agriculture, Economic Research Service.

The product-specific price descriptions are as follows.

United States

Sugar The producer price is the domestic contract no. 12, bulk, duty-paid, spot, New York. The 'world' price is the ISA daily price, f.o.b. and stowed at greater Caribbean ports.

EEC

Sugar The producer price is the EEC intervention price for raw sugar, which applies only to 'quota sugar'.

Beef The producer price is adjusted from live weight prices by a factor of 1.85. Live weight prices are prices received by the producer or first buyer for adult bovine animals.

Soybeans The producer price is the 'guide price', i.e. the minimum price set by the European Commission.

Japan

Sugar The producer price is the price received for refined sugar by the domestic producer. The import price has been converted to refined sugar equivalent at the rate of 0.955 kilograms of refined sugar for each kilogram of raw sugar.

Beef The producer price is measured by the wholesale price for dairy beef. The foreign market price has been measured by the c.i.f. import price.

A.6 Estimation Procedure

When an export market is imperfectly competitive, export demand is less than infinitely elastic with respect to price. Under this condition, estimation of the export demand and supply relationships by the ordinary least squares method will usually not be consistent. The reason is that export supply and demand determine the export price, and the export price variable appears in the export supply and demand relationships as an explanatory variable. The least squares estimator of the coefficient of the price variable will accordingly be correlated with the disturbance term. As a result, random shifts in the demand and supply relationships caused by variations in their disturbance terms will affect the export price, which implies that $\text{cov}(p_{ijt}, v_t)$ is non-zero.

To resolve this problem, the two-stage least squares approach has been used. In the first stage, a set of instruments for the endogenous export price variable is found by estimating the export price relationship. The predetermined variables in the reduced-form equation, obtained from the solution of the export demand and supply equations that determine the export price, serve as the instruments. (As noted by McCarthy 1971: 251–9, the selected instruments must always require that all predetermined variables in the equation be estimated in order to ensure consistency. For suggested instrumental variable sets, see Fair 1970.) The instruments 'purge' the estimated export price variable of its stochastic component. Then, in the second stage, export demand and supply are regressed on the actual values of the exogenous variables and on the estimated values of the endogenous export price variable, which has now been purged of its stochastic component (for details, see Harvey 1981: 331–3).

A.7 Specification Strategy

A systematic approach is used to test the specification of both nested and non-nested hypotheses in the equations of the model. The sequence of hypotheses is ordered and nested in the export demand and foreign import demand equations. The specification strategy begins with the general, or maintained, hypothesis and tests hypotheses in increasing order of restrictiveness. (For an extensive discussion of encompassing, see Mizon 1984.) A simple sequential testing procedure common to all demand relationships follows from the nesting of the activity variable within specifications that include price variables. The activity variable in the export demand relationship is foreign import demand, and the activity variable in the foreign import demand relationship is income. Tests of the significance of the price variable in the demand relationships automatically generate information about the relationship of demand to the activity variable because the specifications encompass simpler ones that include only these variables.

A more complex procedure is used in the sequence of tests of significance for different parameters of the price variable in the demand equations. The test for the short-run relative price effect can be performed only if the first test does not reject the existence of the long-run relative-price effect in the steady-state solution. When the inclusion of the effect of relative price r in the export demand equation (8.17), i.e.

$$\Delta z_t = \alpha_{20} + \beta_{21} \Delta m_{jt} + \beta_{22}^* z_{ij,t-1} + \alpha_{23} \Delta r_{ijt} + \beta_{24} r_{ij,t-1} + v_{2t}^*,$$

is considered, the sequential testing procedure is H_1: $\beta_{24} = 0$ and H_2: $\beta_{24} = 0$, $\alpha_{23} = 0$. The first test is of the null hypothesis H_1: $\beta_{24} = 0$ against the alternative hypothesis $\beta_{24} \neq 0$. If the first test rejects the null hypothesis that $\beta_{24} = 0$, then the succeeding hypothesis H_2: $\beta_{24} = 0$, $\alpha_{23} = 0$ is also rejected. Only if the first test does not reject H_1: $\beta_{24} = 0$ can the second test of $\alpha_{23} = 0$ against $\alpha_{23} \neq 0$ be performed. The same ordering procedure is used to consider the inclusion of the price variables in the import demand equation. Such a backward selection procedure, in contrast to one that starts from the most restrictive hypothesis, has been shown by Anderson (1971) to have optimal properties with regard to power.

A somewhat similar testing procedure has been used for the export supply equation. However, there is no sequential testing procedure for the lags in the explanatory export variable. The test for the export supply equation is whether there is a significant response to the export price of the product lagged one or two periods and whether the shape of the lag structure has been suitably approximated by the selected rational lag structure. Since a test of significance of price lagged one period is not contingent on the test of another lag in the price variable, there is no natural order of hypotheses. In non-nested hypotheses, the optimal sequential test procedure is what Mizon (1977) refers to as the exhaustive test procedure. It consists of sequential tests on all possible orderings of the general rational lag structure selected for each commodity.

As an example, consider the general rational lag structure of the export supply equation for sugar:

$$x_t = \alpha_0 + \alpha_1 x_{t-1} + \beta_1 p_t + \beta_2 p_{t-4}.$$

There are two possible orders of the hypotheses: (1) export supply responds to current and four-year-lagged prices; and (2) the lag coefficients follow a dampened smooth approach to the new equilibrium after the initial response and the fourth-year response. In the first, whether x_t depends only on p_t is examined. A test of significance is performed first on the estimate of β_2 in the maintained hypothesis, and then on the estimate of α_1 on the regression of x_t on p_t and x_{t-1} alone. Its formal purpose is to test the null hypothesis H_0: $\beta_2 = 0$ against the alternative hypothesis H_1: $\beta_2 = 0$, $\alpha_1 = 0$. In the second, the purpose is to determine whether x_t depends only on p_{t-4}. It tests H_0: $\beta_1 = 0$; H_1: $\beta_1 = 0$, $\alpha_1 = 0$. Extension of the exhaustive test procedure to the point where the maintained hypothesis contains more than three explanatory variables is straightforward. However, the extension can become computationally cumbersome since the number of possible orders rises at an exponential rate as the number of explanatory variables in the equation increases.

A.8 Parameter Constancy Test

The aforementioned tests are used to assess whether or not the model specification is internally consistent with the data in the sample. It is also important to test whether the model specification is externally valid when used in policy simulations or in predictions based on post-sample data.

A common procedure is to estimate a model from a subset of the sample and apply a post-sample predictive test to the unused observations at the end of the sample period. The 'mean squared error' (the sum of squares of the prediction errors) is the usual measure of a post-sample goodness of fit. However, there are no statistical criteria for testing the null hypothesis that the model provides an inadequate representation of the data-generating process (Harvey 1981: 187). Thus, the above-mentioned measure can be used to decide whether to accept or reject the model only after comparing it with the post-sample goodness of fit of an alternative specification.

Once a suitable representation of the underlying data-generating process has been selected, the post-sample predictive test becomes a test of parameter constancy. The Chow test provides a measure of parameter constancy (Maddala 1977: 198–201). It is based on the following F-test:[4]

$$F(n_2, n_1 - k - 1) = \frac{RSS - RSS_1}{RSS_1} \frac{n_1 - k - 1}{n_2},$$

where
- RSS = residual sum of squares with $n_1 + n_2$ observations
- RSS_1 = residual sum of squares with n_1 observations
- k = number of constraints
- n_1 = number of observations in first sub-period of the sample
- n_2 = number of observations in second sub-period of the sample

The test is of the null hypothesis of parameter constancy. If the F-test is significant at the 5 per cent level, the hypothesis that the coefficients are stable is not rejected. On the other hand, if the value of the F is not significant at the 5 per cent level the null hypothesis is rejected, which means that the model does not adequately represent the data-generating process. This approach permits the model to be estimated with the entire data sample, while it simultaneously tests for parameter changes that may have occurred as a result of structural change, such as those structural changes that may have ensued after the 1981 world recession.

Notes

1. There are two equations—one that determines import demand and another that determines import price—for each of the 52 principal markets plus one residual market for all ten commodities; and there are two equations—one that determines export demand and another that determines export supply—for each of the 140 bilateral trade flows between the exporters of interest and their export markets; and there are 53 equations used to determine the world market price of the ten commodities.
2. Note that a 'product' constitutes a commodity differentiated by country of origin. Hence there are 49 exporters of the 10 commodities among the 20 Latin American countries.
3. The software performs two edit-checks. The first verifies that the sum of the import data for the countries of destination is equal to the total exports reported by all countries. The other calculates the unit value of exports and checks whether the data for each year are within one standard deviation of their mean.
4. This test is used when the number of observations in one of the two sub-periods in a sample is less than or equal to the number of constraints plus one.

Annex B
Dynamic System of Equations

The first part of this appendix summarizes the system of dynamic equations used to characterize data-generating processes in international commodity trade in Part III, and the second section summarizes the long-run, or steady-state, solution of the system.

Notation

C	Consumption
D_j	General price level of geographic market j
D_i	General price level of country of interest i
K	Stocks
M_j	Import demand of geographic market j
P	World market price
P_j	Import price of geographic market j
P_{ij}	Export price of country of interest i to geographic market j
P^e	Expected world market price
Q	Production
R_{ij}	Export price of country of interest i relative to import price of geographic market j (i.e. P_{ij}/P_j)
T	Secular trend
W_i	Major disturbance in country of interest i
X^d_{ij}	Export demand of country of interest i to geographic market j
X^s_{ij}	Export supply of country of interest i to geographic market j
Y	Economic activity of world
Y_j	Economic activity of geographic market j
Z_{ij}	Export demand of country of interest i relative to export demand of competing suppliers k by geographic markets j

B.1 Dynamic System of Equations
(Lower-case letters denote logarithmic values of variables.)

B.1.1 Foreign Market Imports

(a) Import demand

$$\Delta m_{jt} = \alpha_{10} + \alpha_{11} \Delta y_{jt} + \beta_{12}(m_j - y_j)_{t-1} + \beta_{13} y_{j,t-1} + \beta_{14} \Delta(p_j - d_j)_t$$
$$+ \beta_{15}(p_j - d_j)_{t-1} + v_{1t}, \qquad (8.12)$$

where $\alpha_{11} > 0$, $-1 < \beta_{12} < 0$; $\beta_{13} > \beta_{12}$; and β_{14}, $\beta_{15} < 0$.

(b) *Import price*

$$\Delta p_{jt} = \alpha_{60} + \alpha_{61}\Delta p_t + \beta_{62}(p_j - p)_{t-1} + v_{6t}, \quad (9.50)$$

where $\alpha_{61} > 0$ and $-1 < \beta_{62} < 0$.

B.1.2 Exports

(a) *Export demand*

$$\Delta z_{ijt} = \alpha_{20} + \beta_{21}\Delta m_{jt} + \beta_{22}^* z_{ij,t-1} + \alpha_{23}\Delta r_{ijt} + \beta_{24} r_{ij,t-1} + v_{2t}^*, \quad (8.17)$$

where

$$z_{ij} = \ln\left[\frac{X_{ij}^d/M_j}{1 - (X_{ij}^d/M_j)}\right] = \ln(X_{ij}^d/X_{kj}^d),$$

since $X_{kj}^d = (M_j - X_{ij}^d)$, and where $\beta_{21} > -1$; $-1 < \beta_{22}^* < 0$; and α_{23}, $\beta_{24} < 0$.

(b) *Export supply*

$$\Delta x_{ijt}^s = \alpha_{40} + \beta_{41} x_{ij,t-1}^s + \alpha_{42} x_{ij,t-2}^s + \sum_{k=0}^{n-1} \alpha_{43+k}\Delta(p_{ij} - d_i)_{t-k}$$

$$+ \sum_{k=1}^{n} \gamma_{44+k}(p_{ij} - d_i)_{t-k} + \beta_{45}T + \beta_{46}W_{it} + v_{4t}, \quad (9.41)$$

where $\alpha_{43+k} \leq 0$; $\gamma_{44+k} > 0$; and (a) $-1 < \beta_{41} < 1$, (b) $-1 < \alpha_{42} < 1$, (c) $\beta_{41} + \alpha_{42} < 0$, and (d) $(1 + \beta_{41})^2 \geq \alpha_{42}$.

(c) *Equilibrium condition*

$$X_{ijt}^d = X_{ijt}^s. \quad (5.29)$$

B.1.3 World Market

(a) *Rest-of-world consumption*

$$\Delta c_{kt} = \alpha_{30} + \alpha_{31}\Delta y_{kt} + \beta_{32}(c_k - y_k)_{t-1} + \beta_{33} y_{k,t-1} + \beta_{34}\Delta(p - d)_t$$

$$+ \beta_{35}(p - d)_{t-1} + v_{3t}, \quad (8.29)$$

where $-1 < \beta_{32} < 0$, $\beta_{33} > \beta_{32}$, and β_{34}, $\beta_{35} < 0$.

(b) *Rest-of-world production*

$$\Delta q_{kt} = \alpha_{50} + \beta_{51} q_{k,t-1} + \alpha_{52} q_{k,t-2} + \sum_{m=0}^{n-1} \alpha_{53+m}\Delta(p - d)_{t-m}$$

$$+ \sum_{m=1}^{n} \gamma_{54+m}(p - d)_{t-m} + \beta_{55}T + \beta_{56}W_{kt} + v_{5t}, \quad (9.47)$$

where $\alpha_{53+m} \leq 0$; $\gamma_{54+m} > 0$; $-1 < \beta_{51} < 1$; $-1 < \alpha_{52} < 1$; $\beta_{51} + \alpha_{52} < 0$; and $(1 + \beta_{51})^2 \geq \alpha_{52}$.

(c) Total consumption

$$C = \sum_{j=1}^{m} M_j + C_k. \quad (3.14)$$

(d) Total production

$$Q = \sum_{i=1}^{n} X_i^s + Q_k. \quad (4.24)$$

(e) Demand for stocks

$$\Delta k_t^d = \alpha'_{70} + \alpha_{71}\Delta q_t + \gamma_{72}(k^d - q)_{t-1} + \beta_{73}\psi(L)p_{t-1} + v'_{7t}, \quad (8.38)$$

where $\alpha_{71} > 0$, $-1 < \gamma_{72} < 0$, and $\beta_{73} > 0$.

(f) Supply of stocks

$$K_t = K_{t-1} + Q_t - C_t.$$

(g) Equilibrium condition
  (i) Perishable commodities:

$$C_t = Q_t. \quad (5.32)$$

  (ii) Storable commodities:

$$K_t^d = K_t^s. \quad (5.35)$$

B.2 Steady-State Solutions

B.2.1 Foreign Market Imports

(a) Import demand

$$M_j = k_1^* Y_j^{1-(\beta_{13}/\beta_{12})} \left(\frac{P_j}{D_j}\right)^{-\beta_{15}/\beta_{12}}, \quad (8.13)$$

where $k_1^* = \exp\{[-\alpha_{10} + (1 - \alpha_{11})g_4]/\beta_{12}\}$.

(b) Import price

$$P_j = k_6 P, \quad (9.51)$$

where $k_6 = \exp\{-[\alpha_{60} + (\alpha_{61} - 1)g_6]/\beta_{62}\}$.

B.2.2 Exports

(a) Export demand

$$X_{ij}^d = \frac{M_j}{1 + k_2 R_{ij}^{\beta_{24}/\beta_{22}^*}}, \quad (8.19)$$

where $k_2 = \exp[(\alpha_{20} + \beta_{21}g_3)/\beta_{22}^*]$.

(b) Export supply

$$X^s_{ij} = k_3(P_{ij}/D_i)^{\gamma_1} \exp(\sigma_1 T + \sigma_2 W), \qquad (9.42)$$

where $\gamma_1 > 0$, $\sigma_1 \geq 0$, and $\sigma_2 < 0$.

(c) Equilibrium condition

$$X^d_{ij} = X^s_{ij}. \qquad (5.29)$$

B.2.3 World Market

(a) Rest-of-world consumption

$$C_k = k_4 Y^{\epsilon^y}_k (P/D)^{\epsilon^p}, \qquad (8.30)$$

where $\epsilon^y = 1 - (\beta_{33}/\beta_{32}) > 0$, and $\epsilon^p = -(\beta_{35}/\beta_{32}) < 0$.

(b) Rest-of-world production

$$Q_k = k_3(P/D)^{\gamma_2} \exp(\sigma_1 T + \sigma_2 W), \qquad (9.48)$$

where $k_3 = \exp[-\alpha_{50}/(\beta_{51} + \alpha_{52})]$, $\gamma_2 = -\Sigma^n_1 \gamma_{54+m}/(\beta_{51} + \alpha_{52}) > 0$, $\sigma_1 = -\beta_{55}/(\beta_{51} + \alpha_{52}) \geq 0$, and $\sigma_2 = -\beta_{56}/(\beta_{51} + \alpha_{52}) < 0$.

(c) Total consumption

$$C = \sum_{j=1}^{m} M_j + C_k. \qquad (3.14)$$

(d) Total production

$$Q = \sum_{j=1}^{n} X^s_i + Q_k. \qquad (4.24)$$

(e) Demand for stocks

$$K^d = k^*_7 Q P^{e^{\epsilon_k}}, \qquad (8.39)$$

where $k^*_7 = \exp(-\alpha'_{70}/\gamma_{72} + \kappa_4 g_5)$, and $\epsilon_k = -\beta_{73}/\gamma_{72}$.

(f) Supply of stocks

$$\Delta K^s = Q - C. \qquad (5.33)$$

(g) Equilibrium condition
(i) Perishable commodities:

$$C = Q. \qquad (5.32)$$

(ii) Storable commodities:

$$K^d = K^s. \qquad (5.35)$$

Annex C
Regression Results of Model Equations

A Note on Test Statistics

In the tables that follow, the t-statistic is given below the coefficient estimate. The following standard notations appear in the tables:

$\bar{R}^2$	Adjusted square of the multiple correlation coefficient. In two-stage least squares (2SLS) it is adjusted for degrees of freedom in the first and second stages of the regression.
dw	Durbin–Watson statistic
$\hat{\sigma}$	Coefficient of variation
d.o.f.	Degrees of freedom
Test of parameter constancy	Chow test distributed as $F(m, n)$. The entries are the ratio of the F-test times 0.95 to the 5 per cent F-value for m and n degrees of freedom. If the entry is greater than 0.95, then the null hypothesis of parameter constancy can be rejected with 95 per cent confidence.

Table C1 Regression results of import demand equation (8.12)

$$\Delta \ln M_{jt} = \alpha_{10} + \alpha_{11}\Delta \ln Y_{jt} + \beta_{12}\ln(M/Y)_{j,t-1}\ln Y_{j,t-1} + \beta_{13}\ln Y_{j,t-1} + \beta_{14}\Delta \ln(P/D)_{jt} + \beta_{15}\ln(P/D)_{j,t-1}$$

Product Origin	$\Delta \ln Y_{jt}$	$\ln(M/Y)_{j,t-1}\ln Y_{j,t-1}$	$\Delta \ln(P/D)_{jt}$	$\ln(P/D)_{j,t-1}$	$W(t)$	Const.	$\bar{R}^2$	dw	$\hat{\sigma}$	d.o.f.	Test for parameter constancy
Beef											
France	8.85 (3.2)	−0.93 (9.4)	2.92 (10.1)			1.26 (64) (7.7) 0.85 (65) (7.7)	−17.13	0.86	2.6	0.176	20	0.05 (4,15)
W. Germany	4.05 (6.2)	−0.57 (6.3)			−0.16 (2.5)	0.35 (72) (7.0) 0.27 (75) (4.7)	3.27	0.89	1.6	0.049	13	0.53 (4,9)
Greece	4.27 (3.0)	−0.91 (7.0)	0.72 (4.1)	−1.32 (3.8)	−0.57 (1.7)	1.00 (74,75) (5.14)	1.04	0.78	2.5	0.185	17	0.23 (4,11)
Japan	3.50 (2.5)	−0.35 (3.1)	0.84 (5.8)		−0.52 (2.9)	1.00 (71,73) (5.9) 0.53 (76) (3.5)	−8.95	0.82	1.7	0.142	18	0.22 (4,14)
USA		−0.28 (1.8)	−0.23 (1.6)		−0.28 (2.6)	−0.17 (77) (2.3) −0.17 (81) (2.2)	4.14	0.67	2.5	0.070	14	0.04 (4,13)
World	1.37 (1.7)	−0.99 (5.9)	0.38 (3.8)			0.20 (63) (3.1)	8.28	0.67	2.1	0.058	19	0.43 (4,15)

Table C1 (cont.)

Product Origin	$\Delta \ln Y_{jt}$	$\ln(M/Y)_{j,t-1}$	$\ln Y_{j,t-1}$	$\Delta \ln(P/D)_{jt}$	$\ln(P/D)_{j,t-1}$	$W(t)$	Const.	$\bar{R}^2$	dw	$\hat{\sigma}$	d.o.f.	Test for parameter constancy
Maize												
Italy		−0.41 (5.5)				−0.57 (82) (5.1) −0.17 (69) (3.0) −0.21 (77) (3.4)	1.73	0.76	2.1	0.105	24	0.30 (2,21) 0.88 (4,13)
World	3.60 (4.6)	−0.20 (1.8)	0.36 (3.1)				0.89	0.66	1.4	0.055	17	
Bananas												
W. Germany		−0.31 (4.3)		−0.82 (5.2)	−0.81 (4.4)	−0.16 (74) (3.2) 1.00 (81,82) (3.8)	2.74	0.75	2.0	0.046	12	0.22 (4,8)
Italy	1.71 (1.3)	−0.82 (3.4)			−0.40 (1.5)	1.00 (80,81) (2.3)	1.05	0.53	1.8	0.102	10	0.60 (4,6)
Japan	1.40 (1.5)	−0.34 (3.2)		−0.74 (2.8)	−0.45 (1.4)	−0.36 (73−)[a] (2.9) 0.22 (75) (2.3)	1.06	0.65	2.1	0.082	15	0.47 (4,11)
USA	0.34 (1.4)	−0.96 (8.9)				−0.10 (69) (3.2)	6.93	0.78	1.8	0.029	20	0.76 (4,16)
World	0.94 (4.8)	−0.85 (6.0)	−0.49 (6.6)	−0.14 (1.8)	−0.51 (5.0)	1.00 (78,79) (4.3)	11.98	0.76	1.9	0.013	13	0.02 (4,9)

Sugar										
Japan	1.86 (2.4)	−0.04 (0.6) −0.13 (2.9)		−0.07[b] (1.8) −0.10[b] (3.0)	1.00 (74,77) (5.8) −0.14 (70) (2.8) −0.17 (79) (3.5)	0.31 1.85	0.60 0.60	2.0 2.0	0.090 0.046	20 22	0.36 (4,16) 0.02 (4,12)
UK			−0.13[b] (3.0)								
USA	1.84 (1.9)	−0.76 (5.8)		−0.21 (2.8)	0.22 (74) (2.1) 1.00 (80−)[c] (7.8)	8.65	0.78	1.7	0.092	18	0.12 (4,14)
World		−0.89 (5.5)	−0.44 (5.2)		0.10 (70) (2.4) 0.16 (77) (4.2)	12.69	0.76	1.8	0.037	19	0.51 (3,18)
Coffee											
Belgium	1.56 (2.0)	−0.84 (4.9)	0.10 (1.1) −0.34 (4.9)	−0.22 (3.3)	−0.16 (70) (2.4) −0.18 (74) (2.9)	2.85	0.78	2.0	0.058	12	0.05 (4,10)
Finland	2.62 (2.8)	−0.84 (6.7)	−0.26 (2.3)	−0.13 (1.4)	−0.93 (71) (7.3) −0.31 (85) (2.5)	6.86	0.90	2.2	0.111	18	0.18 (4,15)
France		−0.76 (4.2)	−0.23 (3.9) −0.09 (4.0)	−0.03 (1.3)		5.80	0.73	2.5	0.023	12	0.01 (4,8)

Table C1 (cont.)

Product Origin	$\Delta \ln Y_{jt}$	$\ln(M/Y)_{j,t-1}$	$\ln Y_{j,t-1}$	$\Delta \ln(P/D)_{jt}$	$\ln(P/D)_{jt}$	$\ln(P/D)_{j,t-1}$	$W(t)$	Const.	$\bar{R}^2$	dw	$\hat{\sigma}$	d.o.f.	Test for parameter constancy
Coffee (cont.)													
W. Germany		−0.27 (3.3)		−0.09 (2.9)		−0.04 (1.4)	1.00 (73,74) (4.0) 0.10 (79) (3.2)	1.73	0.69	1.5	0.029	12	0.08 (4,10)
Italy	1.22 (4.4)	−0.89 (5.2)	0.18 (3.8)	−0.14 (3.7)		−0.18 (4.9)	0.09 (79) (2.5)	−1.22	0.75	2.3	0.031	18	0.16 (4,15)
Japan	4.02 (3.4)	−0.92 (4.0)	1.06 (5.8)			−0.22 (2.7)	0.26 (79) (1.7)	−12.48	0.79	1.7	0.111	13	0.27 (4,9)
Netherlands	3.41 (5.1)	−0.99 (5.5)	0.14 (2.1)	−0.36 (4.6)		−0.15 (3.3)	1.00 (69,77) (7.4)	6.01	0.86	2.0	0.043	16	0.11 (4,13)
Sweden		−0.46 (2.7)	−0.44 (3.4)	−0.50 (5.2)		−0.13 (1.8)	0.15 (70) (2.0) 0.36 (76) (4.3)	5.56	0.71	1.8	0.072	18	0.63 (4,16)
Switzerland		−0.68 (3.5)	−0.36 (1.7)	−0.15 (2.0)		−0.15 (3.2)	−0.09 (74) (1.6) −0.12 (82) (2.0)	6.37	0.68	2.3	0.546	10	0.30 (4,7)
USA	0.67 (1.4)	−0.92 (4.7)	−0.75 (2.8)	−0.32 (4.8)		−0.14 (2.8)	0.13 (76) (2.3) 0.18 (81) (3.1)	12.65	0.78	2.5	0.048	13	0.24 (4,11)

World	0.45 (1.4)	−0.89 (5.5)	−0.46 (5.5)		−0.09 (3.2)	−1.00 (65,74) (6.5)	11.71	0.86	2.2	0.025	19	0.13 (4,16)
Cocoa												
W. Germany		−0.77 (6.7)	−0.38 (4.3)	−0.05 (1.3)	−0.05 (1.4)	−0.12 (70) (2.7) 0.14 (82) (3.2)	6.63	0.76	2.8	0.038	13	0.02 (4,10)
Italy	2.84 (5.4)	−0.82 (5.9)	−0.68 (4.5)	−0.30 (5.9)	−0.24 (4.4)	−0.29 (65) (2.8)	7.52	0.78	2.5	0.055	17	0.33 (4,13)
Japan	1.98 (2.4)	−0.88 (7.1)	−0.60 (6.3)	−0.14 (1.7)	−0.38 (5.7)	1.00 (71,77) (8.4)	6.73	0.80	2.1	0.092	17	0.13 (4,13)
Netherlands	0.17 (0.8)	−0.27 (5.7)		−0.27 (9.0)	−0.12 (5.6)	0.28 (68) (2.6)	2.10	0.83	1.8	0.024	18	0.28 (4,14)
Spain		−0.99 (6.6)	−0.58 (5.2)	−0.13 (1.3)	−0.09 (1.2)	−0.38 (71) (3.5)	7.09	0.74	2.6	0.102	16	0.03 (4,14)
USA	1.98 (1.4)	−0.30 (2.6)		−0.60 (4.8)	−0.15 (1.2)	−0.26 (70) (1.6) −0.33 (80) (1.9)	1.81	0.56	2.5	0.149	17	0.42 (4,14)
USSR		−0.69 (5.9)	−0.28 (2.5)		−0.11 (1.5)	0.22 (71) (2.1) −0.61 (77) (5.9)	5.98	0.82	2.2	0.100	12	0.24 (4,14)
World		−0.75 (3.7)	−0.46 (3.5)	−0.26 (8.1)	−0.20 (4.1)	0.11 (78) (2.7)	9.62	0.80	1.3	0.033	19	0.11 (4,15)

Table C1 (cont.)

Product Origin	$\Delta \ln Y_{jt}$	$\ln(M/Y)_{j,t-1}$	$\ln Y_{j,t-1}$	$\Delta \ln(P/D)_{jt}$	$\ln(P/D)_{j,t-1}$	$W(t)$	Const.	$\bar{R}^2$	dw	$\hat{\sigma}$	d.o.f.	Test for parameter constancy
Soybeans												
W. Germany	0.55 (0.6)	−0.75 (5.2)	0.73 (4.7)			−0.25 (2.7) (69) 0.24 (2.7) (74)	0.86	0.61	2.4	0.082	19	0.10 (4,15)
Netherlands	5.40 (3.8)	−0.32 (1.5)	0.67 (1.7)	−1.00 (4.2)	−0.22 (1.4)	0.23 (2.0) (72) 1.00 (2.4) (78,79)	−0.73	0.74	2.6	0.111	17	0.03 (4,13)
Spain	0.87 (1.3)	−0.98 (14.4)	0.90 (8.8)	−0.56 (5.0)	−0.53 (6.1)	−0.44 (6.5) (73) 1.00 (5.8) (80,81)	−1.77	0.97	2.0	0.042	12	0.15 (4,8)
Switzerland	4.74 (3.9)	−0.82 (18.7)		−0.75 (2.8)	−0.88 (3.7)		5.51	0.97	2.4	0.182	13	0.10 (4,9)
World	0.87 (1.2)	−0.93 (5.2)	1.29 (4.9)			−0.11 (2.0) (68)	5.30	0.57	2.6	0.055	18	0.08 (4,14)
Cotton												
Belgium	7.94 (5.8)	−0.17 (3.9)		−1.27 (7.0)	−0.73 (4.3)	−0.29 (2.6) (67) 1.00 (4.8) (69,70)	2.01	0.74	2.6	0.106	18	0.31 (4,14)
France	2.63 (3.8)	−0.09 (2.0)		−0.37 (5.4)	−0.13 (2.0)	1.00 (5.8) (73,76)	0.60	0.82	2.1	0.038	13	0.37 (4,9)

W. Germany			−0.09 (2.6)		−0.76 (6.1)	−0.22 (1.9)	1.00 (73,79) (6.1)	1.05	0.77	2.5	0.072	20	0.24 (4,16)
Italy	3.48 (3.5)		−0.99 (5.3)	−0.83 (3.9)	−0.65 (3.5)	−0.46 (2.4)		11.38	0.69	2.1	0.096	18	0.50 (4,14)
Japan			−0.54 (3.3)	−0.53 (3.4)	−0.24 (2.2)	−0.32 (2.9)	−0.17 (69) (2.9) 0.20 (83) (3.2)	8.01	0.66	1.6	0.053	15	0.64 (4,15)
Netherlands	0.85 (1.3)		−0.53 (6.1)		−0.34 (3.3)	−0.14 (1.6)	0.28 (67) (4.9) 0.25 (76) (4.4)	148.54	0.84	2.2	0.053	14	0.02 (4,10)
UK	7.41 (6.2)		−0.07 (1.8)				0.46 (75) (3.7)	0.27	0.65	2.6	0.117	19	0.54 (4,15)
USA			−0.27 (4.2)				1.00 (77,78) (7.2) −1.32 (83) (3.2)	0.69	0.74	2.1	0.040	21	0.20 (3,21)
World	0.72 (1.5)		−0.94 (5.7)	−0.69 (5.4)			−0.07 (69) (2.0) 0.16 (80) (4.2)	13.18	0.70	2.0	0.035	17	0.21 (4,13)

Iron ore

France	6.01 (3.6)		−0.47 (3.1)	0.76 (3.1)			0.34 (63) (2.5) 0.29 (74) (2.7)	−1.49	0.69	2.4	0.102	17	0.71 (4,13)
W. Germany	4.38 (6.0)		−0.57 (4.9)	−0.11 (1.3)			1.00 (73,74) (4.6)	6.71	0.72	2.0	0.073	20	0.48 (4,16)

Table C1 (cont.)

Product Origin	$\Delta \ln Y_{jt}$	$\ln(M/Y)_{j,t-1}$	$\ln Y_{j,t-1}$	$\Delta \ln(P/D)_{jt}$	$\ln(P/D)_{j,t-1}$	$W(t)$	Const.	\multicolumn{4}{c	}{Summary statistics}	Test for parameter constancy
								$\bar{R}^2$	dw	$\hat{\sigma}$	d.o.f.	
Iron ore (cont.)												
Italy	1.48 (1.9)	−0.83 (6.1)	0.24 (1.9)			0.21 (74) (2.7) −0.20 (83) (2.4)	1.09	0.74	1.6	0.077	15	0.17 (4,11)
Japan	0.85 (1.8)	−0.91 (5.8)	0.39 (2.9)	−0.77 (3.0)	−0.52 (2.0)	−0.13 (72) (2.5) −0.14 (83) (3.0)	0.75	0.89	2.0	0.043	16	0.10 (4,13)
UK	2.32 (1.9)	−0.60 (5.2)	−0.41 (1.9)			−0.65 (80) (5.0) −0.41 (82) (3.4)	9.13	0.78	1.4	0.115	19	0.57 (4,15)
USA	2.58 (2.5)	−0.08 (1.4) −0.49 (4.9)			−0.47 (2.9)	1.00 (81,82) (3.7) 0.21 (74) (3.3) −0.18 (82,83) (3.8)	−0.05	0.64	2.6	0.114	19	0.31 (4,15)
World							7.35	0.68	2.3	0.058	15	0.00 (4,11)
Copper												
Belgium	1.93 (3.9)	−0.58 (4.4)		−0.12 (2.0)	−0.18 (3.8)	0.16 (68) (3.0) 1.00 (76,77) (4.1)	3.51	0.72	2.7	0.050	18	0.29 (4,14)

W. Germany	2.02 (4.4)	−0.94 (5.8)	−0.18 (3.0)			−0.11 (81) (2.3)	7.34	0.68	2.1	0.045	20	0.29 (4,16)
Japan	5.78 (3.9)	−0.47 (4.9)	0.25 (1.8)			0.39 (67) (2.5) −0.35 (80) (2.3)	−2.83	0.74	2.6	0.146	14	0.68 (4,15)
USA	5.71 (3.1)	−0.79 (5.9)		−0.63[b] (3.2)	−1.18[b] (5.2)	0.81 (74) (3.9) 0.78 (80) (3.8)	12.89	0.76	2.0	0.167	13	0.18 (4,10)
World	1.15 (2.4)	−0.92 (5.2)	−0.39 (4.5)			−0.08 (71) (2.3)	11.51	0.75	2.4	0.032	17	0.00 (4,13)

[a] Shift variable for 1973: zero before 1973, one thereafter.
[b] Nominal price.
[c] Shift variable in the 1980s to account for the introduction of high fructose corn syrup.

Note: See first page of annex for explanation of test statistics.

Table C2 Regression results of import price equation (9.50)
$\Delta \ln P_{jt} = \alpha_{60} + \alpha_{61}\Delta \ln P_t + \beta_{62}\ln(P_j/P)_{t-1}$

Product Origin	$\Delta \ln P_t$	$\ln(P_j/P)_{t-1}$	$W(t)$	Const.	Summary statistics $\bar{R}^2$	dw	$\hat{\sigma}$	d.o.f.	Test for parameter constancy
Beef									
France	1.06 (10.2)	−0.27 (2.7)	0.13 (76) (3.4) 0.12 (77) (3.3)	−0.01	0.91	1.8	0.035	11	0.10 (4,16)
W. Germany	1.06 (8.7)	−0.43 (1.4)		−0.01	0.91	1.1	0.038	13	0.06 (3,10)
Greece	0.85 (7.8)	−0.23 (2.8)		−0.02	0.83	1.8	0.059	20	0.02 (4,16)
Japan	0.76 (6.9)	−0.13 (2.0)	−0.29 (75) (4.9) 0.27 (81) (4.2) −0.30 (69) (4.0) 1.00 (73,74) (5.6)	0.04	0.86	1.5	0.073	18	0.02 (4,14)
USA	0.85 (9.7)	−0.53 (3.6)		−0.07	0.84	2.2	0.062	22	0.07 (3,10)
World	0.55 (5.3)	−0.21 (1.9)		0.00	0.53	1.6	0.075	22	0.09 (3,10)
Maize									
Italy	0.65 (7.2)	−0.41 (2.7)		0.09	0.73	1.5	0.070	22	0.98 (3,19)

Regression Results of Model Equations

World	0.62 (10.6)	−0.76 (6.4)	0.15	0.87	1.3	0.050	22	0.03 (3,19)
Bananas								
W. Germany	0.81 (5.5)	−0.18 (1.8)	0.04	0.58	2.3	0.064	19	0.10 (3,16)
Italy	1.02 (7.9)	−0.25 (1.6)	0.02	0.75	1.5	0.058	19	0.44 (3,16)
Japan	0.49 (3.6)	−0.30 (2.8)	−0.01	0.59	1.9	0.062	19	0.56 (3,16)
USA	0.27 (3.4)	−0.34 (5.6)	−0.15	0.72	2.3	0.037	16	0.13 (3,16)
World	0.65 (7.7)	−0.41 (3.3)	−0.03	0.73	1.7	0.034	19	0.35 (3,16)
Sugar								
Japan	0.64 (12.6)	−0.58 (3.7)	0.14	0.88	1.6	0.137	22	0.19 (3,19)
UK	1.43 (9.5)	−0.65 (4.0)	0.05	0.82	2.4	0.107	19	0.05 (3,19)
USA	0.74 (8.8)	−0.85 (4.8)	−0.11	0.84	1.2	0.122	22	0.34 (3,19)
World	0.41 (7.4)	−0.26 (4.0)	0.17	0.75	2.2	0.134	19	0.72 (3,19)
Coffee								
Belgium	0.83 (12.3)	−0.45 (3.2)	0.01	0.87	2.1	0.075	22	0.19 (3,19)
Finland	0.80 (12.7)	−0.51 (4.0)	0.03	0.89	1.7	0.070	22	0.18 (3,19)
France	0.84 (11.8)	−0.84 (4.6)	−0.11	0.88	1.8	0.078	22	0.24 (3,19)

Table C2 (cont.)

Product				Summary statistics				Test for parameter constancy
Origin	$\Delta \ln P_t$	$\ln(P_j/P)_{t-1}$	$W(t)$	Const.	$\bar{R}^2$	dw	$\hat{\sigma}$	d.o.f.
Coffee (cont.)									
W. Germany	0.79 (16.6)	−0.84 (5.6)		0.03	0.94	1.5	0.051	22	0.64 (3,19)
Italy	0.65 (8.0)	−0.66 (4.6)		−0.10	0.78	1.4	0.089	22	0.14 (3,19)
Japan	0.86 (6.9)	−0.33 (2.1)		−0.06	0.69	1.9	0.138	22	0.08 (3,19)
Netherlands	0.86 (13.8)	−0.54 (3.4)		−0.02	0.90	2.0	0.069	22	0.06 (3,19)
Sweden	0.84 (14.6)	−0.89 (5.0)		0.01	0.91	1.7	0.063	22	0.26 (3,19)
Switzerland	0.72 (14.1)	−0.86 (6.1)		0.01	0.90	1.4	0.061	22	0.01 (3,19)
USA	0.80 (15.3)	−0.95 (5.8)		−0.12	0.93	1.6	0.056	22	0.10 (3,19)
World	0.79 (13.8)	−0.90 (5.5)		−0.09	0.91	1.5	0.062	22	0.04 (3,19)
Cocoa									
W. Germany	0.34 (5.5)	−0.78 (10.0)		−0.04	0.88	1.4	0.083	22	0.21 (3,19)
Italy	0.43 (6.2)	−0.84 (8.2)		0.01	0.86	2.2	0.093	22	0.16 (3,19)

Regression Results of Model Equations

Japan	0.67 (8.7)	−0.87 (5.5)		0.01	0.88	1.9	0.097	22	0.50 (3,19)
Netherlands	0.47 (6.3)	−0.86 (7.6)		−0.07	0.87	1.4	0.096	22	0.02 (3,19)
Spain	0.49 (5.2)	−0.53 (4.5)		0.01	0.68	1.8	0.131	22	0.25 (3,19)
USA	0.56 (8.8)	−0.84 (8.9)	−0.35 (76) (4.3)	−0.13	0.92	1.7	0.077	21	0.26 (3,19)
USSR	0.33 (4.5)	−0.88 (9.3)		−0.06	0.86	1.6	0.098	22	0.35 (3,19)
World	0.47 (7.2)	−0.80 (8.1)		−0.06	0.88	1.6	0.085	22	0.62 (3,19)
Soybeans									
W. Germany	0.56 (10.4)	−0.89 (8.2)		−0.02	0.88	2.8	0.049	22	0.44 (3,19)
Netherlands	0.70 (17.5)	−0.96 (8.8)	0.07 (75) (1.8) 0.12 (77) (3.3)	−0.03	0.95	2.5	0.034	22	0.52 (3,19)
Spain	0.70 (17.7)	−0.94 (8.3)		0.01	0.94	2.4	0.035	20	0.48 (3,19)
Switzerland	0.67 (10.3)	−0.91 (8.9)		0.06	0.92	2.1	0.057	14	0.86 (3,19)
World	0.59 (12.6)	−0.89 (8.4)		0.01	0.91	2.9	0.043	22	0.51 (3,19)
Cotton									
Belgium	0.77 (4.5)	−0.37 (2.3)		−0.26	0.50	2.0	0.125	22	0.23 (3,19)

Table C2 (cont.)

Product Origin	$\Delta \ln P_t$	$\ln(P_{ij}/P)_{t-1}$	$W(t)$	Const.	Summary statistics $\bar{R}^2$	dw	$\hat{\sigma}$	d.o.f.	Test for parameter constancy
Cotton (cont.)									
France	0.70 (5.3)	−0.23 (1.9)		−0.16	0.56	2.2	0.098	22	0.15 (3,19)
W. Germany	0.75 (5.8)	−0.25 (1.9)		−0.16	0.60	2.2	0.095	22	0.18 (3,19)
Italy	0.70 (5.5)	−0.32 (2.5)		−0.19	0.63	2.5	0.091	22	0.08 (3,19)
Japan	0.58 (4.5)	−0.32 (2.7)		−0.20	0.56	2.1	0.092	22	0.50 (3,19)
Netherlands	0.76 (5.3)	−0.28 (2.0)		−0.18	0.58	2.2	0.103	22	0.52 (3,19)
UK	0.78 (6.5)	−0.24 (2.0)		−0.15	0.67	2.2	0.087	22	0.22 (3,19)
USA	0.51 (2.8)	−0.60 (4.9)	0.50 (77) (3.3) −0.60 (80) (3.9)	0.04	0.68	1.8	0.146	20	0.91 (3,19)
World	0.42 (5.7)	−0.88 (7.6)		−0.01	0.76	1.4	0.062	22	0.92 (3,19)
Iron ore									
France	0.54 (3.7)	−0.53 (3.6)		−0.03	0.67	1.5	0.068	12	0.51 (3,19)
W. Germany	0.48 (4.5)	−0.36 (4.1)		0.03	0.69	2.5	0.050	12	0.61 (3,19)

Regression Results of Model Equations

Italy	0.58 (4.2)	−0.43 (3.2)		−0.02	0.68	1.5	0.063	12	0.19 (3,19)
Japan	0.24 (3.9)	−0.29 (5.2)		−8.35	0.72	1.6	0.034	18	0.61 (3,19)
UK	0.77 (6.2)	−0.91 (5.6)		0.09	0.81	2.2	0.057	12	0.00 (3,19)
USA	0.49 (2.6)	−0.23 (3.1)	−0.28 (72) (3.0)	0.11	0.66	1.8	0.085	11	0.29 (3,19)
World	0.39 (4.3)	−0.36 (3.9)		0.01	0.67	1.4	0.042	12	0.94 (3,19)
Copper									
Belgium	0.74 (5.7)	−0.32 (2.3)		0.01	0.65	1.6	0.141	22	0.18 (3,19)
W. Germany	0.77 (12.7)	−0.25 (2.3)		0.01	0.88	1.8	0.068	22	0.13 (3,19)
Japan	0.67 (7.8)	−0.42 (3.3)		0.00	0.77	1.7	0.096	22	0.01 (3,19)
USA	0.64 (4.5)	−0.45 (2.7)		0.02	0.50	2.1	0.160	22	0.16 (3,19)
World	0.73 (8.4)	−0.10 (1.2)		0.04	0.76	2.0	0.100	22	0.10 (3,19)

Note: See first page of annex for explanation of test statistics.

Table C3 Regression results of export demand equation (8.17)

$$\Delta \ln Z_{ijt} = \alpha_{20} + \beta_{21}\Delta \ln M_{jt} + \beta_{22}^* \ln Z_{ij,t-1} + \alpha_{23}\Delta \ln R_{ijt} + \beta_{24}\ln R_{ij,t-1}$$

Product Origin Destination	$\Delta \ln M_{jt}$	$\ln Z_{ij,t-1}$	$\Delta \ln R_{ijt}$	$\ln R_{ij,t-1}$	$W(t)$	Const.	$\bar{R}^2$	dw	$\hat{\sigma}$	d.o.f.	Test for parameter constancy
Beef											
Argentina											
France[a]	−0.46 (1.9)	−0.13 (1.8)	−1.02 (2.6)	−1.40 (3.6)	2.02 (63) (3.8) 1.30 (67) (2.7)	−1.09	0.73	1.9	0.440	17	0.19 (4,13)
W. Germany[a]	1.11 (1.9)	−0.55 (3.9)	−1.55 (2.5)	−1.80 (4.1)	1.00 (67,68) (5.2)	−1.11	0.66	2.2	0.428	18	0.03 (4,14)
Greece		−0.96 (3.8)		−1.64 (1.6)	1.13 (69) (2.3) 0.74 (72) (1.5)	−1.40	0.56	1.7	0.477	10	0.87 (4,14)
Rest-of-world	1.47 (2.7)	−0.21 (2.1)		−0.91 (3.1)	1.00 (76,77) (5.2)	−0.86	0.67	2.0	0.243	16	0.75 (4,11)
Costa Rica											
USA[a]	−0.75 (2.4)	−0.50 (5.6)	−1.11 (2.8)	−2.53 (5.3)	1.00 (76,78) (7.0)	−1.74	0.80	2.7	0.144	14	0.22 (4,10)
Honduras											
USA		−0.07 (1.3)			0.42 (65) (3.0) 0.40 (69) (4.5)	−0.25	0.75	2.3	0.134	20	0.06 (4,17)

Regression Results of Model Equations

Nicaragua									
USA	−0.76 (2.5)	−0.36 (3.1)			−1.16	0.60	1.9	0.239	19	0.78 (4,15)
Uruguay									
W. Germany	−0.60 (3.5)			−1.28 (77) (2.7) 2.70 (74) (4.7) −1.49 (77) (2.6)	−2.28	0.61	1.9	0.443	12	0.26 (4,8)
Greece	−0.13 (1.2)			−0.90 (63) (3.6) 1.00 (80,81) (3.8)	−0.60	0.68	2.1	0.543	13	0.17 (4,12)
Rest-of-world	−0.90 (6.2)		−0.55 (1.5)	1.00 (78,79) (3.7)	−3.25	0.68	2.2	0.211	17	0.59 (4,12)
Maize									
Argentina									
Italy	−0.26 (2.4)			1.00 (74,75) (4.3) −1.77 (80) (4.9)	−0.05	0.73	2.1	0.352	16	0.24 (4,12)
Rest-of-world	−0.92 (4.7)	−1.79 (1.6)	−2.37 (1.5)	−0.86 (72) (2.3) −1.02 (76) (2.8)	−2.93	0.74	2.2	0.327	13	0.07 (4,9)
Bananas									
Costa Rica									
USA[a]	−0.78 (3.9)	−0.48 (2.3)	−0.70 (2.3)	1.00 (75,76) (4.1)	−1.11	0.62	2.5	0.217	10	0.40 (4,6)

Table C3 (cont.)

Product Origin Destination	$\Delta \ln M_{jt}$	$\ln Z_{ij,t-1}$	$\Delta \ln R_{ijt}$	$\ln R_{ij,t-1}$	$W(t)$	Const.	$\bar{R}^2$	dw	$\hat{\sigma}$	d.o.f.	Test for parameter constancy
Bananas											
Costa Rica (cont.)											
Rest-of-world		−0.49 (4.7)	−0.43 (1.5)	−0.72 (1.7)	1.00 (75,76) (2.6) 0.45 (81) (1.8)	−1.59	0.79	2.6	0.225	10	0.02 (4,8)
Ecuador											
W. Germany	1.79 (1.9)	−0.22 (2.0)		−1.50 (1.6)	1.00 (66,67) (3.4) 1.68 (74) (4.1)	−1.90	0.59	2.1	0.349	15	0.06 (4,11)
Japan		−0.29 (3.2)		−2.41 (3.3)	1.00 (65,68) (4.4)	−2.67	0.56	2.4	0.837	20	0.91 (4,16)
Rest-of-world	−0.77 (1.4)	−0.39 (3.1)			1.00 (68,69) (4.9) 1.00 (73,74) (3.7)	−0.65	0.72	2.0	0.126	20	0.94 (4,16)
Honduras											
W. Germany	1.85 (1.9)	−0.58 (3.9)			−0.67 (73) (1.9) 0.78 (80) (2.1)	−1.11	0.64	2.6	0.341	12	0.11 (4,8)

Regression Results of Model Equations

USA	−0.69 (1.1)	−0.93 (5.7)	−0.73 (3.2)	0.48 (71) (2.8) −1.19 (75) (5.7)	−0.85	0.74	1.5	0.164	19	0.82 (4,15)
Rest-of-world		−0.49 (6.0)		1.00 (68,69) (3.0) −0.87 (75) (4.6)	−1.85	0.74	1.4	0.181	16	0.28 (4,12)
Panama										
W. Germany		−0.97 (11.5) −0.75 (7.0)	−0.23 (1.5)		−1.13	0.90	1.5	0.245	14	0.09 (4,10)
Italy				−1.16 (79) (2.7) 0.76 (82) (1.7)	−0.92	0.81	1.2	0.436	11	0.28 (4,9)
USA		−0.26 (4.4)	−0.48 (3.1)	1.00 (76,77) (6.9) −0.54 (80) (3.1)	−0.73	0.86	1.7	0.167	12	0.55 (4,8)
Rest-of-world		−0.69 (11.0)	−0.27 (1.9)	−0.34 (79) (2.0)	−2.63	0.91	2.1	0.162	12	0.12 (4,9)
Sugar Barbados UK		−0.59 (8.0)		−3.62 (76) (11.1)	−1.78	0.88	1.5	0.319	22	0.32 (3,19)
Brazil USA	0.86 (1.9)	−0.93 (14.2)		−6.41 (76) (15.3)	−1.72	0.95	1.2	0.401	17	0.24 (3,13)

Table C3 (cont.)

Product Origin Destination	$\Delta \ln M_{jt}$	$\ln Z_{ij,t-1}$	$\Delta \ln R_{ijt}$	$\ln R_{ij,t-1}$	$W(t)$	Const.	$\bar{R}^2$	dw	$\hat{\sigma}$	d.o.f.	Test for parameter constancy
Sugar											
Brazil *(cont.)*											
Rest-of-world		−0.16 (3.8)			1.06 (72) (3.8) 1.00 (81,82) (4.6)	−0.42	0.68	2.2	0.271	15	0.15 (4,11)
Costa Rica											
USA	−0.57 (5.2)	−0.66 (6.0)			0.73 (84) (7.2)	−2.69	0.86	2.3	0.096	15	0.38 (4,13)
Dom. Republic											
USA		−0.37 (4.2)			1.00 (75,76) (6.6) −0.28 (84) (3.2)	−0.69	0.76	2.5	0.085	17	0.79 (3,15)
Rest-of-world		−0.59 (9.9)				−2.33	0.86	1.7	0.530	15	0.24 (4,11)
El Salvador											
USA	−0.84 (2.3)	−0.18 (1.2)			−1.86 (80) (5.1) −0.81 (84) (2.3)	−0.53	0.73	1.9	0.320	13	0.12 (4,9)

Regression Results of Model Equations

Guatemala									
USA	1.31 (3.2)	−0.32 (2.8)		1.44 (76) (3.7) 1.96 (83) (4.9)	−1.33	0.79	2.0	0.369	17	0.91 (4,16)
Guyana									
USA	0.74 (1.6)	−0.93 (5.7)		−1.68 (77) (3.8)	−3.94	0.72	2.0	0.428	18	0.05 (4,14)
Jamaica									
UK		−0.63 (4.9)		0.71 (71) (4.2) 1.00 (74,75) (3.0)	−1.38	0.67	2.1	0.163	18	0.75 (4,15)
USA		−0.59 (5.6)		0.92 (77) (3.6) 1.00 (81,82) (5.3)	−2.42	0.70	1.8	0.249	18	0.86 (4,8)
Panama									
USA	−0.96 (5.2)	−0.18 (2.3)		−0.40 (72) (2.2) −0.37 (84) (2.0)	−0.57	0.79	2.0	0.162	10	0.19 (4,7)
Peru									
USA		−0.28 (4.3)		2.16 (80) (8.3)	−0.76	0.78	2.1	0.254	21	0.89 (3,18)

Table C3 (cont.)

Product / Origin / Destination	$\Delta \ln M_{jt}$	$\ln Z_{ij,t-1}$	$\Delta \ln R_{ijt}$	$\ln R_{ij,t-1}$	$W(t)$	Const.	$\bar{R}^2$	dw	$\hat{\sigma}$	d.o.f.	Test for parameter constancy
Coffee											
Brazil											
Italy		−0.31 (3.2)	−1.40 (2.3)	−2.43 (2.1)	1.63 (66) (3.8) 1.00 (69,71) (4.8)	−0.54	0.73	2.5	0.382	15	0.22 (4,11)
USA	1.46 (14.7)	−0.18 (6.9)	−0.14 (1.9)	−0.48 (4.7)	1.00 (72,73) (11.9) 1.00 (79,80) (20.6)	−0.28	0.99	2.4	0.040	10	0.41 (4,6)
Rest-of-world	0.78 (2.2)	−0.59 (6.3)	−0.20 (1.4)	−0.44 (2.2)	1.00 (73,74) (3.9)	−1.08	0.87	1.9	0.065	9	0.18 (4,5)
Colombia											
W. Germany		−0.30 (3.0)		−1.16 (1.7)	1.00 (79,80) (5.6) 0.28 (83) (1.8)	−0.35	0.65	2.7	0.144	16	0.13 (4,15)
USA	0.51 (2.9)	−0.21 (2.5)	−0.77 (3.6)	−0.41 (2.5)	0.16 (67) (2.0) −0.27 (76) (3.0)	−0.34	0.82	1.9	0.076	17	0.17 (4,15)

Regression Results of Model Equations

Rest-of-world[a]		−0.33 (4.0)	−0.72 (2.7)	−1.24 (3.2)	1.00 (76,78) (6.9) 0.22 (81) (2.7)	−0.83	0.78	1.6	0.074	18	0.27 (4,14)
Costa Rica											
Finland	−0.79 (7.7)	−0.53 (7.6)		−0.78 (1.5)	0.41 (74) (2.4) 0.33 (81) (1.8)	−1.10	0.91	2.5	0.164	13	0.11 (4,10)
W. Germany[a]		−0.78 (4.1)	−1.79 (1.3)	−2.83 (1.9)	0.45 (78) (2.3) 1.00 (82,83) (4.6)	−2.66	0.77	2.1	0.187	13	0.10 (4,10)
Sweden[a]		−0.80 (6.2)	−1.84 (1.4)	−3.06 (2.3)	0.67 (74) (2.9) −0.52 (80) (2.1)	−2.74	0.69	1.8	0.192	15	0.52 (4,11)
USA	−0.72 (1.6)	−0.76 (4.4)			−0.62 (76) (3.3) 1.00 (78,79) (3.9)	−3.24	0.66	1.6	0.162	18	0.39 (4,14)
Rest-of-world		−0.68 (4.9)		−5.83 (5.2)	1.00 (73,76) (6.7) 1.00 (80,82) (7.2)	−2.97	0.87	2.5	0.151	11	0.56 (4,7)

Table C3 (cont.)

Product Origin Destination	$\Delta \ln M_{jt}$	$\ln Z_{ij,t-1}$	$\Delta \ln R_{ijt}$	$\ln R_{ij,t-1}$	$W(t)$	Const.	$\bar{R}^2$	dw	$\hat{\sigma}$	d.o.f.	Test for parameter constancy
Coffee (cont.)											
Dom. Republic											
USA	0.83 (1.5)	−0.65 (6.1)			0.41 (65) (1.9) 1.00 (77,79) (5.3)	−2.53	0.80	1.4	0.206	19	0.77 (4,15)
Rest-of-world	1.76 (2.1)	−0.21 (1.4)			0.69 (74) (2.8) 0.89 (81) (3.3)	−1.59	0.62	2.3	0.232	11	0.25 (4,7)
Ecuador											
W. Germany		−0.25 (2.4)			1.00 (72,73) (3.9) 1.00 (76,77) (5.5)	−1.47	0.75	2.9	0.332	11	0.24 (4,7)
USA		−0.56 (2.1)	−0.62 (2.5)	−0.99 (2.7)	0.47 (70) (3.4) 1.00 (76,78) (5.8)	−2.23	0.83	2.5	0.127	9	0.93 (4,10)
Rest-of-world	3.06 (2.7)	−0.82 (5.2)	−0.93 (3.3)		−0.77 (81) (3.3)	−4.60	0.72	1.8	0.221	10	0.85 (4,12)

Regression Results of Model Equations

El Salvador									
W. Germany	−0.34 (3.8)			−0.63 (73) (3.4) 0.42 (78) (2.6)	−0.58	0.63	2.4	0.183	20	0.29 (4,16)
Netherlands[a]	−0.43 (2.7)	−2.76 (1.2)	−3.36 (1.8)	1.32 (74) (2.8) 1.00 (81,82) (4.2)	−1.49	0.68	2.2	0.449	18	0.15 (4,15)
USA	−0.88 (5.1)	−0.90 (4.4)	−0.63 (1.9)	0.84 (72) (2.8) 1.18 (81) (3.7)	−2.69	0.80	1.3	0.290	9	0.04 (4,7)
Rest-of-world	−0.49 (3.7)	−4.99 (3.5)	−6.60 (4.1)	1.00 (69,71) (3.6) −1.45 (79) (4.4)	−2.54	0.61	2.3	0.403	19	0.07 (4,15)
Guatemala										
W. Germany	−0.22 (2.7)			1.00 (72,74) (3.3) 1.00 (77,79) (3.3)	−0.64	0.67	1.5	0.115	12	0.42 (4,9)
USA	−0.80 (5.1)	−1.35 (2.0)	−2.84 (3.4)	−0.51 (76) (3.0) −0.52 (81) (3.4)	−2.41	0.73	2.2	0.149	12	0.19 (4,9)
Rest-of-world[a]	−0.64 (4.9)	−1.20 (1.9)	−1.20 (1.5)	−0.61 (73) (3.5) 0.45 (82) (3.1)	−2.84	0.75	2.3	0.138	11	0.30 (4,8)

Table C3 (*cont.*)

Product Origin Destination	$\Delta \ln M_{jt}$	$\ln Z_{ij,t-1}$	$\Delta \ln R_{ijt}$	$\ln R_{ij,t-1}$	$W(t)$	Const.	$\bar{R}^2$	dw	$\hat{\sigma}$	d.o.f.	Test for parameter constancy
Coffee (*cont.*) Haiti											
Belgium		−0.23 (1.7)			1.00 (80,81) (3.0) −0.47 (84) (2.3)	−0.77	0.61	2.2	0.199	10	0.35 (4,7)
France[a]		−0.47 (3.4)	−1.04 (1.3)	−3.17 (3.7)	1.00 (64,65) (3.6) 1.00 (82,83) (2.5)	−2.36	0.66	1.8	0.193	17	0.28 (4,14)
Italy		−0.85 (4.4)			1.00 (80,81) (4.7)	−3.50	0.88	1.5	0.157	12	0.30 (4,8)
Netherlands		−0.34 (3.2)			1.00 (76,79) (3.5) 1.00 (81,84) (4.4)	−1.73	0.75	1.5	0.230	10	0.24 (4,6)
Rest-of-world	1.88 (1.7)	−0.74 (4.7)		−1.08 (1.5)	0.57 (71) (2.0) −1.46 (79) (4.7)	−4.84	0.75	2.3	0.268	13	0.60 (4,9)

Regression Results of Model Equations

Honduras											
W. Germany	1.55 (1.5)	−0.22 (2.8)		−2.47 (3.4)	1.00 (67,68) (4.1) 1.00 (83,84) (3.6)	−1.11	0.69	2.5	0.205	18	0.05 (4,14)
Japan	0.94 (2.4)	−0.21 (1.4)		−2.04 (2.6)	1.00 (72,73) (6.7) −0.71 (76) (1.9)	−0.71	0.87	2.2	0.348	9	0.10 (4,7)
USA		−0.49 (3.8)	−2.81 (3.7)	−2.27 (2.1)	−0.39 (72) (2.0) 0.92 (78) (4.7)	−2.12	0.74	2.3	0.176	9	0.17 (4,7)
Rest-of-world		−0.22 (2.6)			−1.75 (68) (5.6) 1.00 (71,73) (4.4)	−1.05	0.72	2.5	0.303	16	0.29 (4,12)
Mexico											
W. Germany		−0.38 (3.5)			−0.60 (71) (3.4) 0.76 (81) (4.3)	−1.31	0.80	1.4	0.171	12	0.36 (4,8)
Switzerland	−0.86 (2.0)	−0.29 (3.1)			1.00 (73,75) (7.1) 0.43 (83) (2.7)	−0.63	0.86	2.2	0.150	12	0.04 (4,9)

Table C3 (cont.)

Product / Origin / Destination	$\Delta \ln M_{jt}$	$\ln Z_{ij,t-1}$	$\Delta \ln R_{ijt}$	$\ln R_{ij,t-1}$	$W(t)$	Const.	$\bar{R}^2$	dw	$\hat{\sigma}$	d.o.f.	Test for parameter constancy
Coffee											
Mexico (cont.)											
USA		−0.32 (3.2)	−0.48 (2.0)	−0.34 (1.8)	1.00 (75,76) (5.8) 0.53 (83) (6.0)	−0.87	0.87	2.6	0.074	9	0.10 (4,6)
Rest-of-world	1.79 (1.7)	−0.11 (1.4)			0.62 (70) (1.8) 1.00 (79,80) (6.1)	−0.73	0.70	1.8	0.328	14	0.27 (4,10)
Nicaragua											
Belgium	−0.65 (2.2)	−0.53 (4.7)		−2.06 (4.1)	1.00 (80,81) (5.6) 1.00 (84,85) (5.5)	−1.78	0.72	2.8	0.164	19	0.06 (4,14)
W. Germany[a]		−0.49 (4.4)	−1.47 (4.3)	−3.44 (7.9)	0.44 (78) (4.1) 1.00 (81,83) (6.0)	−2.31	0.87	1.5	0.084	8	0.01 (4,4)
Netherlands	−0.75 (3.0)	−0.63 (4.9)		−0.62 (2.6)	0.32 (66) (2.4) 1.00 (79,81) (4.8)	−2.16	0.75	1.7	0.133	16	0.09 (4,17)

Regression Results of Model Equations 335

USA	1.81 (3.3)	−0.56 (9.9)				1.00 (76,77) (3.4) 1.00 (82,83) (13.0)	−2.81	0.94	1.5	0.205	13	0.18 (4,10)
Rest-of-world		−0.19 (2.8)		−0.29 (2.7)		−1.14 (78) (4.5) 1.00 (81,83) (3.7)	−1.06	0.77	2.1	0.220	10	0.13 (4,6)
Cocoa												
Brazil												
Spain	1.01 (1.7)	−0.78 (6.9)				1.00 (78,79) (3.7) 1.12 (82) (3.1)	−0.63	0.86	1.8	0.311	9	0.31 (4,5)
USSR	1.40 (2.6)	−0.99 (6.5)				−1.19 (72) (2.4)	−1.56	0.76	1.6	0.487	13	0.36 (4,9)
Rest-of-world[a]		−0.96 (6.3)	−0.47 (1.4)	−0.62 (1.3)		1.00 (73,75) (4.6) −0.46 (84) (3.2)	−2.41	0.76	2.3	0.139	12	0.06 (4,9)
Dom. Republic												
USA	−0.82 (5.3)	−0.31 (3.9)				0.49 (82) (3.2)	−0.64	0.80	2.3	0.148	20	0.04 (4,16)
Ecuador												
W. Germany		−0.54 (3.7)		−0.54 (2.8)		0.76 (78) (3.5) 1.00 (72,73) (4.0)	−2.40	0.75	2.0	0.199	12	0.11 (4,8)

Table C3 (cont.)

Product Origin Destination	$\Delta \ln M_{jt}$	$\ln Z_{ij,t-1}$	$\Delta \ln R_{ijt}$	$\ln R_{ij,t-1}$	$W(t)$	Const.	$\bar{R}^2$	dw	$\hat{\sigma}$	d.o.f.	Test for parameter constancy
Cocoa											
Ecuador (cont.)											
Italy		−0.64 (4.5)		−0.20 (1.4)	−0.25 (71) (1.7) 0.73 (78) (5.2)	−1.60	0.75	1.3	0.137	14	0.19 (4,10)
Japan	−0.65 (1.5)	−0.76 (3.1)		−0.60 (2.2)	0.63 (68) (1.5) −0.64 (73) (1.9)	−2.02	0.74	2.1	0.300	12	0.00 (4,8)
Netherlands[a]		−0.95 (6.2)	−0.68 (2.1)	−0.98 (2.8)	1.13 (72) (4.7) 0.90 (76) (3.1)	−4.73	0.78	2.1	0.213	8	0.04 (4,4)
USA	−0.36 (2.3)	−0.74 (6.2)			−0.87 (69) (5.1) 0.59 (78) (3.5)	−2.05	0.83	2.1	0.161	16	0.27 (4,12)
Rest-of-world		−0.73 (5.8)		−0.31 (2.3)	0.37 (76) (2.1) 0.66 (78) (3.6)	−3.28	0.79	1.9	0.170	12	0.17 (4,8)

Regression Results of Model Equations

Soybeans

Brazil

W. Germany	2.81 (2.2)	−0.29 (2.2) −0.22 (2.0)			1.00 (68,69) (4.9) −3.35 (70) (3.5) 1.00 (72,73) (4.2)	−1.08	0.65	1.7	0.731	16	0.41 (4,14) 0.28 (4,12)
Netherlands			−4.76 (4.5)	−5.85 (3.3)		−0.63	0.70	1.8	0.909	14	
Spain		−0.80 (4.7) −0.33 (2.3)			−2.92 (82) (4.3) −2.45 (78) (5.1)	−1.61	0.87	1.7	0.650	9	0.06 (4,5) 0.57 (4,7)
Rest-of-world						−0.87	0.75	1.7	0.448	11	

Paraguay

W. Germany		−0.48 (2.7)			0.70 (76) (2.9) 0.66 (82) (2.7)	−2.34	0.71	1.7	0.226	8	0.09 (4,7)
Netherlands		−0.67 (6.3)			−0.91 (78) (4.5) 1.00 (80,81) (5.4)	−1.62	0.92	2.0	0.178	6	0.01 (4,2)
Switzerland	1.96 (1.8)	−0.99 (3.9) −0.24 (1.5)		−4.96 (1.2)		−2.25	0.61	2.9	0.720	5	0.00 (4,1) 0.00 (4,1)
Rest-of-world					−0.69 (80) (2.7) 0.44 (82) (1.7)	−1.04	0.63	2.8	0.238	4	

Table C3 (cont.)

Product Origin Destination	$\Delta \ln M_{jt}$	$\ln Z_{ij,t-1}$	$\Delta \ln R_{ijt}$	$\ln R_{ij,t-1}$	$W(t)$	Const.	Summary statistics $\bar{R}^2$	dw	$\hat{\sigma}$	d.o.f.	Test for parameter constancy
Cotton											
Brazil											
Belgium	0.81 (1.3)	−0.16 (1.5)			1.00 (69,72) (2.6) 1.00 (83,84) (5.1)	−0.60	0.67	1.9	0.597	15	0.31 (4,8)
W. Germany	−0.94 (2.5)	−0.31 (8.3)			1.00 (76,77) (9.4) 2.35 (83) (9.7)	−1.17	0.95	1.7	0.231	11	0.09 (4,7)
Japan		−0.53 (7.9)			1.00 (76,79) (12.0) −4.28 (84) (6.3)	−1.66	0.91	1.9	0.668	21	1.20 (4,18)
Rest-of-world		−0.45 (6.2)			−3.90 (76) (8.5) −2.44 (80) (5.2)	−1.53	0.86	1.7	0.451	20	0.27 (4,17)
El Salvador											
Japan		−0.42 (5.8)		−2.22 (3.6)	−0.91 (74) (3.1) 1.00 (80,84) (10.4)	−1.58	0.89	1.8	0.285	16	0.30 (4,12)

Regression Results of Model Equations

Rest-of-world	3.48 (3.3)	−0.45 (8.8)			1.00 (73,74) (6.7) 1.00 (84,85) (6.8)	−2.43	0.94	2.3	0.234	10	0.05 (4,7)
Guatemala											
W. Germany[a]	−0.97 (1.5)	−0.38 (2.7)	−1.96 (2.6)	−2.40 (2.8)	−0.71 (67) (2.0) −0.77 (81) (1.8)	−1.91	0.67	1.8	0.346	13	0.41 (4,12)
Italy[a]		−0.71 (5.4)	−0.98 (1.3)	−2.90 (3.7)	1.00 (76,78) (3.1) −1.24 (81) (2.8)	−2.94	0.71	1.7	0.402	17	0.31 (4,13)
Japan	−0.50 (1.9)	−0.43 (6.6)			1.00 (79,80) (7.7) −0.40 (83) (2.8)	−1.33	0.81	2.2	0.132	20	0.06 (4,16)
Rest-of-world[a]		−0.72 (6.0)	−1.16 (1.3)	−1.35 (1.5)	1.00 (71,79) (2.7) 1.00 (81,83) (4.9)	−4.05	0.71	2.0	0.419	17	0.27 (4,13)
Mexico											
Italy		−0.70 (4.9)		−1.71 (2.9)	−1.38 (64) (2.4) −1.33 (71) (2.3)	−2.54	0.62	1.9	0.550	18	0.16 (4,14)
Rest-of-world		−0.57 (4.0)		−0.36 (1.5)	1.00 (82,83) (5.4)	−2.03	0.66	1.9	0.212	15	0.00 (4,14)

Table C3 (cont.)

Product Origin Destination	$\Delta \ln M_{jt}$	$\ln Z_{ij,t-1}$	$\Delta \ln R_{ijt}$	$\ln R_{ij,t-1}$	$W(t)$	Const.	Summary statistics				Test for parameter constancy
							$\bar{R}^2$	dw	$\hat{\sigma}$	d.o.f.	
Cotton (cont.) Nicaragua											
W. Germany	−0.77 (1.4)	−0.74 (5.7)			1.00 (74,75) (3.4) 1.00 (80,82) (3.4)	−2.55	0.79	1.3	0.324	20	0.15 (4,16)
Italy		−0.81 (4.8)		−2.95 (2.3)	1.00 (76,77) (4.2) −1.36 (82) (2.5)	−4.26	0.67	1.5	0.521	19	0.31 (4,17)
Japan	−0.96 (1.7)	−0.43 (5.2)			1.00 (79,80) (7.9)	−1.08	0.80	2.1	0.298	21	0.08 (4,17)
Rest-of-world	2.62 (3.3)	−0.33 (3.6)		−0.71 (2.1)	−0.74 (70) (3.5) 1.04 (74) (4.5)	−1.84	0.69	2.1	0.195	11	0.13 (4,9)
Paraguay Belgium	0.86 (1.7)	−0.12 (1.5)			1.00 (70,72) (6.2) 1.00 (74,75) (4.7)	−0.38	0.77	2.4	0.490	19	0.16 (4,15)
France		−0.92 (9.5)			1.00 (71,72) (8.3) −3.64 (78) (6.0)	−4.33	0.93	1.6	0.573	12	0.72 (4,8)

Regression Results of Model Equations

W. Germany	0.88 (1.9)	−0.06 (1.5)			1.00 (62,64) (4.5) 1.33 (72) (3.9)	0.01	0.78	1.8	0.327	19	0.18 (4,15)
UK	1.25 (1.7)	−0.77 (5.4)			−2.32 (71) (3.3) 1.00 (81,82) (3.7)	−3.77	0.75	1.8	0.671	14	0.14 (4,10)
Rest-of-world		−0.08 (1.5)			2.26 (61) (5.7) 1.00 (75,76) (4.2)	−0.37	0.77	1.8	0.320	20	0.65 (4,16)
Peru Belgium[a]	−0.81 (2.7)	−0.28 (4.4)	−1.13 (1.5)	−3.28 (3.0)	−1.03 (73) (3.5) 1.00 (81,82) (7.5)	0.01	0.86	1.8	0.278	17	0.07 (4,14)
France		−0.47 (3.8)			1.00 (79,81) (4.0) 1.00 (83,84) (3.8)	−1.98	0.64	1.7	0.447	16	0.04 (4,12)
W. Germany	−0.63 (1.6)	−0.53 (5.2)		−1.68 (3.2)	0.63 (74) (2.0) 1.00 (83,84) (6.1)	−1.57	0.68	2.6	0.271	19	0.30 (4,16)
UK[a]		−0.34 (2.6)	−1.75 (3.6)	−1.83 (2.8)	0.40 (66) (2.2) −0.35 (71) (2.0)	−0.84	0.62	2.1	0.167	9	0.05 (4,7)

Table C3 (cont.)

Product Origin Destination	$\Delta \ln M_{jt}$	$\ln Z_{ij,t-1}$	$\Delta \ln R_{ijt}$	$\ln R_{ij,t-1}$	$W(t)$	Const.	Summary statistics $\bar{R}^2$	dw	$\hat{\sigma}$	d.o.f.	Test for parameter constancy
Iron ore											
Brazil											
France	−0.51 (3.3)	−0.43 (5.5)	−0.57 (1.7)	−0.83 (3.1)	−0.24 (66) (2.4) 0.24 (78) (2.3)	−0.75	0.81	2.0	0.087	10	0.01 (4,8)
W. Germany	−0.80 (3.0)	−0.19 (2.4)			0.25 (73) (1.7) 1.00 (80,82) (3.2)	−0.16	0.63	2.6	0.135	12	0.25 (4,8)
Italy	−0.35 (1.4)	−0.18 (2.4)			−0.78 (66) (4.7) 1.00 (80,82) (4.9)	−0.20	0.70	2.2	0.163	16	0.05 (4,12)
UK[a]		−0.15 (3.4)	−0.80 (2.0)	−0.92 (2.2)	−0.32 (73) (2.2) 1.00 (68,69) (5.2)	−0.62	0.67	2.3	0.140	15	0.07 (4,11)
USA		−0.43 (3.0)		−1.69 (2.9)	1.79 (78) (3.3) 1.00 (73,74) (4.0)	−1.90	0.60	1.6	0.481	14	0.68 (4,10)

Regression Results of Model Equations

	C1	C2	C3	C4	R²	DW	#	Last
Rest-of-world	0.50 (1.8)	−0.14 (2.5)		0.43 (75) (3.8) 0.50 (82) (1.8)	−0.29	0.71	2.7	0.101	11	0.02 (4,8)
Peru										
Japan		−0.11 (2.3)	−0.24 (1.4)	0.56 (71) (3.3) −0.65 (75) (3.7)	−0.65	0.66	2.3	0.158	12	0.25 (4,8)
USA	1.35 (4.7)	−0.38 (7.5)		−1.68 (81) (6.5) 1.00 (83,84) (6.4)	−1.42	0.81	2.1	0.248	19	0.15 (4,16)
Rest-of-world	−0.90 (1.8)	−0.11 (1.3)		1.00 (74,75) (4.6) 1.07 (78) (4.7)	−0.64	0.72	1.7	0.217	17	0.36 (4,13)
Copper										
Chile										
W. Germany		−0.25 (3.1) −0.27 (1.3)	−0.79 (3.9)	0.31 (70) (2.8) 1.00 (73,74) (5.4)	−0.59	0.69	2.0	0.102	16	0.14 (4,12)
Japan		−0.81 (2.9) −3.13 (5.7)	−3.74 (3.8)	−1.99 (77) (2.5)	−1.69	0.82	2.2	0.480	13	0.21 (4,9)
USA	−0.62 (4.0)	−0.17 (2.1)		−0.94 (71) (4.2) 1.00 (74,75) (3.4)	−0.16	0.63	2.0	0.213	18	0.07 (4,14)

Table C3 (*cont.*)

Product Origin Destination	$\Delta\ln M_{jt}$	$\ln Z_{ij,t-1}$	$\Delta\ln R_{ijt}$	$\ln R_{ij,t-1}$	$W(t)$	Const.	Summary statistics $\bar{R}^2$	dw	$\hat{\sigma}$	d.o.f.	Test for parameter constancy
Copper											
Chile (*cont.*)											
Rest-of-world		−0.31 (3.1)	−0.62 (1.9)	−0.77 (1.9)	−2.04 (77) (2.5)	−0.71	0.86	2.1	0.178	18	0.89 (4,14)
Peru											
Belgium		−0.75 (11.9)			−4.63 (73) (11.9) −0.97 (79) (2.5)	−2.33	0.95	1.4	0.378	16	0.70 (4,14)
USA	−0.32 (1.3)	−0.80 (8.7)			−3.25 (73) (8.7) −0.97 (76) (2.5)	−1.34	0.89	1.8	0.365	19	0.46 (4,16)
Rest-of-world[a]		−0.47 (4.9)	−0.88 (1.2)	−1.27 (2.6)	−1.58 (73) (7.2) 0.59 (77) (2.7)	−1.94	0.84	2.3	0.209	12	0.60 (4,10)

[a] Estimated using 2SLS.

Notes:
Estimated using OLS unless otherwise specified.
See first page of annex for explanation of test statistics.

Table C4 Regression results of export supply equation (9.41)

$$\Delta \ln X^s_{ijt} = \alpha_{40} + \beta_{41}\ln X^s_{ij,t-1} + \alpha_{42}\ln X^s_{ij,t-2} + \sum_{k=0}^{n-1} \alpha_{43+k}\Delta\ln(P_{ij}/D_i)_{t-k} + \sum_{k=1}^{n} \gamma_{44+k}\ln(P_{ij}/D_i)_{t-k} + \beta_{45}T + \beta_{46}W_{it}$$

Product Origin Destination	$\ln X^s_{ij,t-1}$	$\ln X^s_{ij,t-2}$	$\Delta\ln(P_{ij}/D_i)_{t-n}$	$\ln(P_{ij}/D_i)_{t-n}$	$\Delta\ln(P_{ij}/D_i)_{t-n}$	$\ln(P_{ij}/D_i)_{t-n}$	T	T^2	$W(t)$	Const.	\multicolumn{4}{c}{Summary statistics}	Test for parameter constancy
											$\bar{R}^2$	dw	$\hat{\sigma}$	d.o.f.	
Beef															
Argentina															
France	−0.70 (3.8)	−0.25 (1.5)	0.62(n=5) (2.0)	0.72(n=6) (2.3)			−0.13 (5.2)		1.16(72) (3.6) −0.59(75) (1.7)	268.0	0.74	2.4	0.289	10	0.39 (4,7)
W. Germany	−0.97 (6.7) −0.46 (3.0)	−0.33 (2.7)	3.66(n=6) (5.9)	0.87(n=5) (2.3) 1.38(n=7) (2.0)			−0.07 (2.5)	0.01 (4.8)	1.00(72,73) (4.2) −2.23(71) (4.6) 1.32(76) (3.0)	147.3 2.17	0.77 0.81	2.0 2.7	0.359 0.422	12 11	0.11 (4,8) 0.15 (4,8)
Rest-of-world	−0.90 (6.6)			0.76(n=8) (4.0)			−0.08 (5.5)		1.00(74,75) (5.6)	175.0	0.79	1.4	0.232	12	0.38 (4,8)
Costa Rica															
USA	−0.68 (4.5)	−0.33 (2.7)	0.63(n=1) (2.3)	1.31(n=2) (4.7)	0.76(n=7) (2.1)	1.48(n=8) (4.1)			1.00(76,78) (5.3)	2.56	0.80	2.8	0.141	10	0.52 (4,6)
Honduras															
USA	−0.32 (1.4)	−0.32 (1.7)	2.38(n=0) (5.0)	1.62(n=1) (2.8)	2.30(n=7) (3.0)	1.92(n=8) (2.2)			1.00(74,75) (4.5) −0.74(80) (3.3)	−2.72	0.76	1.7	0.163	7	0.04 (4,7)

Table C4 (cont.)

Product Origin Destination	$\ln X^s_{ij,t-1}$	$\ln X^s_{ij,t-2}$ $\Delta\ln(P_{ij}/D_i)_{t-n}$	$\ln(P_{ij}/D_i)_{t-n}$ $\Delta\ln(P_{ij}/D_i)_{t-n}$	$\ln(P_{ij}/D_i)_{t-n}$	T	T^2	$W(t)$	Const.	Summary statistics $\bar{R}^2$	dw	$\hat{\sigma}$	d.o.f.	Test for parameter constancy
Beef (cont.)													
Nicaragua													
USA	−0.65 (6.2)	0.99(n=6) (4.5)				−0.01 (5.1)	−0.50(74) (4.2) −0.43(81) (3.1)	4.12	0.87	2.3	0.112	12	0.13 (4,8)
Uruguay													
W. Germany	−0.54 (4.5)	1.06(n=1) (2.7)					1.00(69,70) (5.1) 1.00(83,84) (3.3)	1.73	0.77	2.7	0.393	13	0.07 (4,9)
Greece	−0.58 (4.7)	0.41(n=2) (1.5)		−0.08 (4.3)			1.00(73,77) (4.1) 1.00(80,81) (3.2)	154.7	0.80	2.0	0.360	13	0.27 (4,9)
Rest-of-world	−0.90 (6.3)	0.49(n=2) (2.5)					−0.62(75) (3.2)	9.14	0.79	1.5	0.163	12	0.02 (3,10)
Maize													
Argentina													
Italy[a]	−0.22 (2.4)	0.77(n=0) (2.1)					−0.92(75) (3.3) −1.56(80) (6.4)	3.43	0.72	2.5	0.222	19	0.90 (4,15)

Rest-of-world	−0.50 (4.6)		0.75(n=4) (1.8)		−0.94(68) (3.1) −1.04(72) (3.3)	7.55	0.73	1.8	0.296	13	0.49 (4,9)
Bananas										
Costa Rica										
USA	−0.22 (1.4)		0.13(n=1) (1.7)		1.00(71,73) (3.9) 1.00(75,77) (5.3)	2.91	0.89	2.1	0.080	11	0.04 (4,8)
Rest-of-world	−0.89 (11.6)		0.84(n=2) (5.5)		1.00(73,74) (4.8) 1.00(83,84) (4.1)	11.4	0.92	2.3	0.124	11	0.10 (4,8)
Ecuador										
W. Germany	−0.97 (5.2)	5.63(n=0) (5.7)	4.42(n=1) (4.2)	−0.09 (3.7)	1.00(74,76) (5.7) −0.51(80) (2.7)	12.6	0.85	2.9	0.170	9	0.51 (4,5)
Japan[a]	−0.74 (6.5)		8.88(n=0) (5.7)	−0.07 (4.6)	1.00(63,65) (4.0) 1.37(81) (2.5)	143.3	0.79	2.4	0.492	19	0.18 (4,15)
Rest-of-world	−0.24 (2.5)	1.57(n=0) (2.7)	1.67(n=1) (2.3)		1.00(73,74) (6.6) −0.42(83) (4.1)	3.22	0.84	1.9	0.090	10	0.20 (4,7)
Honduras										
W. Germany	−0.86 (7.3)	1.10(n=1) (3.0)	2.02(n=2) (4.1)		0.71(72) (3.3) 0.59(80) (2.7)	8.97	0.83	2.5	0.209	10	0.18 (4,7)

Table C4 (cont.)

Product / Origin / Destination	$\ln X^s_{ij,t-1}$	$\ln X^s_{ij,t-2}$	$\Delta\ln(P_{ij}/D_i)_{t-n}$	$\Delta\ln(P_{ij}/D_i)_{t-n}$	$\ln(P_{ij}/D_i)_{t-n}$	T	T^2	$W(t)$	Const.	Summary statistics $\bar{R}^2$	dw	$\hat{\sigma}$	d.o.f.	Test for parameter constancy
Bananas														
Honduras (cont.)														
USA	−0.46 (3.1)	0.94(n=1) (4.1)	0.76(n=2) (3.4)					0.41(71) (2.8) 0.30(73) (2.0)	5.76	0.74	1.9	0.145	15	0.41 (4,13)
Rest-of-world	−0.45 (4.0)	0.32(n=1) (2.2)	0.38(n=2) (2.7)					−0.58(75) (3.4) 0.53(79) (3.6)	5.18	0.79	1.8	0.139	10	0.35 (4,7)
Panama														
W. Germany	−0.59 (5.3)	3.99(n=0) (3.3)	3.30(n=1) (6.7)				0.23 (6.1)	1.00(66,67) (7.4)	−450.5	0.82	2.2	0.528	18	0.06 (4,14)
Italy	−0.46 (3.8)	3.85(n=0) (3.1)	3.50(n=1) (3.8)				0.17 (3.7)	1.00(69,70) (6.9)	−337.7	0.78	1.8	0.438	19	0.23 (4,15)
USA	−0.75 (10.5)	0.75(n=0) (1.9)					0.05 (6.9)	1.00(74,75) (9.3) −0.52(80) (4.5)	−93.1	0.95	1.4	0.105	8	0.07 (4,4)
Rest-of-world	−0.33 (6.8)	2.46(n=1) (3.5)						−0.43(85) (1.8)	4.17	0.83	2.1	0.223	14	0.07 (4,11)

348

Sugar													
Barbados UK	−0.57 (3.1)	3.15(n=3) (4.0)		3.90(n=5) (3.9)	4.39(n=6) (4.5)	−0.13 (3.5)	1.21(73) (1.9) −3.22(81) (3.4)	257.4	0.67	2.1	0.584	10	0.07 (4.7)
Brazil USA	−0.98 (14.9)	0.40(n=0) (1.6)	0.82(n=1) (2.3)				−1.94(75) (3.7) −6.60(76) (15.3)	12.3	0.97	1.2	0.373	9	0.42 (4,5)
Rest-of-world	−0.89 (8.8)	0.22(n=0) (1.7)		0.27(n=2) (2.5)	−0.21(n=3) (2.1)		0.003 1.00(72,73) (6.6) (5.8) 1.00(79,81) (3.7)	11.5	0.86	2.3	0.204	13	0.36 (4.9)
Costa Rica USA	−0.99 (7.7)	0.15(n=5) (2.2)	0.22(n=6) (3.1)				−0.001 0.11(73) (7.2) (1.7) 0.70(84) (8.1)	11.2	0.95	2.0	0.060	8	0.46 (4,5)
Dom. Republic USA	−0.26 (2.7)		0.35(n=1) (3.5)			−0.01 (2.3)	0.47(79) (3.5) −0.72(82) (5.0)	25.2	0.73	2.2	0.125	15	0.77 (4,11)
Rest-of-world	−0.55 (5.1)	1.44(n=0) (2.5)		0.94(n=1) (1.6)	1.26(n=2) (1.8)		1.00(68,70) (6.7)	5.8	0.77	1.5	0.978	18	0.39 (4,14)
El Salvador USA	−0.99 (7.9)		1.13(n=1) (6.3)				1.00(78,79) (6.2) 0.85(83) (3.5)	9.93	0.85	2.5	0.223	11	0.13 (4,8)
Guatemala USA	−0.99 (6.0)	1.89(n=2) (4.1)		2.01(n=3) (3.3)	2.35(n=4) (3.5)		0.002 −1.95(82) (3.4) (5.2)	8.67	0.88	1.7	0.334	11	0.34 (4,8)

Table C4 (*cont.*)

Product Origin Destination	$\ln X^s_{ij,t-1}$	$\ln X^s_{ij,t-2}$	$\Delta\ln(P_{ij}/D_i)_{t-n}$	$\ln(P_{ij}/D_i)_{t-n}$	$\Delta\ln(P_{ij}/D_i)_{t-n}$	$\ln(P_{ij}/D_i)_{t-n}$	T	T^2	$W(t)$	Const.	Summary statistics				Test for parameter constancy
											$\bar{R}^2$	dw	$\hat{\sigma}$	d.o.f.	
Sugar (*cont.*)															
Guyana USA	−0.95 (2.7)	0.98(n=0) (2.7)		0.99(n=1) (2.8)	0.63(n=2) (1.6)		−0.03 (1.7)		−1.56(73) (3.5) −2.42(85) (5.1)	68.5	0.81	2.0	0.409	13	0.13 (4,10)
Jamaica UK	−0.46 (4.6)	0.52(n=4) (4.6)	0.44(n=5) (4.0)				−0.02 (3.0)		1.00(71,73) (6.4) −1.02(79) (6.2)	46.5	0.82	1.9	0.119	14	0.41 (4,6)
USA	−0.48 (4.1)	0.40(n=0) (1.6)	0.46(n=2) (1.4)						−1.02(77) (2.5) 0.72(84) (2.4)	4.85	0.73	2.0	0.276	14	0.07 (4,11)
Panama USA	−0.58 (3.9)	0.54(n=3) (4.4)		0.15(n=4) (1.5)	0.70(n=5) (3.9)		0.04 (2.8)		1.00(70,72) (2.7) 0.25(75) (2.1)	−81.8	0.69	2.2	0.113	10	0.93 (4,11)
Peru USA	−0.54 (5.2)	0.58(n=0) (1.6)	0.82(n=2) (2.4)				−0.05 (3.8)		−1.83(80) (6.6)	97.1	0.85	2.2	0.232	15	0.30 (4,11)

Coffee

Brazil

Importer											
Italy	−0.32 (3.4)		0.12($n=4$) (1.6)	−0.02 (2.4)	1.00(75,76) (6.1)	41.2	0.77	2.5	0.097	11	0.24 (4,7)
USA	−0.70 (7.1)	0.19($n=7$) (1.8)	0.38($n=8$) (2.9)	−0.03 (3.2)	−0.51(74) (4.7) 1.00(77,79) (8.0)	74.1	0.91	2.9	0.103	10	0.76 (4,6)
Rest-of-world	−0.67 (4.8)	0.22($n=7$) (2.4)	0.23($n=8$) (2.6)		0.22(73) (2.2) 1.00(77,79) (4.5)	8.25	0.73	1.7	0.097	10	0.10 (4,6)

Colombia

Importer											
W. Germany	−0.20 (2.6)	0.26($n=4$) (2.3)	0.24($n=5$) (1.8)		0.29(75) (2.6) 1.00(78,79) (5.6)	1.62	0.66	2.0	0.106	14	0.26 (4,10)
USA	−0.44 (3.7)	0.34($n=1$) (2.5)	0.34($n=2$) (2.7)		−0.47(77) (3.2) −0.61(81) (4.7)	4.31	0.73	2.4	0.121	15	0.94 (4,11)
Rest-of-world	−0.08 (1.6)	0.26($n=2$) (3.2)	0.09($n=3$) (1.2)		1.00(76,77) (5.7)	0.79	0.78	1.6	0.079	18	0.54 (4,14)

Costa Rica

Importer											
Finland	−0.57 (7.9)	0.10($n=8$) (1.3)	0.74($n=9$) (5.0)		0.26(74) (3.5) 1.00(76,77) (3.7)	2.97	0.89	2.0	0.071	11	0.39 (4,7)
W. Germany	−0.67 (3.4)	0.29($n=4$) (1.7)	0.77($n=5$) (4.5)		0.33(78) (1.9) −0.76(83) (3.1)	4.34	0.81	2.0	0.167	10	0.88 (4,7)

Table C4 (cont.)

Product Origin Destination	$\ln X^s_{ij,t-1}$	$\ln X^s_{ij,t-2}$	$\Delta \ln(P_{ij}/D_i)_t$	$\ln(P_{ij}/D_i)_{t-n}$ $\Delta\ln(P_{ij}/D_i)_{t-n}$	$\ln(P_{ij}/D_i)_{t-n}$	T	T^2	$W(t)$	Const.	$\bar{R}^2$	dw	$\hat{\sigma}$	d.o.f.	Test for parameter constancy
Coffee														
Costa Rica (cont.)														
Sweden	−0.44 (2.9)			0.40(n=8) (2.3)		0.001 (2.8)		0.68(74) (3.9) 0.51(81) (2.7)	2.78	0.69	2.4	0.167	12	0.31 (4,8)
USA	−0.52 (3.7)	0.23(n=7) (1.2)	0.42(n=8) (2.4)			−0.01 (1.3)		0.31(70) (1.8) −0.30(75) (1.8)	29.9	0.67	2.8	0.158	10	0.38 (4,6)
Rest-of-world	−0.93 (5.5)	0.39(n=7) (2.0)	1.03(n=8) (5.1)					0.51(79) (2.9) 1.00(81,83) (4.9)	6.72	0.82	2.4	0.165	12	0.34 (4,8)
Dom. Republic														
USA	−0.99 (6.1)			0.39(n=1) (1.7)				1.03(77) (4.6) 0.28(84) (1.3)	8.94	0.76	1.7	0.205	14	0.19 (4,11)
Rest-of-world	−0.46 (3.8)	0.74(n=4) (2.8)	0.65(n=5) (2.7)			−0.02 (1.8)		−1.00(77) (5.1) −0.68(80) (2.7)	41.2	0.75	1.6	0.186	14	0.95 (4,10)

Country												
Ecuador												
W. Germany	−0.59 (4.4)	1.15(n=7) (2.0)	2.39(n=8) (2.3)			−0.79(71) (2.2) 1.00(75,77) (3.7)	−1.13	0.74	1.3	0.329	10	0.13 (4,7)
USA	−0.62 (4.5)	0.44(n=6) (2.7)	0.60(n=7) (3.7)			−0.32(71) (1.9) 0.41(81) (2.4)	4.90	0.73	2.5	0.159	12	0.05 (4,8)
Rest-of-world	−0.78 (4.0)		1.16(n=12) (2.7)	−0.04 (2.5)		1.00(79,81) (2.9)	78.6	0.75	1.8	0.216	9	0.47 (4,5)
El Salvador												
W. Germany	−0.96 (6.7)	0.52(n=7) (3.8)	0.53(n=8) (4.1)	−0.04 (5.7)		−0.39(73) (3.2) 1.00(81,82) (5.4)	93.3	0.87	2.0	0.111	11	0.12 (4,7)
Netherlands	−0.14 (1.5)		1.26(n=6) (2.2)		−0.002 (3.7)	1.00(74,75) (5.3) 1.00(81,82) (7.0)	−1.90	0.90	1.9	0.285	12	0.18 (4,8)
USA	−0.38 (2.4)		0.27(n=7) (1.8)			0.91(73) (3.1) 1.00(78,79) (5.4)	3.28	0.81	2.4	0.275	12	0.87 (4,8)
Rest-of-world	−0.98 (7.4)		1.08(n=4) (4.8)	0.09 (5.7)		1.00(68,71) (4.5) 0.60(78) (2.2)	5.42	0.86	2.4	0.252	16	0.35 (4,12)
Guatemala												
W. Germany	−0.28 (2.9)	0.69(n=9) (3.4)	0.68(n=10) (2.7)			−0.26(73) (2.7) −0.39(83) (4.2)	0.97	0.75	2.4	0.089	9	0.09 (4,6)

Table C4 (*cont.*)

Product Origin Destination	$\ln X^s_{ij,t-1}$	$\Delta \ln X^s_{ij,t-2}$	$\Delta \ln(P_{ij}/D_i)_{t-n}$	$\ln(P_{ij}/D_i)_{t-n}$	$\ln(P_{ij}/D_i)_{t-n}$	T	T^2	$W(t)$	Const.	$\bar{R}^2$	dw	$\hat{\sigma}$	d.o.f.	Test for parameter constancy
Coffee														
Guatemala (*cont.*)														
USA	−0.58 (3.5)		0.26(n=5) (2.1)					0.35(73) (2.3) −0.69(81) (4.7)	5.52	0.73	2.0	0.140	16	0.14 (4,12)
Rest-of-world	−0.59 (4.3)	0.63(n=5) (3.5)	0.28(n=6) (1.6)					−0.54(67) (3.0) 0.28(72) (1.6)	5.47	0.68	1.8	0.168	12	0.10 (4,9)
Haiti														
Belgium	−0.91 (4.3)	0.41(n=5) (2.6)	0.67(n=6) (4.7)			−0.06 (6.2)		0.56(80) (3.0)	128.6	0.80	2.2	0.124	11	0.15 (4,7)
France	−0.49 (3.4)	0.26(n=2) (1.5)	0.85(n=3) (4.3)					0.72(65) (3.7) −0.55(81) (2.6)	1.63	0.69	1.9	0.178	16	0.24 (4,12)
Italy	−0.99 (5.8)	0.81(n=6) (3.0)	0.47(n=7) (2.0)					0.90(80) (4.1)	6.71	0.77	1.8	0.214	14	0.31 (4,10)
Netherlands	−0.40 (3.4)		2.71(n=9) (2.9)					1.00(71,72) (3.8) 0.57(76) (2.0)	−4.53	0.64	1.8	0.267	11	0.26 (4,7)
Rest-of-world	−0.98 (5.0)	0.53(n=6) (2.1)	0.83(n=7) (2.7)				0.002 (2.2)	−0.96(79) (3.3)	6.71	0.76	2.1	0.266	12	0.21 (4,8)

Honduras											
W. Germany	−0.34 (5.2)		0.22(n=2) (1.4)		1.00(67,68) (3.3) / 1.00(73,75) (3.3)	2.45	0.69	1.7	0.202	18	0.52 (4,15)
Japan	−0.89 (7.8)	0.39(n=4) (1.8)	0.29(n=5) (1.4)	0.17 (7.2)	1.00(71,72) (7.9) / −0.90(74) (3.4)	−333.7	0.92	2.1	0.251	14	0.14 (4,10)
USA	−0.90 (6.9)	0.72(n=1) (4.1)	0.94(n=2) (5.2)		1.00(67,68) (3.7) / 1.00(75,78) (4.1)	6.08	0.80	2.2	0.185	18	0.19 (4,14)
Rest-of-world	−0.93 (5.4)	0.49(n=5) (1.8)	0.82(n=6) (2.7)	0.003 (4.3)	1.00(67,68) (3.7) / 1.00(71,72) (2.8)	5.03	0.77	2.0	0.264	13	0.08 (4,9)
Mexico											
W. Germany	−0.38 (3.9)		0.27(n=6) (2.5)		−0.62(71) (3.6) / 1.00(77,78) (5.0)	2.91	0.81	1.3	0.160	11	0.22 (4,7)
Switzerland	−0.21 (1.8)	0.52(n=10) (1.5)	1.23(n=11) (3.0)		0.93(73) (6.0) / −0.47(77) (2.8)	−1.30	0.82	2.0	0.142	9	0.92 (4,5)
USA	−0.90 (5.0)		1.17(n=11) (3.5)	−0.001 (2.1)	1.00(76,78) (1.9) / −0.37(81) (2.5)	7.29	0.66	2.6	0.126	9	0.14 (4,5)
Rest-of-world	−0.43 (3.5)	0.44(n=4) (3.3)	0.40(n=5) (2.1)	−0.001 (1.4)	1.00(80,82) (7.2)	2.19	0.83	2.1	0.250	12	0.06 (4,8)

Table C4 (cont.)

Product Origin Destination	ln $X^s_{ij,t-1}$	ln $X^s_{ij,t-2}$	Δln$(P_{ij}/D_i)_{t-n}$	ln$(P_{ij}/D_i)_{t-n}$	Δln$(P_{ij}/D_i)_{t-n}$	ln$(P_{ij}/D_i)_{t-n}$	T	T^2	W(t)	Const.	$\bar{R}^2$	dw	$\hat{\sigma}$	d.o.f.	Test for parameter constancy
Coffee (cont.)															
Nicaragua															
Belgium	−0.25 (3.5)		0.83(n=10) (3.2)						1.00(80,81) (5.8) −0.62(84) (4.7)	−0.01	0.80	2.4	0.123	11	0.14 (4,8)
W. Germany	−0.65 (4.9)		0.37(n=10) (1.7)						−0.44(71) (3.3) 0.28(78) (2.1)	5.37	0.73	1.4	0.122	10	0.10 (4,6)
Netherlands	−0.23 (1.7)		0.49(n=10) (1.7)						1.00(79,80) (4.4) 1.00(83,84) (6.4)	0.68	0.84	2.2	0.146	11	0.01 (4,8)
USA	−0.76 (7.0)		0.51(n=2) (1.7)				−0.05 (2.5)	0.17 (6.3)	1.00(82,83) (8.0)	109.5	0.85	2.1	0.322	14	0.20 (4,11)
Rest-of-world	−0.92 (6.4)		0.17(n=2) (1.5)						−0.86(78) (4.1) 0.63(83) (2.9)	−318.4	0.77	1.4	0.199	16	0.22 (4,13)
Cocoa															
Brazil															
Spain	−0.95 (12.9)	0.39(n=3) (2.3)	0.50(n=4) (3.4)						−2.71(71) (13.1)	7.59	0.95	2.5	0.202	16	0.56 (4,12)

Country										
USSR	−0.97 (6.8)				−1.01(72) (2.2) −1.57(77) (3.5)	8.26	0.80	2.2	0.425	12	0.11 (4,8)
Rest-of-world	−0.99 (6.4)	0.31($n=8$) (2.7)	0.61($n=8$) (2.1)		0.49(75) (3.9) −0.42(84) (3.2)	10.8	0.80	2.4	0.118	11	0.02 (4,8)
Dom. Republic USA	−0.64 (8.7)	0.10($n=8$) (1.7)	0.20($n=9$) (2.5)		1.00(70,72) (6.4) 0.31(82) (3.3)	6.08	0.85	2.3	0.248	13	0.02 (4,8)
Ecuador W. Germany	−0.99 (7.4)		0.22($n=9$) (1.6)		1.00(71,72) (5.3) 1.00(73,75) (2.7)	7.03	0.84	1.4	0.159	12	0.33 (4,8)
Italy	−0.89 (5.3)		0.16($n=9$) (1.9)		1.00(76,77) (5.2)	6.70	0.89	1.7	0.102	13	0.30 (4,9)
Japan	−0.89 (6.8)		0.17($n=9$) (1.4)		1.00(70,71) (6.5) 0.46(78) (2.5)	6.12	0.83	2.5	0.166	12	0.15 (4,8)
Netherlands	−0.77 (3.9)		0.38($n=9$) (1.6)	−0.001 (2.4)	1.00(72,73) (3.0) −0.57(77) (2.3)	4.97	0.71	1.5	0.238	11	0.09 (4,7)
USA	0.93 (6.3)		0.22($n=9$) (1.7)	0.01 (2.3)	1.00(77,78) (5.2) 0.60(71) (4.1)	8.66	0.86	1.6	0.130	13	0.28 (4,7)
Rest-of-world	−0.69 (5.1)		0.15($n=9$) (1.7)		1.00(77,78) (3.6)	6.10	0.79	1.6	0.184	12	0.11 (4,7)

Table C4 (cont.)

Product Origin Destination	$\ln X^s_{ij,t-1}$	$\ln X^s_{ij,t-2}$ $\Delta\ln(P_{ij}/D_i)_{t-n}$	$\ln(P_{ij}/D_i)_{t-n}$ $\Delta\ln(P_{ij}/D_i)_{t-n}$	$\ln(P_{ij}/D_i)_{t-n}$	T	T^2	$W(t)$	Const.	Summary statistics $\bar{R}^2$	dw	$\hat{\sigma}$	d.o.f.	Test for parameter constancy
Soybeans													
Brazil													
W. Germany	−0.69 (4.0)	2.32(n=0) (2.3)		3.67(n=1) (2.9)	3.55(n=2) (2.4)			−2.69(68) (3.7) 1.00(83,85) (4.1)	5.54	0.70	1.5	0.706	14	0.06 (4,11)
Netherlands	−0.48 (4.7)	1.08(n=0) (1.3)	1.89(n=1) (4.1)					1.00(70,72) (5.8) 1.00(81,82) (5.7)	−3.94	0.80	1.6	0.702	9	0.30 (4,15)
Spain	−0.90 (6.0)	0.88(n=1) (1.4)	1.83(n=2) (2.5)				0.004 (3.0)	−2.56(73) (3.5) −2.70(82) (3.7)	8.72	0.85	1.9	0.677	10	0.06 (4,7)
Rest-of-world	−0.36 (2.8)		0.70(n=1) (1.3)				0.001 (1.4)	1.00(68,69) (3.8) 0.90(72) (1.9)	4.15	0.76	2.0	0.457	12	0.30 (4,8)
Paraguay														
W. Germany	−0.49 (8.1)		0.85(n=1) (3.1)				0.002 (4.3)	0.59(73) (2.7) 0.77(76) (3.6)	3.97	0.84	2.6	0.203	11	0.42 (4,7)

Netherlands	−0.99 (5.8)		0.49(n=0) (1.8)		0.004 (3.2)	0.35(82) (1.2)	10.3	0.83	2.9 0.249 5	0.04 (4.2)
Switzerland	−0.96 (2.0)		1.08(n=2) (1.1)		0.003 (1.4)	−0.68(81) (1.4)	8.59	0.69	2.4 0.318 3	0.00 (3.1)
Rest-of-world	−0.52 (1.4)		1.83(n=0) (2.4)		0.004 (1.8)	−0.62(80) (2.2) −0.73(84) (2.5)	4.12	0.72	3.0 0.191 5	0.00 (3,3)
Cotton										
Brazil										
Belgium	−0.27 (2.3)	−0.12 (4.3)	1.31(n=5) (2.0)			1.00(67,68) (5.1) 1.00(72,73) (4.9)	230.4	0.83	2.8 0.259 9	0.01 (4,7)
W. Germany	−0.40 (4.5)	−0.08 (2.2)	2.03(n=8) (1.7)			1.00(76,77) (4.2) 2.69(83) (5.8)	148.7	0.84	1.9 0.407 11	0.52 (4,8)
Japan	−0.60 (5.3)	−0.06 (2.0)	2.76(n=1) (2.5)			−3.51(76) (3.9) −7.09(79) (7.7)	123.9	0.85	1.7 0.863 19	0.92 (4,15)
Rest-of-world	−0.41 (4.2)		1.21(n=1) (1.8)			−2.13(80) (3.7) 1.09(83) (1.8)	2.18	0.78	2.1 0.558 19	0.94 (4,16)
El Salvador										
Japan	−0.66 (5.7)			1.35(n=0) (1.7)	0.002 (3.1)	0.73(78) (1.9) −2.62(84) (6.9)	7.16	0.80	2.1 0.336 19	0.08 (4,16)
Rest-of-world	−0.54 (5.5)	0.07 (3.0)		1.36(n=0) (2.4)		1.70(n=1) (2.2) 1.45(74) (4.2) −0.99(81) (2.9)	−145.0	0.88	1.7 0.310 8	0.02 (4,5)

Table C4 (cont.)

Product Origin Destination	$\ln X^s_{ij,t-1}$	$\ln X^s_{ij,t-2}$ $\Delta\ln(P_{ij}/D_i)_{t-n}$	$\ln(P_{ij}/D_i)_{t-n}$ $\Delta\ln(P_{ij}/D_i)_{t-n}$	$\ln(P_{ij}/D_i)_{t-n}$	T	T^2	$W(t)$	Const.	Summary statistics				Test for parameter constancy d.o.f.
									$\bar{R}^2$	dw	$\hat{\sigma}$	d.o.f.	
Cotton (cont.) Guatemala													
W. Germany	−0.36 (3.6)	1.13(n=0) (1.5)			0.05 (3.5)		1.50(74) (5.8) −1.02(81) (4.1)	−88.8	0.81	2.0	0.226	11	0.25 (4.8)
Italy	−0.84 (5.7)	2.35(n=1) (2.8)	3.29(n=2) (3.4)		0.08 (4.0)		−0.87(71) (2.2) 1.15(75) (2.6)	−163.4	0.75	2.0	0.376	15	0.65 (4,11)
Japan	−0.35 (5.8)	0.48(n=0) (1.7)	0.56(n=1) (3.0)				0.30(72) (2.4) 0.53(76) (2.1)	2.41	0.83	2.2	0.119	18	0.73 (4,14)
Rest-of-world[a]	−0.67 (4.5)	2.32(n=0) (1.7)	3.19(n=1) (2.2)		0.06 (2.0)		1.00(81,83) (4.6)	−112.9	0.80	1.2	0.377	9	0.40 (4,7)
Mexico Italy	−0.75 (3.9)		0.65(n=2) (1.3)		−0.12 (3.4)		−1.03(71) (2.4) 0.82(79) (1.9)	242.6	0.67	2.8	0.378	10	0.27 (4,6)

Rest-of-world	−0.30 (2.1)	0.63(n=0) (3.2)	0.42(n=1) (1.8)	−0.04 (4.2)	0.41(69) (2.2) 0.47(78) (2.4)	82.3	0.71	2.4	0.175	17	0.26 (4,13)
Nicaragua											
W. Germany	−0.46 (5.8)		1.27(n=1) (3.7)		1.34(74) (6.1) 1.00(80,82) (8.2)	1.31	0.89	1.6	0.207	16	0.07 (4,12)
Italy	−0.53 (4.6)		1.45(n=1) (1.8)		1.35(74) (2.5) 1.00(82,83) (4.1)	1.29	0.67	2.4	0.502	19	0.16 (4,15)
Japan	−0.81 (8.2)		0.70(n=5) (1.7)	0.001 (2.6)	0.47(75) (1.7) 1.00(79,80) (7.7)	7.53	0.86	2.0	0.247	15	0.51 (4,11)
Rest-of-world	−0.27 (4.2)	1.17(n=1) (5.1)	1.63(n=2) (5.4)		1.00(72,73) (6.7) 1.00(76,77) (3.4)	−0.89	0.80	2.2	0.186	18	0.12 (4,14)
Paraguay											
Belgium	−0.16 (1.5)		1.00(n=1) (2.2)		1.00(70,72) (5.2) 1.00(75,77) (4.5)	−0.67	0.76	2.5	0.517	17	0.12 (4,14)
France	−0.38 (2.1)		2.05(n=2) (1.8)		1.00(70,71) (4.0) −3.92(78) (3.8)	−1.41	0.75	1.8	0.948	16	0.18 (4,11)
W. Germany	−0.73 (6.3)		0.89(n=1) (3.2)	0.19 (6.1)	1.00(72,73) (5.6) 1.00(77,78) (4.5)	−368.4	0.83	2.1	0.288	15	0.23 (4,11)

Table C4 (cont.)

Product Origin Destination	$\ln X^s_{ij,t-1}$	$\ln X^s_{ij,t-2}$ $\Delta\ln(P_{ij}/D_i)_{t-n}$	$\ln(P_{ij}/D_i)_{t-n}$	$\Delta\ln(P_{ij}/D_i)_{t-n}$	$\ln(P_{ij}/D_i)_{t-n}$	T	T^2	$W(t)$	Const.	Summary statistics $\bar{R}^2$	dw	$\hat{\sigma}$	d.o.f.	Test for parameter constancy
Cotton														
Paraguay (cont.)														
UK	−0.97 (5.1)		1.23(n=2) (1.8)					1.00(70,71) (3.2) 1.00(74,75) (2.9)	4.08	0.81	2.5	0.624	12	0.20 (4,8)
Rest-of-world	−0.62 (7.8)	0.96(n=0) (7.1)	1.70(n=1) (7.5)			0.11 (6.9)		1.00(75,76) (5.9)	−205.9	0.85	2.5	0.168	13	0.41 (4,11)
Peru														
Belgium	−0.99 (11.1)		0.55(n=1) (1.7)			−0.13 (8.8)		1.00(78,81) (8.2) −1.07(84) (4.4)	271.4	0.91	1.6	0.219	19	0.36 (4,16)
France	−0.64 (4.8)		1.95(n=1) (2.9)					1.00(75,77) (4.2) 1.00(79,81) (5.7)	0.39	0.76	1.9	0.378	12	0.76 (4,10)
W. Germany	−0.99 (6.3)	1.21(n=0) (3.9)	0.64(n=1) (2.0)			−0.08 (4.8)		−0.61(68) (2.4) −0.53(77) (2.1)	172.9	0.82	1.6	0.221	11	0.15 (4,8)
UK	−0.76 (6.4)		0.33(n=1) (1.6)			−0.03 (3.4)		0.33(70) (2.3) −0.63(74) (4.7)	57.7	0.85	1.4	0.129	12	0.63 (4,8)

Iron ore

Origin/Dest.										
Brazil										
France	−0.19 (7.4)		0.16(n=2) (1.4)		1.00(73,74) (5.6)	3.20	0.86	2.8	0.058	12	0.30 (4,9)
W. Germany	−0.16 (2.8)		0.50(n=2) (2.1)		0.17(69) (1.6) 0.28(73) (2.7)	3.68	0.64	2.2	0.100	13	0.21 (4,9)
Italy	−0.15 (3.6)		0.35(n=2) (2.3)		−0.84(66) (5.6) 0.40(74) (2.7)	3.06	0.76	2.6	0.144	18	0.25 (4,14)
UK	−0.34 (3.9)		0.44(n=1) (1.5)		1.00(80,81) (4.6)	5.91	0.79	2.1	0.153	13	0.36 (4,9)
USA	−0.69 (6.2)		0.76(n=1) (2.0)	0.03 (2.3)	1.00(73,74) (5.2) 1.00(80,81) (5.3)	−53.3	0.80	2.8	0.321	13	0.17 (4,9)
Rest-of-world	−0.51 (5.7)	1.26(n=0) (4.8)		0.04 (3.6)	1.00(73,74) (3.2) −0.28(79) (3.0)	−77.9	0.82	2.3	0.089	10	0.10 (4,7)
Peru										
Japan	−0.33 (3.2)		0.13(n=2) (1.4)	−0.05 (4.1)	1.00(70,71) (4.9) −0.56(75) (4.5)	95.3	0.86	2.6	0.097	12	0.17 (4,9)
USA	−0.19 (3.3)		0.51(n=1) (1.8)	−0.06 (5.6)	1.00(74,75) (2.1) 1.00(81, 82) (3.3)	129.4	0.73	1.9	0.246	16	0.02 (4,13)
Rest-of-world	−0.51 (4.6)		0.17(n=1) (1.6)	0.002 (5.1)	1.00(71,72) (6.1) 0.84(78) (4.8)	7.02	0.85	1.4	0.163	13	0.07 (4,9)

Table C4 (cont.)

Product Origin Destination	$\ln X^s_{ij,t-1}$	$\ln X^s_{ij,t-2}$	$\Delta \ln(P_{ij}/D_l)_{t-n}$	$\ln(P_{ij}/D_l)_{t-n}$	T	T^2	$W(t)$	Const.	$\bar{R}^2$	dw	$\hat{\sigma}$	d.o.f.	Test for parameter constancy
Copper													
Chile													
W. Germany	−0.73 (4.6)	0.21(n=9) (2.0)	0.45(n=10) (3.6)		0.02 (2.6)		0.25(74) (3.4) −0.11(79) (1.5)	−39.0	0.79	2.0	0.061	9	0.28 (4,5)
Japan	−0.58 (6.6)		0.23(n=7) (1.9)				1.00(72,74) (4.3) −0.78(80) (3.6)	6.03	0.79	1.7	0.208	14	0.65 (4,10)
USA	−0.47 (3.6)		0.50(n=10) (2.2)				−0.83(71) (4.2) 1.00(74,75) (4.8)	4.10	0.81	2.0	0.181	11	0.10 (4,7)
Rest-of-world	−0.76 (10.7)		0.54(n=2) (2.4)				−1.08(77) (4.1)	8.18	0.89	1.3	0.236	20	0.25 (4,16)
Peru													
Belgium	−0.94 (18.5)		0.42(n=2) (2.4)		−0.05 (5.6)		−4.31(73) (16.9) −2.68(81) (10.3)	100.3	0.97	2.3	0.247	18	0.83 (4,14)
USA	−0.99 (7.3)		1.09(n=1) (2.7)				−2.97(73) (6.0)	8.15	0.81	1.6	0.481	20	0.01 (4,16)

Rest-of-world	−0.40 (3.0)	0.82(n=9) (2.6)	0.91(n=10) (3.2)	−1.42(73) (5.7) −0.58(78) (2.3)	2.35	0.83	1.9	0.220	9	0.32 (4,5)

[a] Estimated using 2SLS.

Notes:
Estimated using OLS unless otherwise indicated.
See first page of annex for explanation of test statistics.

Table C5 Regression results of consumption equation (8.29)
$\Delta \ln C_t = \alpha_{30} + \alpha_{31}\Delta \ln Y_t + \beta_{32}\ln(C/Y)_{t-1} + \beta_{33}\ln Y_{t-1} + \beta_{34}\Delta\ln(P/D)_t + \beta_{35}\ln(P/D)_{t-1}$

Product Market	$\Delta \ln Y_t$	$\ln(C/Y)_{t-1}$	$\ln Y_{t-1}$	$\Delta\ln(P/D)_t$	$\ln(P/D)_{t-1}$	$W(t)$	Const.	$\bar{R}^2$	dw	$\hat{\sigma}$	d.o.f.	Test for parameter constancy
Beef World		−0.32 (3.5)	−0.14 (2.0)	−0.11[a] (5.1)		−0.03(71) (2.0)	2.94	0.67	1.8	0.014	14	0.61 (4,8)
Maize World	1.60 (4.7)	−0.21 (1.6)		−0.11 (2.6)	−0.06 (2.1)	−0.09(75) (3.2) 0.12(85) (3.2)	1.65	0.62	1.9	0.027	19	0.60 (4,14)
Sugar USA	0.76 (3.1)	−0.54 (2.9)			−0.07 (2.8)	−0.09(67) (3.6) 0.08(76) (2.3)	1.25	0.73	2.4	0.026	20	0.15 (4,15)
EEC	0.92 (2.2)	−0.94 (22.0)	−0.84 (20.7)			1.00(74,75) (7.7)	7.90	0.97	2.4	0.019	16	0.19 (3,16)
Rest-of-world		−0.62 (6.3)			−0.04[a] (8.1)	−0.03(85) (3.6)	3.97	0.79	2.6	0.010	14	0.36 (4,8)
Coffee World	1.34 (2.3)	−0.99 (7.5)	−0.38 (4.3)	−0.25 (7.2)	−0.17 (5.1)	−0.09(74) (2.3) −0.20(82) (4.8)	6.08	0.94	2.8	0.032	12	0.34 (4,8)
Cocoa World	1.10 (2.9)	−0.53 (3.7)	−0.20 (3.5)	−0.14 (6.5)	−0.10 (3.7)		2.59	0.68	2.3	0.023	21	0.79 (4,15)

Soybeans World		−0.76 (5.8)	−0.22[a] (4.0)	−0.17(75) (4.9) 0.10(80) (2.9)	5.13	0.81	2.2	0.034	10	0.22 (4,4)
Cotton World		−0.35 (4.0)	−0.26[b] (3.7)	−0.44(84) (6.8)	2.57	0.72	2.4	0.063	23	0.93 (3,19)
Copper World	2.79 (10.4)	−0.25 (2.4)	−0.02 (1.0)	−0.07(85) (2.7)	1.07	0.83	2.0	0.024	20	0.87 (4,16)

[a]Current price.
[b]Egyptian price.

Note: See first page of annex for explanation of test statistics.

Table C6 Regression results of production equation (9.47)

$$\Delta \ln Q_t = \alpha_{50} + \beta_{51} \ln Q_{t-1} + \alpha_{52} \ln Q_{t-2} + \sum_{m=0}^{n-1} \alpha_{53+m} \Delta \ln(P/D)_{t-m} + \sum_{m=1}^{n} \gamma_{54+m} \ln(P/D)_{t-m} + \beta_{55} T + \beta_{56} W_t$$

Product Market	$\ln Q_{t-1}$	$\Delta \ln(P/D)_{t-m}$	$\ln(P/D)_{t-n}$	T	T^2	$W(t)$	Const.	Summary statistics $\bar{R}^2$	dw	$\hat{\sigma}$	d.o.f.	Test for parameter constancy
Beef World	−0.22 (2.9)	−0.08[b] (3.9)	0.03[e] (1.5)			−0.03(71) (2.2) 0.05(74) (3.4)	−2.26	0.70	1.7	0.014	15	0.11 (4,10)
Maize World	−0.93 (6.6)		0.09[c] (1.8)	0.03 (6.5)		−0.09(74) (2.1) 0.10(79) (2.2)	−55.51	0.78	1.6	0.043	19	0.62 (4,15)
Sugar USA	−0.53 (3.6)		0.09[c] (1.5)	0.01 (3.2)		0.24(73) (4.0) 0.16(75) (2.5)	−18.36	0.57	2.2	0.058	18	0.65 (4,12)
EEC	−0.94 (6.7)		0.11[b] (1.4)	0.08 (6.7)	−0.01 (5.5)	−0.19(74) (2.8) −0.16(76) (2.4)	−156.95	0.73	1.9	0.062	17	0.03 (4,12)
Rest-of-world	−0.80 (5.7)	0.03[d] (2.9)	0.02[e] (2.4)	0.02 (5.7)		−0.09(80) (4.1) 0.08(82) (3.6)	−28.23	0.78	1.5	0.020	15	0.02 (4,9)

Coffee										
Brazil	−0.96 (10.23)		0.28[e] (2.7)		1.00(73,76) (8.9)	6.14	0.92	1.9	0.143	17	0.70 (4,12)
Rest-of-world	−0.79 (3.9)		0.08[d] (2.5)	0.01 (3.4)	1.00(74,75) (2.8) −0.08(83) (2.2)	−19.98	0.65	2.5	0.033	15	0.01 (4,10)
Cocoa										
World	−0.91 (5.7)		0.09[g] (1.8)	0.01 (3.0)	−0.17(76) (3.5) 1.00(82,83) (3.9)	−17.05	0.73	1.8	0.048	13	0.20 (4,8)
Soybeans										
World	−0.75 (6.4)		0.25[b] (6.3)	0.04 (6.4)	1.00(75,77) (3.5) −0.13(80) (3.5)	−80.54	0.88	2.4	0.036	15	0.00 (4,8)
Cotton										
World	−0.33 (4.2)		0.13[d] (2.0)		−0.18(75) (3.8)	2.86	0.81	2.5	0.045	15	0.14 (4,10)
Copper										
World	−0.04 (1.3)	0.13[a] (5.4)	0.02[f] (1.5)		0.07(68) (2.7) 0.07(72) (2.8)	0.31	0.69	1.8	0.023	16	0.03 (4,10)

[a]Current price.
[b]Lagged 1 period.
[c]Lagged 2 periods.
[d]Lagged 3 periods.
[e]Lagged 4 periods.
[f]Lagged 5 periods.
[g]Lagged 8 periods.

Note: See first page of annex for explanation of test statistics.

Table C7 Regression results of stock demand equation (8.38)

$$\Delta \ln K_t^d = \alpha_{70} + \alpha_{71}\Delta \ln Q_t + \gamma_{72}\ln(K^d/Q)_{t-1} + \beta_{73}\ln P_{t-n}^e$$

Product	$\Delta \ln(Q)_t$	$\ln(K^d/Q)_{t-1}$	$\ln P_{t-n}^e$	T	$W(t)$	Const.	$\bar{R}^2$	dw	$\hat{\sigma}$	d.o.f.	Test for parameter constancy
Beef	0.78 (1.9)	−0.05 (2.9)	0.10 (n=0) (1.7)		0.08 (69) (2.7) 0.12 (76) (4.1)	−0.77	0.70	2.0	0.028	14	0.29 (4,9)
Maize	1.42 (2.6)	−0.27 (1.5)			−0.84 (83) (4.0) 0.46 (85) (2.4)	−0.76	0.85	2.3	0.161	13	0.30 (4,8)
Sugar	0.56 (2.5)	−0.38 (5.0)	0.05 (n=1)[a] (2.5)		1.00 (73,74) (4.1) −0.09 (80) (1.9)	−0.45	0.76	1.3	0.037	18	0.51 (4,12)
Coffee	0.52 (4.7)	−0.16 (1.6)	0.18 (n=2) (2.1)	−0.02 (3.6)	−0.21 (73) (2.0) 1.00 (76,78) (5.2)	37.17	0.75	2.4	0.089	20	0.50 (4,14)
Cocoa	1.98 (12.1)	−0.09 (1.2)	0.11 (n=2) (4.1)		−0.21 (73) (3.0) 1.00 (83,84) (7.8)	−0.91	0.92	1.9	0.063	16	0.30 (4,11)
Soybeans	1.39 (5.1)	−0.29 (3.9)	0.79 (n=2) (8.6)		1.00 (77, 78) (4.2)	−4.90	0.91	1.7	0.111	12	0.13 (4,6)

Regression Results of Model Equations

Cotton	0.79 (4.7)	−0.45 (6.1)	0.15 (n=2) (2.7)	−0.02 (6.1)	0.19 (74) (2.9) 1.00 (83–85) (8.9)	36.83	0.88	1.4	0.061	19	0.45 (3,14)
Copper[b]		−0.36 (7.6)	1.10 (n=2) (4.3)		−0.39 (71) (2.0) 0.41 (83) (2.3)	−7.39	0.79	2.0	0.169	14	0.57 (4,8)

[a] US price.
[b] Current interest rates, the 3-month US Treasury bill, were also included in the relationship. The coefficient on the variable is −0.62 with a t-statistic of 3.7.

Note: See first page of annex for explanation of test statistics.

Annex D

Derivation of Equations

The purpose of this annex is to show how the equations in the main text were derived and to serve as a summary guide to the interrelationship among the equations in each of the major parts of the book.

Part II: Commodity Trade Theory

Demand

1 Let $Q = (Q_1, \ldots, Q_n)$ be a *commodity type* composed of a vector of products from sources $1, \ldots, n$.

2 It is assumed that each importer j has a most-preferred commodity type, denoted Q_j^*, and is indifferent between the compensated difference, d_j, for a commodity type Q_j and Q_j^*, such that

$$Q_j = d_j Q_j^*, \qquad (3.1)$$

where $d_j \geq 1$.

3 Then the *quantity of imports*, denoted M, can be expressed in terms of unit-equivalents of the most-preferred commodity type:

$$M_j = Q_j/d_j \qquad (3.2)$$

for each importer j.

4 By the transitivity property of equalities, the *quantity of exports*, denoted X, to geographic market j can be expressed in terms of unit-equivalents of the most-preferred commodity type:

$$\begin{aligned} X_{1j} + \ldots + X_{nj} &= (Q_{1j} + \ldots + Q_{nj})/d_j \\ &= \frac{Q_{1j}}{d_j} + \ldots + \frac{Q_{nj}}{d_j}. \end{aligned} \qquad (3.3)$$

Non-existence of a most-preferred product Q_{ij}^*, which would require that there be a d_{ij} such that $X_{ij} = Q_{ij}^*$, follows from the importer's preference for diversity of supply sources. Since the importer's preference structure contains the desire for diversity of supply sources, the most-preferred commodity type is a composite of products $Q_{1j}, \ldots, Q_{nj}$. Consequently, there cannot exist a single most-preferred product Q_{ij}^*.

5 Preference ordering of the importer is described by the utility tree. If M_j is *weakly separable* from the numeraire, denoted $N_{o,j}$, and if substitution takes

place in the *constant elasticity form,* then the indifference curve at the first level of decision is given by

$$U(M_j, N_{o,j}) = [\pi_j M_j^\alpha + (1 - \pi_j) N_{o,j}^\alpha]^{1/\alpha}; \qquad (3.4)$$

and at the second level of decision by

$$U_m(X_1, \ldots, X_n) = (\pi_{ij} X_{ij}^\beta + \sum_{k=1}^{n-1} \pi_{kj} X_{kj}^\beta)^{1/\beta}, \qquad (3.5)$$

where both indifference curves are strictly convex for positive values of the variables and for $\alpha, \beta < 1$. The values of the *distribution parameters* are constrained to $0 < \pi < 0.5$ such that $\pi_{ij} + \Sigma_k \pi_{kj} = 1$.

6 In (3.5), the *marginal rate of substitution* between an exporter of interest i and its competing suppliers k to market j is derived as follows. Let

$$M_j = [\pi_{ij} X_{ij}^\beta + (1 - \pi_{ij}) X_{kj}^\beta]^{1/\beta}.$$

Then

$$\frac{\partial M_j}{\partial X_{ij}} = (\ldots)^{1/\beta - 1} \pi_{ij} X_{ij}^{\beta - 1}.$$

Similarly,

$$\frac{\partial M_j}{\partial X_{kj}} = (\ldots)^{1/\beta - 1} (1 - \pi_{ij}) X_{kj}^{\beta - 1}.$$

Hence the *marginal rate of substitution* is given by

$$\frac{\partial M_j / \partial X_{ij}}{\partial M_j / \partial X_{kj}} = \frac{\partial X_{kj}}{\partial X_{ij}} = \left(\frac{\pi_{ij}}{1 - \pi_{ij}}\right)\left(\frac{X_{ij}}{X_{kj}}\right)^{\beta - 1}, \qquad (3.6)$$

where the importer's indifference curve is decreasing and strictly convex for $\beta < 1$.

N.B. The subscript reference to the importer is implicit in the equations that follow.

7 The utility maximization problem for the first level of decision by the importer, given a level of nominal dollar income Y^n, is

$$\max [\pi M^\alpha + (1 - \pi) N_o^\alpha]^{1/\alpha}$$

$$\text{subject to } PM + N_o = Y^n, \qquad (3.7)$$

where $\alpha < 1$ and $0 < \pi < 1$.

The Lagrangian is

$$\mathcal{L}_1 = [\pi M^\alpha + (1 - \pi) N_o^\alpha]^{1/\alpha} + \mu(Y^n - PM - N_o).$$

The first-order conditions are

$$\frac{\partial \mathcal{L}_1}{\partial M} = [\pi M^\alpha + (1 - \pi) N_o^\alpha]^{(1/\alpha) - 1} \pi M^{\alpha - 1} - \mu P = 0$$

Annex D

$$\frac{\partial \mathcal{L}_1}{\partial N_o} = [\pi M^\alpha + (1-\pi)N_o^\alpha]^{(1/\alpha)-1}(1-\pi)N_o^{\alpha-1} - \mu = 0$$

$$\frac{\partial \mathcal{L}_1}{\partial \mu} = Y^n - PM - N_o = 0.$$

From the first-order conditions for M and N_o,

$$\frac{\pi}{1-\pi} \frac{M^{\alpha-1}}{N_o^{\alpha-1}} = P.$$

Solve for N_o:

$$N_o = \left(\frac{\pi}{1-\pi}\right)^{1/(\alpha-1)} P^{-1/(\alpha-1)} M, \qquad [*]$$

and substitute into the first-order condition for μ:

$$Y^n = \left[P + \left(\frac{\pi}{1-\pi}\right)^{1/(\alpha-1)} P^{-1/(\alpha-1)}\right] M.$$

Solve for M:

$$M = \frac{Y^n}{P + [\pi/(1-\pi)]^{1/(\alpha-1)} P^{-1/(\alpha-1)}}.$$

Divide through by Y^n, and factor out π and P from the expression in the denominator of the right-hand side:

$$\frac{M}{Y^n} = \left(\frac{\pi}{1-\pi}\right)^{-1/(\alpha-1)} \frac{P^{1/(\alpha-1)}}{[\pi/(1-\pi)]^{-1/(\alpha-1)} P^{\alpha/(\alpha-1)} + 1}.$$

Raise the above expression by the power of $\alpha - 1$:

$$\left(\frac{M}{Y^n}\right)^{\alpha-1} = \left(\frac{\pi}{1-\pi}\right)^{-1} \frac{P}{\{[\pi/(1-\pi)]^{-1/(\alpha-1)} P^{\alpha/(\alpha-1)} + 1\}^{\alpha-1}}.$$

Then raise the above expression by the power of $1/\alpha$:

$$\left(\frac{M}{Y^n}\right)^{(\alpha-1)/\alpha} = \left(\frac{\pi}{1-\pi}\right)^{-1/\alpha} \frac{P^{1/\alpha}}{\{[\pi/(1-\pi)]^{-1/(\alpha-1)} P^{\alpha/(\alpha-1)} + 1\}^{(\alpha-1)/\alpha}}$$

$$= \left(\frac{\pi}{1-\pi}\right)^{-1/\alpha} \frac{P^{1/\alpha}}{D},$$

where $D = \{[\pi/(1-\pi)]^{-1/(\alpha-1)} P^{\alpha/(\alpha-1)} + 1\}^{(\alpha-1)/\alpha}$ is the deflator.
Solve to derive the *demand schedule for M*:

$$M^d = k_1 \left(\frac{P^{\epsilon_m^p}}{D^{\epsilon_m^p+1}}\right) Y^n$$

$$= k_1 \left(\frac{P^{\epsilon_m^p}}{D^{\epsilon_m^p}}\right) \frac{Y^n}{D}$$

$$= k_1 \left(\frac{P}{D}\right)^{\epsilon_m^p} Y, \qquad (3.8a)$$

where $\epsilon_m^p = 1/(\alpha - 1)$, $k_1 = [(1 - \pi)/\pi]^{1/(1-\alpha)}$, and $Y = Y^n/D$, the constant dollar income.

Next, substitute the first expression in (3.8a) into the first-order condition for μ:

$$Y^n - k_1\left(\frac{PP^{\epsilon_m^p}}{D^{\epsilon_m^p+1}}\right)Y^n - N_o = 0.$$

Rearrange terms:

$$\left[1 - k_1\left(\frac{P}{D}\right)^{\epsilon_m^p+1}\right]Y^n - N_o = 0.$$

Solve to derive the demand schedule for N_o:

$$N_o = (1 - k_1)\left(\frac{P}{D}\right)^{\epsilon_n^p} Y^n, \tag{3.8b}$$

where $\epsilon_n^p = \alpha/(\alpha - 1)$.

8 The importer's *utility maximization problem for the second level* of decision is as follows:

$$\max [\pi_i X_i^\beta + (1 - \pi_i) X_k^\beta]^{1/\beta}$$

$$\text{subject to } P_i X_i + P_k X_k = Y_m^n, \tag{3.9}$$

where $\beta < 1$ and $0 < \pi_i < 0.5$.

The Lagrangian is

$$\mathcal{L}_m = [\pi_i X_i^\beta + (1 - \pi_i) X_k^\beta]^{1/\beta} + \mu_m(Y_m^n - P_i X_i - P_k X_k),$$

where $Y_m^n = PM$.

The first-order conditions are

$$\frac{\partial \mathcal{L}_m}{\partial X_i} = [\pi_i X_i^\beta + (1 - \pi_i) X_k^\beta]^{(1/\beta)-1} \pi_i X_i^{\beta-1} - \mu_m P_i = 0$$

$$\frac{\partial \mathcal{L}_m}{\partial X_k} = [\pi_i X_i^\beta + (1 - \pi_i) X_k^\beta]^{(1/\beta)-1}(1 - \pi_i) X_k^{\beta-1} - \mu_m P_k = 0$$

$$\frac{\partial \mathcal{L}_m}{\partial \mu_k} = Y_m^n - P_i X_i - P_k X_k = 0.$$

From the first-order conditions for X_i and X_k, the *rate of product substitution* is

$$\frac{X_i}{X_k} = \left(\frac{\pi_i}{1 - \pi_i}\right)^{-1/(\beta-1)}\left(\frac{P_i}{P_k}\right)^{-1/(\beta-1)}.$$

Invert the above expression, then raise it by the power of β and add one to both sides:

$$\frac{X_i^\beta + X_k^\beta}{X_i^\beta} = \left(\frac{1-\pi_i}{\pi_i}\right)^{-\beta/(1-\beta)}\left(\frac{P_i^{\beta/(\beta-1)} + P_k^{\beta/(\beta-1)}}{P_i^{\beta/(\beta-1)}}\right).$$

Raise the expression by the power of $1/\beta$:

$$\frac{(X_i^\beta + X_k^\beta)^{1/\beta}}{X_i} = \left(\frac{1-\pi_i}{\pi_i}\right)^{-1/(\beta-1)}\left(\frac{P_i^{\beta/(\beta-1)} + P_k^{\beta/(\beta-1)}}{P_i^{\beta/(\beta-1)}}\right)^{1/\beta}.$$

or

$$\frac{M}{X_i} = \left(\frac{1 - \pi_i}{\pi_i}\right)^{-1/(\beta-1)} \left(\frac{P}{P_i}\right)^{1/(\beta-1)}$$

where $P = (P_i^{\beta/(\beta-1)} + P_k^{\beta/(\beta-1)})^{(\beta-1)/\beta}$ and, as before, $M = (X_i^\beta + X_k^\beta)^{1/\beta}$.
Solve the above expression to derive the *demand schedule for* X_i:

$$X_i^d = k_2 \left(\frac{P_i}{P}\right)^{\epsilon_x^p} M \qquad (3.10a)$$

where $k_2 = [\pi_i/(1 - \pi_i)]^{1/(\beta-1)}$, so that $0 < k_2 < 1$, and $\epsilon_x^p = 1/(\beta - 1)$.

The *demand schedule for* X_k is derived in the same manner, but without the rate of product substitution being inverted:

$$X_k^d = (1 - k_2)\left(\frac{P_k}{P}\right)^{\epsilon_x^p} M \qquad (3.10b)$$

where, as before, $\epsilon_x^p = 1/(\beta - 1)$.

Supply

Subscript references to the exporter and the importer are implicit in this section.

9 Let the production schedule of the exporter be described by a generalized CES function. Then the quantity of exports, denoted X, is related to labour, A, and capital, B, inputs as follows:

$$X = k_3(A^\theta + B^\theta)^{\tau/\theta}, \qquad (4.15)$$

where $k_3 = \exp(\sigma_0 + \sigma_1 T + \sigma_2 W)$, in which T is an efficiency parameter and W is a shift parameter, and where $\tau > 0$ and $\theta < 1$.

10 If V_1 and V_2 are the constant unit costs of capital and labour respectively, then the cost of production, denoted E, equals

$$E = V_1 A + V_2 B. \qquad (4.16)$$

11 The cost minimization problem is

$$\min V_1 A + V_2 B$$
$$\text{subject to } k_3^{\theta/\tau} A^\theta + k_3^{\theta/\tau} B^\theta = X^{\theta/\tau}.$$

The Lagrangian is

$$\mathcal{L}_2 = V_1 A + V_2 B + \gamma(X^{\theta/\tau} - k_3^{\theta/\tau} A^\theta - k_3^{\theta/\tau} B^\theta).$$

The first-order conditions are

$$\frac{\partial \mathcal{L}_2}{\partial A} = V_1 - \gamma \theta k_3^{\theta/\tau} A^{\theta-1} = 0.$$

$$\frac{\partial \mathcal{L}_2}{\partial B} = V_2 - \gamma \theta k_3^{\theta/\tau} B^{\theta-1} = 0.$$

Derivation of Equations

Hence the rate of technical substitution is

$$\frac{V_1}{V_2} = \frac{A^{\theta-1}}{B^{\theta-1}}.$$

Transform the above expression to

$$\frac{V_1 A}{V_2 B} = \frac{A^\theta}{B^\theta}.$$

Add one to both sides:

$$\frac{V_1 A + V_2 B}{V_2 B} = \frac{A^\theta + B^\theta}{B^\theta}$$

or

$$\frac{E}{V_2 B} = \frac{A^\theta + B^\theta}{B^\theta}.$$

Multiply both sides by $k_3^{\theta/\tau}$:

$$\frac{k_3^{\theta/\tau} E}{V_2 B} = \frac{k_3^{\theta/\tau} A^\theta + k_3^{\theta/\tau} B^\theta}{B^\theta} = \frac{X^{\theta/\tau}}{B^\theta}.$$

Solve for B:

$$B = k_3^{\theta/[\theta(1-\theta)]} X^{\theta/[\tau(\theta-1)]} V_2^{1/(\theta-1)} E^{1/(1-\theta)}.$$

Multiply both sides by V_2:

$$V_2 B = k_3^{\theta/[\tau(1-\theta)]} X^{\theta/[\tau(\theta-1)]} V_2^{\theta/(\theta-1)} E^{1/(1-\theta)}.$$

By symmetry,

$$V_1 B = k_3^{\theta/[\tau(1-\theta)]} X^{\theta/[\tau(\theta-1)]} V_1^{\theta/(\theta-1)} E^{1/(1-\theta)}.$$

Hence

$$\begin{aligned} E &= V_1 A + V_2 B \\ &= k_3^{\theta/[\tau(1-\theta)]} X^{\theta/[\tau(\theta-1)]} E^{1/(1-\theta)} [V_1^{\theta/(\theta-1)} + V_2^{\theta/(\theta-1)}]. \end{aligned}$$

Solve for E to derive the *cost schedule*:

$$E = k_3^{1/\tau} X^{1/\tau} [V_1^{\theta/(\theta-1)} + V_2^{\theta/(\theta-1)}]^{(\theta-1)/\theta}. \tag{4.17}$$

12 The *average cost schedule* is

$$E/X = k_3^{1/\tau} X^{1/\tau - 1} [V_1^{\theta/(\theta-1)} + V_2^{\theta/(\theta-1)}]^{(\theta-1)/\theta}$$

and the *marginal cost schedule* is

$$\frac{\partial E}{\partial X} = 1/\tau \, k_3^{1/\tau} X^{1/\tau - 1} [V_1^{\theta/(\theta-1)} + V_2^{\theta/(\theta-1)}]^{(\theta-1)/\theta}.$$

13 The quantity of the product that the exporting country will supply is determined by the first-order condition of the profit maximization objective of producers. The profit maximization problem is

$$\max P(X)X - E(X).$$

The first-order condition is

$$\left(\frac{\partial P}{\partial X}X + P\right) - \frac{\partial E}{\partial X} = 0$$

or

$$\left(\frac{\partial P}{\partial X}\frac{X}{P} + 1\right)P - 1/\tau\, k_3^{1/\tau} X^{1/\tau-1}[V_1^{\theta/(\theta-1)} + V_2^{\theta/(\theta-1)}]^{(\theta-1)/\theta} = 0.$$

The inverse of $(\partial P/\partial X)(X/P)$ is the price elasticity of export demand, which from (3.10a) equals ϵ_x^p. Substitution of the expression for the price elasticity of export demand ϵ_x^p into the above expression yields

$$\left(1 + \frac{1}{\epsilon_x^p}\right)P - 1/\tau\, k_3^{1/\tau} X^{1/\tau-1}[V_1^{\theta/(\theta-1)} + V_2^{\theta/(\theta-1)}]^{(\theta-1)/\theta} = 0.$$

The first term on the left-hand side is the marginal revenue of the exporter; the second term is the marginal cost.

The solution for X from the above first-order condition yields the *export supply schedule*.

$$X^s = k_4\left(\frac{P}{D}\right)^\gamma \exp(\phi_1 T + \phi_2 W)_1, \qquad (4.18)$$

where T is a trend variable, W is a shift variable, and

$$D = [V_1^{\theta/(\theta-1)} + V_2^{\theta/(\theta-1)}]^{(\theta-1)/\theta}$$

is the deflator. The parameters are defined as

$$\gamma = \frac{\tau}{1-\tau}, \qquad \phi_1 = \frac{\sigma_1}{\tau-1}, \qquad \phi_2 = \frac{\sigma_2}{\tau-1},$$

and

$$k_4 = e^{\sigma_0/(\tau-1)} \tau^{\tau/(1-\tau)} \left[1 + \frac{1}{\epsilon_x^p}\right]^{\tau/(1-\tau)}.$$

14 Let the *import supply schedule* be described by a generalized exponential function that relates the import supply M^s of a market j to the constant dollar price of the commodity P/D, the ratio between the import price P_j and the market price P, and the foreign exchange reserves of the country H_j:

$$M_j^s = k_5(P/D)^{\rho_1}(P_j/P)^{\rho_2} H_j^{\rho_3}, \qquad (4.21)$$

where $k_5 = \exp(\rho_4 + \rho_5 T + \rho_6 W)$ and the variables T and W represent efficiency and shift parameters respectively. Since $\rho_3 \simeq 0$ so that $H_j^{\rho_3} \simeq 1$, (4.21) becomes

$$M_j^s = k_5(P/D)^{\rho_1}(P_j/P)^{\rho_2}.$$

15 Solve for P_j to find the *inverse import supply schedule*:

$$P_j = k_5^{-1/\rho_2} P^{(\rho_2+\rho_1)/\rho_2} D^{\rho_1/\rho_2} M^{-1/\rho_2}.$$

Unless the importer is a monopsonist, the import price will be given and the relative price elasticity of import supply will approach infinity. The fact that $\rho_2 \simeq \infty$ implies that

(a) $p_1/p_2 \simeq 0$, so that $D^{p_1/p_2} \simeq 1$;
(b) $-1/p_2 \simeq 0$, so that $M^{-1/p_2} \simeq 1$;
(c) $(p_2 - p_1)/p_2 \simeq 1$ by L'Hôpital's rule; and
(d) $-[\exp(p_4 + p_5 T + p_6 W)]/p_2$ does not necessarily degenerate to zero.

From whence it can be seen that

$$P_j = k_6 P \tag{4.22}$$

where $k_6 = k_5^{-1/p_2}$.

Part III: Dynamic Specifications

The derivation in this section of the dynamic specification of the equations in Part III for the most part follows the general approach presented in Chapter 7. However, specific derivations sometimes contain unique transformations, such as the logit form used in the export demand equation.

In the notations that follow, lower-case letters denote logarithms of corresponding capital letters.

Import Demand

16 The general stochastic specification of the expression for import demand M of a geographic market j in terms of its economic activity Y_j and the import price P_j, relative to the general price deflator D_j, is

$$m_{jt} = \alpha_{10} + \alpha_{11} y_{jt} + \alpha_{12} y_{j,t-1} + \alpha_{13}(p_j - d_j)_t + \alpha_{14}(p_j - d_j)_{t-1}$$
$$+ \alpha_{15} m_{j,t-1} + v_{1t}, \tag{8.11}$$

where $\alpha_{11}, \alpha_{12} > 0$; $\alpha_{13}, \alpha_{14} < 0$; and $0 < \alpha_{15} < 1$.

17 Equation (3.8a) states that import demand has a proportional response to a change in economic activity, which implies that $\alpha_{11} + \alpha_{12} + \alpha_{15} = 1$. In order to obtain a proportional response of import demand to changes of economic activity in (8.11), first substitute $(1 - \alpha_{11} - \alpha_{15})$ for α_{12} and add $(-m_{j,t-1})$ to both sides:

$$\Delta m_{jt} = \alpha_{10} + \alpha_{11} \Delta y_{jt} + \beta_{12}(m_j - y_j)_{t-1} + \alpha_{13}(p_j - d_j)_t$$
$$+ \alpha_{14}(p_j - d_j)_{t-1} + v_{1t},$$

where $\beta_{12} = (\alpha_{15} - 1)$.

Next, transform the import price variable so that a 'difference' formulation is nested in the levels form of the equation:

$$\Delta m_{jt} = \alpha_{10} + \alpha_{11} \Delta y_{jt} + \beta_{12}(m_j - y_j)_{t-1} + \beta_{14}\Delta(p_j - d_j)_t$$
$$+ \beta_{15}(p_j - d_j)_{t-1} + v_{1t},$$

where $\beta_{14} = \alpha_{13}$ and $\beta_{15} = \alpha_{13} + \alpha_{14}$.

Finally, add $y_{j,t-1}$ as a separate regressor in order to test the hypothesis that import demand has a proportional response to a change in economic activity:

$$\Delta m_{jt} = \alpha_{10} + \alpha_{11}\Delta y_{jt} + \beta_{12}(m_j - y_j)_{t-1} + \beta_{13}y_{j,t-1}$$
$$+ \beta_{14}\Delta(p_j - d_j)_t + \beta_{15}(p_j - d_j)_{t-1} + v_{1t}, \qquad (8.12)$$

such that $\beta_{13} = (\alpha_{11} + \alpha_{12} + \alpha_{15} - 1)$. The expected signs of the coefficients are $\alpha_{11} > 0$; $-1 < \beta_{12} < 0$; $\beta_{13} > \beta_{12}$; and $\beta_{14}, \beta_{15} < 0$.

18   In a steady-state path, $\Delta m_{jt} = g_3$, $\Delta y_{jt} = g_4$, and $\Delta(p - d)_{jt} = 0$:

$$g_3 = \alpha_{10} + \alpha_{11}g_4 + \beta_{12}(m - y)_j + \beta_{13}y_j + \beta_{15}(p - d)_j.$$

From the first difference of (8.11), the relationship between g_3 and g_4 is found to be

$$g_3 = \frac{\alpha_{11} + \alpha_{12}}{-\beta_{12}} g_4.$$

Then the long-run equilibrium relationship is

$$m_j = -\alpha_{10}/\beta_{12} - [(\alpha_{11} + \alpha_{12})/\beta_{12}^2 + \alpha_{11}/\beta_{12}]g_4$$
$$+ [(\beta_{12} - \beta_{13})/\beta_{12}]y_j - \beta_{15}/\beta_{12}(p - d)_j.$$

Since $\alpha_{12} = (1 - \alpha_{11} - \alpha_{15})$ and $\beta_{12} = (\alpha_{12} - 1)$ in the coefficient of g_4, the above expression yields

$$M_j = k_1^* Y_j^{\epsilon_m^y}(P/D)_j^{\epsilon_m^p}, \qquad (8.13')$$

where $k_1^* = \exp\{-\alpha_{10}/\beta_{12} + [(1 - \alpha_{11})/\beta_{12}]g_4\}$; $\epsilon_m^y = [1 - (\beta_{13}/\beta_{12})] > 0$; and $\epsilon_m^p = (-\beta_{15}/\beta_{12}) < 0$.

Export Demand

19   The general stochastic specification of the expression for export demand X^d of a country i by geographic market j as a function of import demand M of the geographic market and relative price R is

$$x^d_{ijt} = \alpha_{20} + \alpha_{21}m_{jt} + \alpha_{22}m_{j,t-1} + \alpha_{23}r_{ijt} + \alpha_{24}r_{ij,t-1} + \alpha_{25}x^d_{ij,t-1} + v_{2t}, \qquad (8.15)$$

where relative price $r_{ij} = \ln(P_{ij}/P_j)$, and $\alpha_{21}, \alpha_{22} > 0$; $\alpha_{23}, \alpha_{24} < 0$; and $0 < \alpha_{25} < 1$.

20   A proportional response of export demand to a change in the demand for imports in the foreign market in (3.10) implies that $\alpha_{21} + \alpha_{22} + \alpha_{25} = 1$. Substitute $(1 - \alpha_{21} - \alpha_{25})$ for α_{22} and add $(-m_{j,t-1})$ to both sides:

$$\Delta x^d_{ijt} = \alpha_{20} + \alpha_{21}\Delta m_{jt} + \beta_{22}(x^d_{ij} - m_j)_{t-1} + \alpha_{23}r_{ijt} + \alpha_{24}r_{ij,t-1} + v_{2t},$$

where $\beta_{22} = (\alpha_{25} - 1)$ with expected sign $-1 < \beta_{22} < 0$. Then transform the relative price variable so that a 'difference' formulation is nested in the levels form of the equation:

$$\Delta x^d_{ijt} = \alpha_{20} + \alpha_{21}\Delta m_{jt} + \beta_{22}(x^d_{ij} - m_j)_{t-1} + \alpha_{23}\Delta r_{ijt} + \beta_{24}r_{ij,t-1} + v_{2t}, \qquad (8.16)$$

where $\beta_{24} = (\alpha_{23} + \alpha_{24}) < 0$.

Derivation of Equations

Finally, apply a logit transformation to the regressand, as well as to the disequilibrium adjustment term, so that $0 < X_{ij}^d/M_j < 1$ for all t. As a first step in the transformation, subtract Δm from both sides:

$$\Delta(x_{ij}^d - m_j)_t = \alpha_{20} + \beta_{21}\Delta m_{jt} + \beta_{22}(x_{ij}^d - m_j)_{t-1} + \alpha_{23}\Delta r_{ijt} + \beta_{24}r_{ij,t-1} + v_{2t}^*,$$

where $\beta_{21} = (\alpha_{21} - 1)$, such that $\beta_{21} > 0$ if $\alpha_{21} > 1$; $\beta_{21} = 0$ if $\alpha_{21} = 1$; and $-1 < \beta_{21} < 0$ if $0 < \alpha_{21} < 1$. A logit transformation can then be applied to the regressand and the disequilibrium adjustment term:

$$\Delta z_{ijt} = \alpha_{20} + \beta_{21}\Delta m_{jt} + \beta_{22}^* z_{ij,t-1} + \alpha_{23}\Delta r_{ijt} + \beta_{24}r_{ij,t-1} + v_{2t}^*, \quad (8.17)$$

where

$$z_{ij} = \ln\left[\frac{X_{ij}^d/M_j}{1 - (X_{ij}^d/M_j)}\right] = \ln(X_i^d/X_k^d)_j,$$

since $X_{kj}^d = (M - X_i^d)_j$, with expected signs $\beta_{21} > -1$; $-1 < \beta_{22}^* < 0$; and α_{23}, $\beta_{24} < 0$.

21 The *steady-state solution* of (8.17) has constant import growth $g_3 = \Delta m_{jt}$ and $\Delta r_{ijt} = 0 = \Delta z_{ijt}$, so that it converges over time to

$$0 = \alpha_{20} + \beta_{21}g_3 + \beta_{22}^* z_{ij} + \beta_{24}r_{ij}$$

or

$$z_{ij} = -\alpha_{20}/\beta_{22}^* - (\beta_{21}/\beta_{22}^*)g_3 - (\beta_{24}/\beta_{22}^*)r_{ij}. \quad (8.18)$$

For simplicity of exposition, let $z = z_{ij}$ and let $s = X_{ij}/M_j$, so that $z = \ln[s/(1-s)]$. Then $e^z = s/(1-s)$. Solve for s in this expression by first dividing through by $1 - s$:

$$s = e^z(1 - s),$$

adding se^z to both sides:

$$e^z = s(1 + e^z),$$

and dividing through by $(1 + e^z)$:

$$s = \frac{1}{1 + e^{-z}},$$

or, in terms of original notations and after multiplying both sides of the expression by M_j,

$$X_{ij}^d = \frac{M_j}{1 + k_2 R_{ij}^{\beta_{24}/\beta_{22}^*}}, \quad (8.19)$$

where $k_2 = \exp[(\alpha_{20} + \beta_{21}g_3)/\beta_{22}^*]$.

22 The *price elasticity of export demand* can be derived from (8.19) when it is expressed as

$$\ln X_i^d = \ln M - \ln[1 + \exp(k_2 + \lambda \ln P_i - \lambda \ln P)] \quad (8.19')$$

where, for notational simplicity, subscript j is implicit and $\lambda = \beta_{24}/\beta_{22}^*$. Then the elasticity is given by

$$\epsilon_x^p = \frac{\partial \ln X_i^d}{\partial \ln P_i}$$

$$= \frac{\partial \ln M}{\partial \ln P_i} - \frac{\exp(k_2 + \lambda \ln P_i - \lambda \ln P)}{1 + \exp(k_2 + \lambda \ln P_i - \lambda \ln P)} \frac{\partial(k_2 + \lambda \ln P_i - \lambda \ln P)}{\partial \ln P_i}$$

$$= -\frac{e^{-z}}{1 + e^{-z}} \lambda$$

$$= -\frac{1}{1 + s/(1-s)} \lambda$$

$$= -\lambda(1-s). \tag{8.20}$$

23 The *import growth elasticity* of export demand is calculated by first multiplying the numerator and denominator of (8.19) by e^z, and then taking the partial derivative with respect to g (where, for notational simplicity, subscripts are implicit):

$$\frac{\partial}{\partial g}\left(\frac{e^z M}{1 + e^z}\right) = \frac{f_1 hM - fh_1 M}{h^2},$$

such that $f(g,r) = e^z$ and $h(g,r) = 1 + e^z$. Also, for simplicity of exposition, let $\gamma = -\beta_{21}/\beta_{22}^*$. The solution to the above expression is

$$\frac{\gamma fhM - \gamma f^2 M}{h^2} = \frac{\gamma fM(h-f)}{h^2}$$

$$= \frac{\gamma fM}{h^2}.$$

Finally, multiply both sides by $1/X_{ij}^d$ to obtain the import growth elasticity:

$$\kappa_2 = \frac{\partial X_{ij}^d}{\partial g_3} \frac{1}{X_{ij}^d} = \frac{\gamma e^z}{(1+e^z)^2} \frac{1}{s}$$

$$= \gamma(1-s). \tag{8.21}$$

24 The *elasticity of substitution for market shares* is derived in a manner similar to that for the price elasticity of export demand. First, divide both sides of (8.19) by M and express the relationship in the form

$$\ln(X_i^d/M) = -\ln[1 + \exp(k_2 + \gamma \ln R_i)], \tag{8.18''}$$

where, as before, subscript j is implicit and $\lambda = \beta_{24}/\beta_{22}^*$. Then this elasticity is given by

$$\epsilon_x^s = \frac{\partial \ln(X_i^d/M)}{\partial \ln R_i}$$

$$= -\frac{\exp(k_2 + \lambda \ln R_i)}{1 + \exp(k_2 + \lambda \ln R_i)} \frac{\partial(k_2 + \lambda \ln R_i)}{\partial \ln R_i}$$

$$= -\frac{1}{1 + s/(1-s)} \lambda$$

$$= -\lambda(1-s). \tag{8.25}$$

Export Supply

25 The stochastic difference equation for export supply X^s of country i to its geographic market j as a function of its export price P, relative to the general price level D, as well as of major disturbances W and of a secular trend T, is given by

$$x^s_{ijt} = \alpha_{40} + \alpha_{41}x^s_{ij,t-1} + \alpha_{42}x^s_{ij,t-2} + \sum_{k=0}^{n}\alpha_{43+k}(P_{ij} - d_i)_{t-k}$$
$$+ \beta_{45}T + \beta_{46}W_{it} + v_{4t}, \qquad (9.40)$$

where $\alpha_{43+k} > 0$, $\beta_{45} \leq 0$ and $\beta_{46} < 0$. If $\alpha_{42} = 0$, then $0 < \alpha_{41} < 1$; if $\alpha_{42} \neq 0$, then $0 < \alpha_{41} < 2$; $-1 < \alpha_{42} < 1$; $(1 - \alpha_{41} - \alpha_{42}) > 0$; and $\alpha^2_{41} \geq -4\alpha_{42}$.

Transform the above equation in such a way that 'differences' formulations of the variables are nested in the levels form of the equation:

$$\Delta x^s_{ijt} = \alpha_{40} + \beta_{41}x^s_{ij,t-1} + \alpha_{42}x^s_{ij,t-2} + \sum_{k=0}^{n-1}\alpha_{43+k}\Delta(P_{ij} - d_i)_{t-k}$$
$$+ \sum_{k=1}^{n}\gamma_{44+k}(P_{ij} - d_i)_{t-k} + \beta_{45}T + \beta_{46}W_{it} + v_{4t}, \qquad (9.41)$$

where $\beta_{41} = (\alpha_{41} - 1)$, $\sum_{k=1}^{n}\gamma_{44+k} = \sum_{k=0}^{n-2}\alpha_{43+k}$ and $\gamma_{44+n} = \alpha_{43+n-1} + \alpha_{43+n}$. The expected signs are $\alpha_{43+k} \leq 0$, $\gamma_{44+k} > 0$, such that $(\alpha_{43+k} + \gamma_{44+k}) > 0$, and for the lagged dependent variable $-1 < \beta_{41} < 1$; $-1 < \alpha_{42} < 1$; $\beta_{41} + \alpha_{41} < 0$; and $(1 + \beta_{41})^2 \geq \alpha_{42}$.

Since $\Delta \ln(P_{ij}/D_i) = 0$ implies $\Delta x_{ij} = 0$, the long-run relationship in (9.41) is

$$X^s_{ij} = k_3(P_{ij}/D_i)^{\gamma_1} \exp(\sigma_1 T + \sigma_2 W), \qquad (9.42)$$

where $k_3 = \exp(-\alpha_{40}/\beta_{41} + \alpha_{42})$, $\gamma_1 = -\sum_{k=1}^{n}\gamma_{44+k}/(\beta_{41} + \alpha_{42}) > 0$, $\sigma_1 = \beta_{45}/(\beta_{41} + \alpha_{42}) \geq 0$, and $\sigma_2 = -\beta_{46}/(\beta_{41} + \alpha_{42}) < 0$.

Import Price

26 The first-order stochastic difference equation for the relationship between the import price of a geographic market P_j and the world market price P is given by

$$p_{jt} = \alpha_{60} + \alpha_{61}p_t + \alpha_{62}p_{t-1} + \alpha_{63}p_{j,t-1} + v_{6t}, \qquad (9.49)$$

where $\alpha_{61}, \alpha_{62} > 0$ and $0 < \alpha_{63} < 1$.

Since a proportional relationship between P_j and P requires that $(\alpha_{61} + \alpha_{62} + \alpha_{63}) = 1$, first substitute $(1 - \alpha_{61} - \alpha_{63})$ for α_{62}:

$$p_{jt} = \alpha_{60} + \alpha_{61}p_t + (1 - \alpha_{61} - \alpha_{63})p_{t-1} + \alpha_{63}p_{j,t-1} + v_{6t}.$$

Then add $(-p_{j,t-1})$ to both sides to obtain the following ECM specification:

$$\Delta p_{jt} = \alpha_{60} + \alpha_{61}\Delta p_t + \beta_{62}(p_j - p)_{t-1} + v_{6t}, \qquad (9.50)$$

where $\beta_{62} = (\alpha_{63} - 1)$, with expected signs $\alpha_{61} > 0$ and $-1 < \beta_{62} < 0$.

27 In a steady-state path $\Delta p_j = g_6 = \Delta p$,

$$g_6 = \alpha_{60} + \alpha_{61}g_6 + \beta_{62}p_j - \beta_{62}p,$$

which yields the long-run equilibrium solution
$$P_j = k_6 P, \tag{9.51}$$
where $k_6 = \exp\{-[\alpha_{60} + (\alpha_{61} - 1)g_6]/\beta_{62}\}$.

Part V: Commodity Trade Policies

28 The *foreign income multiplier* for a country's exports X is derived from the equilibrium identity $X^d = X^s = X$. In order to simplify notation, the subscripts for the exporter and its geographic market are implicit. The equilibrium identity for the export demand and supply functions f and g, respectively, can be expressed as

$$f\{M(Y), R[M(Y)]\} - X = 0$$
$$g[P(Y)] - X = 0,$$

where, for example, the country's relative export price R is shown to depend on foreign market imports M, which in turn depends on income Y of the foreign market.

Total differentiation of the two identities with respect to income, after rearrangement of terms, yields

$$\frac{\partial X^d}{\partial P}\frac{dP}{dY} - \frac{dX}{dY} = -\frac{\partial X^d}{\partial Y}$$

$$\frac{\partial X^s}{\partial P}\frac{dP}{dY} - \frac{dX}{dY} = 0$$

which in matrix notation may be written as

$$\begin{bmatrix} \frac{\partial X^d}{\partial P} - 1 \\ \frac{\partial X^s}{\partial P} - 1 \end{bmatrix} \begin{bmatrix} \frac{dP}{dY} \\ \frac{dX^d}{dY} \end{bmatrix} = \begin{bmatrix} -\frac{\partial X^d}{\partial Y} \\ 0 \end{bmatrix}.$$

By Cramer's rule, the foreign income multiplier is found to be

$$\frac{dX}{dY} = \frac{\frac{\partial X^s}{\partial P}\frac{\partial X^d}{\partial Y}}{\frac{\partial X^s}{\partial P} - \frac{\partial X^d}{\partial P}},$$

whose expected sign can be evaluated when the expression is expanded:

$$\frac{dX}{dY} = \frac{\left(\frac{\partial X^s}{\partial P}\frac{dP}{dY}\right)\left(\frac{\partial X^d}{\partial M}\frac{\partial M}{\partial Y} + \frac{\partial X^d}{\partial R}\frac{\partial R}{\partial M}\frac{\partial M}{\partial Y}\right)}{\frac{\partial X^s}{\partial P}\frac{dP}{dY} - \frac{\partial X^d}{\partial R}\frac{\partial R}{\partial M}\frac{\partial M}{\partial P}\frac{dP}{dY} - \frac{\partial X^d}{\partial M}\frac{\partial M}{\partial P}\frac{dP}{dY}} > 0$$

since $\partial R/\partial M < 0$.

29 The impact of an *export subsidy* on the marginal cost of an exporter is

$$\frac{\partial C}{\partial X} = (1/\tau)k_3^{1/\tau}X^{(1/\tau)-1}D - S.$$

The country will export at the level at which marginal revenue is equal to marginal cost:

$$(1 + 1/\epsilon_x^p)P - (1/\tau)k_3^{1/\tau}X^{(1/\tau)-1}D + S = 0.$$

Solve for X to obtain the amount of exports that will be supplied with a subsidy:

$$X_i^s = k_4' \left[\frac{(1 + 1/\epsilon_x^p)P_i + S}{D_i} \right]^\gamma t^{\phi_1} w^{\phi_2}, \tag{19.13a}$$

where $k_4' = k_4/(1 + 1/\epsilon_x^p)^\gamma$.

In order to find the effect of a change in a subsidy on export volume and price, set the export demand equal to the export supply, and solve for price. Then substitute the solution for price into the export demand or supply function to obtain the amount exported. Taking the partial derivative of X with respect to s and dividing by X/s yields

$$\frac{\partial X}{\partial s}\frac{s}{X} = -\frac{s\epsilon_x^p}{(1 - \epsilon_x^p/\gamma)(1 + 1/\epsilon_x^p + s)}, \tag{19.14a}$$

and taking the partial derivative of P with respect to s and dividing by P/s yields

$$\frac{\partial P}{\partial s}\frac{s}{P} = -\frac{s}{(1 - \epsilon_x^p/\gamma)(1 + 1/\epsilon_x^p + s)}. \tag{19.14b}$$

The same procedure is followed to find the effect on exports produced by a tax.

References

ABAELU, J. N., and MANDERSCHEID, L. V. (1968), 'US Import Demand for Green Coffee by Variety', *American Journal of Agricultural Economics* 50: 232–42.

ADAMS, F. G. (1978), 'Primary Commodity Markets in a World Model System', in F. G. Adams and S. A. Klein (eds.), *Stabilizing World Commodity Markets: Analysis, Practice, and Policy*. Lexington, Mass.: D. C. Heath.

—— (1979), 'Integrating Commodity Models into LINK', in J. A. Sawyer (ed.), *Modelling the International Transmission Mechanism*. Amsterdam: North-Holland.

—— and BEHRMAN, J. R. (1976), *Econometric Models of World Agricultural Commodity Markets*. Cambridge, Mass.: Ballinger.

—— and —— (1982), *Commodity Exports and Economic Development*. Lexington, Mass.: D. C. Heath.

ADLER, F. M. (1970), 'The Relationship Between the Income and Price Elasticities of Demand for United States Exports', *Review of Economics and Statistics* 52: 313–19.

AKIYAMA, T., and DUNCAN, R. C. (1982), *Analysis of the World Coffee Market*. World Bank Staff Commodity Working Paper no. 7, Commodities and Export Projections Division, International Bank for Reconstruction and Development, Washington, DC.

ALLEN, R. G. D. (1956), *Mathematical Economics*. London: St Martin's Press.

ANDERSON, T. W. (1971), *The Statistical Analysis of Time Series*. New York: John Wiley.

ARMINGTON, P. S. (1969), 'A Theory of Demand for Products Distinguished by Place of Production', *IMF Staff Papers* 16: 159–78.

ARTUS, J. R., and SOSA, S. C. (1978), 'Relative Price Effects on Export Performance: The Case of Nonelectric Machinery', *IMF Staff Papers* 25: 25–47.

ASKARI, H., and CUMMINGS, J. T. (1977), 'Estimating Agricultural Supply Response with the Nerlove Model: A Survey', *International Economic Review* 18(2): 257–92.

ATKIN, M. J. (1983), *Commodities and Financial Futures Yearbook 1983/4*. London: Landell Mills Commodities Studies Ltd in association with International Commodities Clearing House Ltd.

AUGUSTO, S., and POLLAK, P. K. (1981), 'Structure and Prospects of the World Fats and Oils Economy', in *World Bank Commodity Models*, 1. Commodity and Export Projections Division, International Bank for Reconstruction and Development, Washington, DC.

BALASSA, B. (1978), 'Export Incentives and Export Performance in Developing Countries: A Comparative Analysis', *Weltwirtschaftliches Archiv* 114: 24–61.

—— (1979), 'Export Composition and Export Performance in the Industrialized Countries 1953–71', *Review of Economics and Statistics* 61: 604–7.

—— (1982), *Development Strategies in Semi-Industrial Economies*. Baltimore: Johns Hopkins University Press.
BALDWIN, R. E. (1962), 'Implications of Structural Changes in Commodity Trade', in *Factors Affecting the United States Balance of Payments*, US Congress Joint Economic Committee. Washington, DC: US Government Printing Office.
—— and KRUGMAN, P. R. (1988), 'Market Access and International Competition: A Simulation Study of 16K Random Access Memories', in R. C. Feenstra (ed.), *Empirical Methods for International Trade*. Cambridge, Mass.: MIT Press.
BALE, M. D., and LUTZ, E. (1979), 'The Effects of Trade Intervention on International Price Instability', *American Journal of Agricultural Economics* 61: 512–16.
—— and RYAN, M. E. (1977), 'Wheat Protein Premiums and Price Differentials', *American Journal of Agricultural Economics* 59: 530–2.
BANERJI, R. (1974), 'The Export Performance of Less Developed Countries: A Constant Market Share Analysis', *Weltwirtschaftliches Archiv* 110(3): 447–81.
BARKER, T. S. (1969), 'Aggregation Error and Estimation of the UK Import Demand Function', in K. Hilton and D. F. Heathfield (eds.), *The Econometric Study of the United Kingdom*. London: Macmillan.
BECKER, G. S. (1965), 'A Theory of the Allocation of Time', *Economic Journal* 75: 493–517.
BEHRMAN, J. R. (1977), *International Commodity Agreements: An Evaluation of the UNCTAD Integrated Commodity Programme*. Washington, DC: Overseas Development Council.
BHAGWATI, J. N. (1978), *Foreign Trade Regimes and Economic Development: Anatomy and Consequences of Exchange Control Regimes*. Lexington, Mass.: Ballinger for the National Bureau of Economic Research.
BISMUT, C., and OLIVEIRA-MARTINS, J. (1987), *Market Shares, Price Competitiveness and Product Differentiation*, Working Paper no. 87.07, Centre d'Etudes Prospectives et d'Informations Internationales, Paris.
BOND, M. E. (1987), 'An Econometric Study of Primary Commodity Exports from Developing Country Regions to the World', *IMF Staff Papers* 34(2): 191–227.
BOYLAN, T. A., CUDDY, M. P., and O'MUIRCHEARTAIGH, I. O. (1980), 'The Functional Form of the Aggregate Import Demand Equation', *Journal of International Economics* 10: 561–6.
BOX, G. E. P., and JENKINS, G. M. (1976), *Time Series Analysis*. San Francisco: Holden-Day.
BRANDER, J. A., and SPENCER, B. J. (1983), 'Export Subsidies and International Market Share Rivalry', unpublished manuscript.
—— and —— (1984), 'Tariff Protection and Imperfect Competition', in H. Kierzkowski (ed.), *Monopolistic Competition and International Trade*. Oxford: Clarendon Press.
—— and —— (1985), 'Export Subsidies and International Market Share Rivalry', *Journal of International Economics* 18: 83–100.
BRORSEN, B. W., CHAVAS, J., and GRANT, W. R. (1984), 'Dynamic Relationships of Rice Import Prices in Europe', *European Review of Agricultural Economics* 11(1): 29–42.

BROWN, M., and HEIEN, D. (1972), 'The S-Branch Utility Tree: A Generalization of the Linear Expenditure System', *Econometrica* 40(4): 737–47.

CAIRNCROSS, A. K. (1961), 'International Trade and Economic Development', *Economica* 28(3): 235–51.

CHAMBERLIN, E. H. (1933), *The Theory of Monopolistic Competition*. Cambridge, Mass.: Harvard University Press.

CHANG, H. (1977), 'Functional Forms and the Demand for Meat in the United States', *Review of Economics and Statistics* 59(3): 355–9.

CHIANG, A. C. (1984), *Fundamental Methods of Mathematical Economics*. New York: McGraw-Hill.

Chicago Board of Trade (CBT) (1982), *Grains: Production, Processing, Marketing*. Chicago.

CHU, K. (1978), 'Short-Run Forecasting of Commodity Prices: An Application of Autoregressive Moving Average Models', *IMF Staff Papers* 25(1): 90–111.

—— and KRISHNAMURTY, K. (1978), 'Temporal Relationship Between International Prices and Export Unit Values of Primary Commodities', Working Paper DM/78/32, International Monetary Fund, Washington, DC.

—— and MORRISON, T. K. (1984), 'The 1981–82 Recession and Non-Oil Primary Commodity Prices: An Econometric Analysis', *IMF Staff Papers* 31(1): 93–140.

CLINE, W. R., et al. (1978), *Trade Negotiations in the Tokyo Round: A Quantitative Assessment*. Washington, DC: Brookings Institution.

COHEN, B. I., and SISLER, D. (1971), 'Exports of Developing Countries in the 1960s', *Review of Economics and Statistics* 53: 354–62.

COLLINS, K. J. (1977), 'An Economic Analysis of Export Competition in the World Coarse Grain Market: A Short-Run Constant Elasticity of Substitution Approach', Ph.D. thesis, North Carolina State University.

Committee of Twenty (1974), *Outline of International Monetary Reform*, Washington, DC.

COOPER, R. D., and LAWRENCE, R. Z. (1975), 'The 1972–75 Commodity Boom', *Brookings Papers on Economic Activity* 3: 671–723.

COPPOCK, J. D. (1962), *International Economic Instability*. New York: McGraw-Hill.

CROUHY-VEYRAC, L., CROUHY, M., and MELITZ, J. (1982), 'More About the Law of One Price', *European Economic Review* 18: 325–44.

CURRIE, D. (1981), 'Some Long-Run Features of Dynamic Time Series Models', *Economic Journal* 91: 704–15.

CURTIS, B. N., et al. (1987), *Cocoa: A Trader's Guide*. Geneva: International Trade Centre UNCTAD/GATT.

DAS, S. P. (1982), 'Economies of Scale, Imperfect Competition, and the Pattern of Trade', *Economic Journal* 92: 684–93.

DAVIDSON, J. E. H., HENDRY, D. F., SRBA, F., and YEO, S. (1978), 'Econometric Modelling of the Aggregate Time-Series Relationship Between Consumers' Expenditure and Income in the United Kingdom', *Economic Journal* 88: 661–92.

—— and KEIL, M. (1981), 'An Econometric Model of the Money Supply and

Balance of Payments in the United Kingdom', International Centre for Economics and Related Disciplines, London School of Economics.

DEARDORFF, A. V. (1984), 'Testing Trade Theories and Predicting Trade Flows', in R. W. Jones and P. B. Kenen (eds.), *Handbook of International Economics*, 1. Amsterdam: North-Holland.

—— and STERN, R. M. (1987), 'Current Issues in Trade Policy: An Overview', in R. M. Stern (ed.), *US Trade Policies in a Changing World Economy*. Cambridge, Mass.: MIT Press.

DEATON, A., and MUELLBAUER, J. (1980), *Economics and Consumer Behavior*. Cambridge: University Press.

DEGRAFF, H. (1960), *Beef: Production and Distribution*. Norman, Okla.: University of Oklahoma Press.

DEMLER, F. R., and TILTON, J. E. (1980), 'Modeling International Trade Flows in Mineral Markets, with Implications for Latin America's Trade Policies', in W. C. Labys, M. I. Nadiri, and J. Núñez del Arco (eds.), *Commodity Markets and Latin American Development: A Modeling Approach*. Cambridge, Mass.: Ballinger Publishing Company for National Bureau of Economic Research.

DENT, W. T. (1967), 'Application of Markov Analysis to International Wool Flows', *Review of Economics and Statistics* 49: 613-16.

DIAZ-ALEJANDRO, C. F. (1975), 'Trade Policies and Economic Development', in P. B. Kenen (ed.), *International Trade and Finance*. Cambridge: University Press.

DIXIT, A. K. (1984), 'International Trade Policy for Oligopolistic Industries', *Economic Journal Conference Papers* 94: 1-16.

—— (1987), 'Issues of Strategic Trade Policy for Small Countries', *Scandinavian Journal of Economics* 89(3): 349-67.

—— (1988), 'Optimal Trade and Industrial Policies for the US Automobile Industry', in R. C. Feenstra (ed.), *Empirical Methods for International Trade*. Cambridge, Mass.: MIT Press.

—— and GROSSMAN, G. M. (1984), 'Targeted Industrial Policy with Several Oligopolistic Industries', unpublished manuscript, Princeton University.

—— and NORMAN, V. (1980), *Theory of International Trade*. Welwyn, Herts: James Nisbet and Cambridge University Press.

—— and STIGLITZ, J. E. (1977), 'Monopolistic Competition and Optimum Product Diversity', *American Economic Review* 67(3): 297-308.

DOMOWITZ, I., and ELBADAWI, I. (1987), 'An Error-Correction Approach to Money Demand: The Case of Sudan', *Journal of Development Economics* 26(2): 257-74.

DONNENFELD, S. (1984), 'Imperfect Information and Trade in Differentiated Products', unpublished manuscript, University of Haifa and New York University.

EATON, J., and GROSSMAN, G. M. (1986), 'Optimal Trade and Industrial Policy Under Oligopoly', *Quarterly Journal of Economics* 101(2): 383-406.

EDWARDS, R., and PARIKH, A. (1976), 'A Stochastic Policy Simulation of the World Coffee Economy', *American Journal of Agricultural Economics* 156: 152-60.

ENGLE, R. F., and GRANGER, C. W. J. (1987), 'Co-Integration and Error

Correction: Representation, Estimation, and Testing', *Econometrica* 55(2): 251–76.

ERB, G. F., and FISHER, B. S. (1977), 'US Commodity Policy: What Response to Third World Initiative?', *Law and Policy in International Business* 9(2): 479–513.

FAIR, R. (1970), 'The Estimation of Simultaneous Equation Models with Lagged Endogenous Variables and First Order Serially Correlated Errors', *Econometrica* 38(3): 507–16.

FEENSTRA, R. C. (ed.) (1988), *Empirical Methods for International Trade*. Cambridge, Mass.: MIT Press.

FINGER, M., and DeROSA, D. (1978), 'Commodity Price Stabilization and the Ratchet Effect', *The World Economy* 1(2): 195–204.

FISHER, F. M., COOTNER, P. H., and BAILY, M. (1972), 'An Econometric Model of the World Copper Industry', *Bell Journal of Economics and Management Science* 3: 568–609.

FLAM, H., and HELPMAN, E. (1987), 'Industrial Policy under Monopolistic Competition', *Journal of International Economics* 32(1/2): 79–102.

FLEMING, J. M., and TSIANG, S. C. (1956), 'Changes in Competitive Strength and Export Shares in Major Industrial Countries', *IMF Staff Papers* 5: 218–48.

FRANZ, J., STENBERG, B., and STRONGMAN, J. (1986), *Iron Ore: Global Prospects for the Industry, 1985–95*, Industry and Finance Series Vol. 12. Washington, DC: International Bank for Reconstruction and Development.

FRY, J. (1985), 'Sugar: Aspects of a Complex Commodity Market', Division Working Paper no. 1985-1, Commodities and Export Projections Division, International Bank for Reconstruction and Development, Washington, DC.

GANA, J. L., et al. (1979), 'Alternative Approaches to Linkage of National Econometric Models', in J. A. Sawyer (ed.), *Modelling the International Transmission Mechanism*. Amsterdam: North-Holland.

GANNON, C. A. (1977), 'Product Differentiation and Locational Competition in Spatial Markets', *International Economic Review* 18(2): 293–322.

GARDINER, W. H., and DIXIT, P. M. (1987), *Price Elasticity of Export Demand: Concepts and Estimates*, Foreign Agricultural Economic Report no. 228, Economic Research Service. Washington, DC: US Department of Agriculture.

GARDNER, B. (1985), 'Estimating Effects of Commodity Policy and Trade Liberalization in Agriculture', background paper for *World Development Report 1986*. Washington, DC: International Bank for Reconstruction and Development.

GERACI, V., and PREWO, W. (1982), 'An Empirical Demand and Supply Model of Multilateral Trade', *Review of Economics and Statistics* 64(3): 432–41.

GHOSH, S., GILBERT, C. L., and HUGHES HALLETT, A. J. (1987), *Stabilizing Speculative Commodity Markets*. Oxford: Clarendon Press.

GILBERT, C. L., and DI MARCHI, N. (eds.) (1989), *The History and Methodology of Econometrics*. Oxford: University Press.

—— and PALASKAS, T. B. (1990), 'Modelling Expectations Formation in Primary Commodity Markets', in D. Sapsford and A. L. Winters (eds.),

Primary Commodity Prices: Economic Models and Economic Policy. Cambridge: University Press.
GINSBURG, A. L. (1969), *American and British Regional Export Determinants*. Amsterdam: North-Holland.
—— and STERN, R. M. (1965), 'The Determination of the Factors Affecting American and British Exports in the Inter-War and Post-War Period', *Oxford Economic Papers* 17: 263–78.
GLEZAKOS, C. (1973), 'Export Instability and Economic Growth: A Statistical Verification', *Economic Development and Cultural Change* 21(4): 670–8.
GOLDBERGER, A. S. (1964), *Econometric Theory*. New York: John Wiley.
GOLDSTEIN, M., and KHAN, M. S. (1976), 'Large Versus Small Price Changes and the Demand for Imports', *IMF Staff Papers* 23: 200–25.
—— and —— (1978), 'The Supply and Demand for Exports: A Simultaneous Approach', *Review of Economics and Statistics* 60(2): 275–86.
—— and —— (1982), 'Effects of Slowdown in Industrial Countries on Growth in Non-Oil Developing Countries', Occasional Paper no. 12, International Monetary Fund, Washington, DC.
—— and —— (1985), 'Income and Price Effects in Foreign Trade', in R. W. Jones and P. B. Kenen (eds.), *Handbook of International Economics*, 2. Amsterdam: North-Holland.
GOODWIN, T. H., and SHEFFRIN, S. M. (1982), 'Testing the Rational Expectations Hypothesis in an Agricultural Market', *Review of Economics and Statistics* 64(4): 658–67.
GORMAN, W. M. (1959), 'Separable Utility and Aggregation', *Econometrica* 27: 469–81.
—— (1980), 'A Possible Procedure for Analyzing Quality Differentials in the Egg Market', *Review of Economic Studies* 47: 843–56.
GRANGER, C. W. J. (1986), 'Developments in the Study of Cointegrated Economic Variables', *Oxford Bulletin of Economics and Statistics* 48(3): 213–28.
—— and NEWBOLD, P. (1974), 'Spurious Regressions in Econometrics', *Journal of Econometrics* 2: 111–20.
—— and WEISS, A. A. (1983), 'Time Series Analysis of Error-Correcting Models', in S. Karlin, T. Amemiya, and L. A. Goodman (eds.), *Studies in Econometrics, Time Series, and Multivariate Statistics*. New York: Academic Press.
GRAY, H. P., and MARTIN, J. J. (1980), 'The Meaning and Measurement of Product Differentiation in International Commodity Trade', *Weltwirtschaftliches Archiv* 10(1): 141–56.
GREENAWAY, D. (1984), 'The Measurement of Product Differentiation in Empirical Studies of Trade Flows', in H. Kierzkowski (ed.), *Monopolistic Competition and International Trade*. Oxford: Clarendon Press.
GREENHUT, M. L., NORMAN, G., and HUNG, C.-S. (1987), *The Economics of Imperfect Competition: A Spatial Approach*. Cambridge: University Press.
GREENWALD, B. C., and STIGLITZ, J. E. (1988), 'Financial Market Imperfections and Business Cycles', Working Paper no. 2494, National Bureau of Economic Research, New York.
GRENNES, T., JOHNSON, P. R., and THURSBY, M. (1978), *The Economics of*

World Grain Trade. New York: Praeger.
GRIFFITH, G. and MEILKE, K. (1982), *A Structural Econometric Model of the World Markets for Rapeseed, Soybeans and Their Products*. Guelph, Ontario: School of Agricultural Economics and Extension Education, University of Guelph.
GRILICHES, Z. (1967), 'Distributed Lags: A Survey', *Econometrica* 35(1): 16–49.
GRILLI, E., and YANG, C. (1988), 'Primary Commodity Prices, Manufactured Goods Prices, and the Terms of Trade of Developing Countries: What the Long Run Shows', *World Bank Economic Review* 2(1): 1–47.
GRISSA, A. (1976), *Structure of the International Sugar Market and Its Impact on Developing Countries*. Paris: Organisation for Economic Co-operation and Development.
GROSSMAN, G. M. (1982), 'Import Competition from Developed and Developing Countries', *Review of Economics and Statistics* 64(2): 271–81.
GRUBEL, H. G., and LLOYD, P. J. (1975), *Intra-Industry Trade: The Theory and Measurement of International Trade in Differentiated Products*. New York: John Wiley.
GRUNWALD, J., and MUSGROVE, P. (1970), *Natural Resources in Latin American Development*. Baltimore: Johns Hopkins Press for Resources for the Future.
GUSTAFSON, R. L. (1958), *Carryover Levels for Grains*, Technical Bulletin no. 1178. Washington DC: US Department of Agriculture.

HANSON, J. R. (1977), 'More on Trade as a Handmaiden of Growth', *Economic Journal* 87: 554–7.
HARBERGER, A. C. (1957), 'Some Evidence on the International Price Mechanism', *Journal of Political Economy* 65(6): 506–21.
HART, O. D. (1985*a*), 'Monopolistic Competition in the Spirit of Chamberlin: Special Results', *Economic Journal* 95: 889–908.
—— (1985*b*), 'Monopolistic Competition in the Spirit of Chamberlin: A General Model', *Review of Economic Studies* 52: 529–46.
HARVEY, A. C. (1981), *The Econometric Analysis of Time Series*. Oxford: Philip Allan.
HAYNES, S. E., and STONE, J. A. (1983), 'Specification of Supply Behavior in International Trade', *Review of Economics and Statistics* 65(4): 626–94.
HAZILLA, M., and KOPP, R. J. (1984), 'Assessing US Vulnerability to Raw Material Supply Disruptions', *Southern Economic Journal* 51(2): 341–55.
HELLIWELL, J. F., and PADMORE, T. (1985), 'Empirical Studies of Macroeconomic Interdependence', in R. W. Jones and P. B. Kenen (eds.), *Handbook of International Economics*, 2. Amsterdam: North-Holland.
HELPMAN, E. (1981), 'International Trade in the Presence of Product Differentiation, Economics of Scale, and Monopolistic Competition: A Chamberlin–Heckscher–Ohlin Approach', *Journal of International Economics* 11(3): 305–40.
—— (1984), 'Increasing Returns, Imperfect Markets, and Trade Theory', in R. W. Jones and P. B. Kenen (eds.), *Handbook of International Economics*, 1. Amsterdam: North-Holland.
—— and KRUGMAN, P. R. (1985), *Market Structure and Foreign Trade: Increasing Returns, Imperfect Competition, and the International Economy*.

Cambridge, Mass.: MIT Press.

HENDLER, R. (1975), 'Lancaster's New Approach to Consumer Demand and Its Limitations', *American Economic Review* 65: 194–9.

HENDRY, D. F. (1979), 'Predictive Failure and Econometric Modelling in Macro-economics: The Transactions Demand for Money', in P. Ormerod (ed.), *Economic Modelling*. London: Heinemann Educational Books.

—— (1980), 'Econometrics: Alchemy or Science?' *Economica* 47: 387–406.

—— (1983), 'Econometric Modelling: The "Consumption Function" in Retrospect', *Scottish Journal of Political Economy* 30(3): 193–219.

—— (1986), 'Econometric Modelling with Cointegrated Variables: An Overview', *Oxford Bulletin of Economics and Statistics* 48(3): 201–12.

—— and MIZON, G. E. (1978), 'Serial Correlation as a Convenient Simplification not a Nuisance: A Comment on a Study of the Demand for Money by the Bank of England', *Economic Journal* 88: 549–63.

—— PAGAN, A. R., and SARGAN, J. D. (1984), 'Dynamic Specification', in Z. Griliches and M. Intrilingator (eds.), *Handbook of Econometrics*, 2. Amsterdam: North-Holland.

—— and RICHARD, J.-F. (1981), 'The Econometric Analysis of Economic Time Series', CORE Discussion Paper no. 8122, Université Catholique de Louvain-la-Neuve, Belgium.

—— and VON UNGERN-STERNBERG, T. (1980), 'Liquidity and Inflation Effects on Consumers' Behavior', in A. S. Deaton (ed.), *Essays in the Theory and Measurement of Consumers' Behavior*. Cambridge: University Press.

—— HENRY, S. G. B., and ORMEROD, P. (1979), 'Rational Expectations in a Wage-Price Model of the UK, 1972–79', paper presented to the SSRC Sussex Conference on Rational Expectations.

HICKMAN, G., and LAU, L. J. (1973), 'Elasticities of Substitution and Export Demands in a World Trade Model', *European Economic Review* 4: 347–80.

HICKS, J. R. (1932), *The Theory of Wages*. London: Macmillan.

HITIRIS, T., and PETOUSSIS, E. (1984), 'Price and Tariff Effect in a Dynamic Specification of the Demand for Imports', *Applied Economics* 16(1): 15–24.

HORN, H. (1984), 'Product Diversity, Trade, and Welfare', in H. Kierzkowski (ed.), *Monopolistic Competition and International Trade*. Oxford: Clarendon Press.

HOTELLING, H. (1929), 'Stability in Competition', *Economic Journal* 39(153): 41–57.

HOUCK, J. P., and MANN, J. S. (1968), *An Analysis of Domestic and Foreign Demand for U.S. Soybeans and Soybean Products*, Technical Bulletin no. 256, Agricultural Experiment Station. Minneapolis: University of Minnesota Press.

HOUTHAKKER, H. S., and MAGEE, S. P. (1969), 'Income and Price Elasticities in World Trade', *Review of Economics and Statistics* 51(2): 111–25.

HUFBAUER, G. C. (1970), 'The Impact of National Characteristics and Technology of the Commodity Composition of Trade in Manufactured Goods', in R. Vernon (ed.), *The Technology Factor in International Trade*. New York: Columbia Press for National Bureau of Economic Research.

HUMPHREY, D. H. (1976), 'Disaggregated Import Functions for the UK, West Germany and France', *Oxford Bulletin of Economics and Statistics* 38(4): 281–97.

HUNTER, L. C., and MARKUSEN, J. R. (1988), 'Per-Capita Income as a Determinant of Trade', in R. C. Feenstra (ed.), *Empirical Methods for International Trade*. Cambridge, Mass.: MIT Press.

HUSTED, S., and KOLLINTZAS, T. (1984), 'Import Demand with Rational Expectations: Estimates for Bauxite, Cocoa, Coffee, and Petroleum', *Review of Economics and Statistics* 66: 608-18.

HWA, E. C. (1985), 'A Model of Price and Quantity Adjustments in Primary Commodity Markets', *Journal of Policy Modeling* 7(2): 305-38.

HYLLEBERG, S., and MIZON, G. E. (1989), 'Cointegration and Error Correction Mechanisms', *Economic Journal Conference Papers* 99: 114-25.

International Bank for Reconstruction and Development (IBRD) (1981), *Beef Handbook*, Commodities and Export Projections Division, Economic Analysis and Projections Department, Washington, DC.

—— (1989), *World Debt Tables*, Washington, DC.

International Monetary Fund (IMF) (1989), *International Financial Statistics: Yearbook* 42(2), Washington, DC.

IRELAND, N. J. (1987), *Product Differentiation and Non-Price Competition*. Oxford: Basil Blackwell.

ISARD, P. (1977), 'How Far Can We Push the "Law of One Price"?', *American Economic Review* 67(5): 942-8.

JARVIS, L. S. (1974), 'Cattle as Capital Goods and Ranchers as Portfolio Managers: An Application to the Argentine Cattle Sector', *Journal of Political Economy* 80: 489-520.

JEPMA, C. J. (1986), *Extensions of the Market Shares Analysis with an Application to Long-Term Export Data of Developing Countries*. New Delhi: Indian Economic Association.

JONES, L. E. (1984), 'A Competitive Model of Commodity Differentiation', *Econometrica* 52(2): 507-30.

JUNG, W. S., and LEE, G. (1986), 'The Effectiveness of Export Promotion Policies: The Case of Korea', *Weltwirtschaftliches Archiv* 122(2): 340-57.

JUNZ, H. B., and RHOMBERG, R. R. (1965), 'Prices and Export Performance of Industrial Countries, 1953-63', *IMF Staff Papers* 12: 224-69.

—— and —— (1973), 'Price Competitiveness in Export Trade Among Industrial Countries', *American Economic Review* 63: 412-18.

KALDOR, N. (1976), 'Inflation and Recession in the World Economy', *Economic Journal* 86: 703-14.

KATRAK, H. (1975), 'An Application of Lancaster's Consumer Demand Theory to Some Recent Hypotheses of International Trade', in M. Parkin and A. R. Nobay (eds.), *Contemporary Issues in Economics*. Manchester: University Press.

KATZNER, D. W. (1968), 'A Note on the Differentiability of Consumer Demand Functions', *Econometrica* 36(2): 415-18.

KAWATA, G. (1975), 'The World Banana Industry: Current Situation and Prospects', unpublished manuscript, Inter-American Development Bank, Washington, DC.

KHAN, M. S. (1974), 'Import and Export Demand in Developing Countries', *IMF Staff Papers* 21: 678–93.
—— and Ross, K. Z. (1975), 'Cyclical and Secular Income Elasticities of the Demand for Imports', *Review of Economics and Statistics* 57: 357–61.
—— and —— (1977), 'The Functional Form of the Aggregate Import Demand Equation', *Journal of International Economics* 7: 149–60.
KIRMANI, N., MOLAJONI, L., and MAYER, T. (1984), 'Effects of Increased Market Access on Exports of Developing Countries', *IMF Staff Papers* 31: 661–84.
KLEIN, L. R. (1974), *A Textbook of Econometrics*. Englewood Cliffs, NJ: Prentice-Hall.
—— and RUBIN, H. (1948), 'A Constant Utility Index of the Cost of Living', *Review of Economic Studies* 15: 84–7.
KNUDSEN, O., and PARNES, A. (1975), *Trade Instability and Economic Development*. Lexington, Mass.: Lexington Books.
KOESTER, U., and SCHMITZ, P. M. (1982), 'The EC Sugar Market Policy and Developing Countries', *European Review of Agricultural Economics* 9(2): 183–204.
KRAVIS, I. B. (1970*a*), 'Trade as the Handmaiden of Growth: Similarities between the Nineteenth and Twentieth Centuries', *Economic Journal* 80: 850–72.
—— (1970*b*), 'External Demand and Internal Supply Factors in LDC Export Performance', *Banca Nazionale del Lavoro: Quarterly Review* 23: 157–79.
—— and LIPSEY, R. E. (1971), *Price Competitiveness in World Trade*. New York: Columbia University Press for National Bureau of Economic Research.
—— and —— (1972), 'The Elasticity of Substitution as a Variable in World Trade', in D. J. Daly (ed.), *International Comparison of Prices and Real Incomes*. New York: Columbia University Press for National Bureau of Economic Research.
—— and —— (1974), 'International Trade Prices and Price Proxies', in N. D. Ruggles (ed.), *The Role of the Computer in Economic and Social Research in Latin America*. New York: Columbia University Press.
—— and —— (1982), 'Prices and Market Shares in the International Machinery Trade', *Review of Economics and Statistics* 64(1): 110–16.
KREININ, M. E. (1967), 'Price Elasticities in International Trade', *Review of Economics and Statistics* 49(4): 510–16.
—— (1973), 'Disaggregated Import Demand Functions: Further Results', *Southern Economic Journal* 40: 19–25.
—— (1977), 'The Effect of Exchange Rate Changes on the Prices and Volume of Foreign Trade', *IMF Staff Papers* 24(2): 297–329.
KRUEGER, A. (1978), *Foreign Trade Regimes and Economic Development: Liberalization Attempts and Consequences*. Lexington, Mass.: Ballinger for National Bureau of Economic Research.
KRUGMAN, P. R. (1979), 'Increasing Returns, Monopolistic Competition and International Trade', *Journal of International Economics* 9(4): 469–80.
—— (1981), 'Intraindustry Specialization and the Gains from Trade', *Journal of Political Economy* 89(5): 959–73.
—— (1982), 'Trade in Differentiated Products and the Political Economy of

Trade Liberalization', in J. N. Bhagwati (ed.), *Import Competition and Response*, National Bureau of Economic Research Conference Report Series. Chicago: University Press.

—— (1987), 'Strategic Sectors and International Competition', in R. M. Stern (ed.), *US Trade Policies in a Changing World Economy*. Cambridge, Mass.: MIT Press.

LABYS, W. C. (1973), *Dynamic Commodity Models: Specification, Estimation and Simulation*. Lexington, Mass.: Lexington Books.

—— (1980), 'Commodity Price Stabilization Models: A Review and Appraisal', *Journal of Policy Modeling* 2(1): 121–36.

—— (1987), *Primary Commodity Markets and Models: An International Bibliography*. Farnborough, Hants: Gower.

—— and GRANGER, C. W. J. (1970), *Speculation, Hedging and Commodity Price Forecasts*. Lexington, Mass.: D. C. Heath.

—— and HUNKELER, J. (1974), 'Survey of Commodity Demand and Supply Elasticities', Research Memorandum no. 48, UNCTAD/RD/70, Research Division, Geneva.

—— and LORD, M. J. (1990), 'Portfolio Optimization and the Design of Latin American Export Diversification Policies', *Journal of Development Studies* 26(2): 260–77.

—— and POLLAK, P. K. (1984), *Commodity Models for Forecasting and Policy Analysis*. London: Croom Helm.

LADD, G. W. (1983), 'Survey of Promising Developments in Demand Analysis: Economics of Product Characteristics', in G. C. Rausser (ed.), *New Directions in Econometric Modeling and Forecasting in US Agriculture*. New York: North-Holland.

—— and SUVANNUNT, V. (1976), 'A Model of Consumer Goods Characteristics', *American Journal of Agricultural Economics* 58: 504–10.

—— and ZOBER, Z. (1977), 'Models of Consumer Reaction to Product Characteristics', *Journal of Consumer Research* 4: 89–101.

LANCASTER, K. J. (1966), 'A New Approach to Consumer Theory', *Journal of Political Economy* 74: 132–57.

—— (1971), *Consumer Demand: A New Approach*. New York: Columbia University Press.

—— (1975), 'Socially Optimal Product Differentiation', *American Economic Review* 65(4): 567–85.

—— (1979), *Variety, Equity, and Efficiency*. Oxford: Basil Blackwell.

—— (1980), 'Intra-Industry Trade under Perfect Monopolistic Competition', *Journal of International Economics* 10: 151–75.

LARSEN, F., LLEWELLYN, J., and POTTER, S. (1983), 'International Economic Linkages', in *OECD Economic Studies*. Paris: Organisation for Economic Co-operation and Development.

LASAGA, M. (1981), *The Copper Industry in the Chilean Economy: An Econometric Analysis*. Lexington, Mass.: Lexington Books, D. C. Heath.

LAWRENCE, C., and SPILLER, P. T. (1983), 'Product Diversity, Economies of Scale, and International Trade', *Quarterly Journal of Economics* 73(2): 63–83.

LEAMER, E. E., and STERN, R. M. (1970), *Quantitative International Economics*. Boston: Allyn and Bacon.

LEONTIEF, W. (1947), 'Introduction to a Theory of the Internal Structure of Functional Relationships', *Econometrica* 15: 361–73.

LEUTHOLD, R. M., MACCORMICK, A. J. A., SCHMITZ, A., and WATTS, D. G. (1970), 'Forecasting Daily Hog Prices and Quantities: A Study of Alternative Forecasting Techniques', *Journal of the American Statistical Association* 65: 90–107.

LEWIS, W. A. (1980), 'The Slowing Down of the Engine of Growth', *American Economic Review* 70(4): 555–64.

LINNEMANN, H. (1966), *An Econometric Study of International Trade Flows*. Amsterdam: North-Holland.

LITTLE, I. M. D. (1982), *Economic Development: Theory, Policy, and International Relations*. New York: Basic Books.

——, SCITOVSKY, T., and SCOTT, M. (1970), *Industry and Trade in Some Developing Countries*. London: Oxford University Press for the Development Centre of the Organisation for Economic Co-operation and Development.

LLEWELLYN, G. E. J., and PESARAN, M. H. (1976), 'The Determination of United Kingdom Import Prices: A Note', *Economic Journal* 86: 315–20.

LORD, M. J. (1989a), 'Product Differentiation in International Commodity Trade', *Oxford Bulletin of Economics and Statistics* 51(1): 35–55.

—— (1989b), 'International Commodity Trade and Latin American Exports: An Empirical Investigation', in P. L. Brock, M. B. Connolly, and C. Gonzalez-Vega (eds.), *Latin American Debt and Adjustment: External Shocks and Macroeconomic Policies*. New York: Praeger.

—— (1989c), 'Primary Commodities as an Engine for Export Growth of Latin America', in N. Islam (ed.), *The Balance between Industry and Agriculture in Economic Development*, 5. London: Macmillan in association with the International Economic Association.

—— (1990), 'Post-Recession Commodity Price Formation', in O. Guvenen, W. C. Labys, and J. B. Lesourd (eds.), *International Commodity Market Modelling: Advances in Methodology and Applications*. London: Chapman Hall.

—— and BOYE, G. R. (1987), *Commodity Export Prospects of Latin America*. Washington, DC: Inter-American Development Bank.

—— and —— (1991), 'The Determinants of International Trade in Latin America's Commodity Exports', in M. Urrutia (ed.), *Long Term Trends in Latin American Economic Development*. Washington, DC: Inter-American Development Bank.

LUCAS, R. E. (1975), 'Hedonic Price Functions', *Economic Inquiry* 13: 157–78.

MACBEAN, A. (1966), *Export Instability and Economic Development*. London: George Allen & Unwin.

MACDOUGALL, G. D. A. (1952), 'British and American Exports: A Study Suggested by the Theory of Comparative Costs, Part I', *Economic Journal* 61: 697–724.

MADDALA, G. S. (1977), *Econometrics*. New York: McGraw-Hill.

MAGEE, S. P. (1975), 'Prices, Incomes, and Foreign Trade', in P. B. Kenen (ed.), *International Trade and Finance*. Cambridge: University Press.

MAIZELS, A. (1963), *Industrial Growth and World Trade*. Cambridge: University Press.
—— (1968), *Exports and Economic Growth of Developing Countries*. Cambridge: University Press.
MARQUEZ, J., and MCNEILLY, C. (1988), 'Income and Price Elasticities for Exports of Developing Countries', *Review of Economics and Statistics* 70(2): 306–14.
MARSHALL, A. (1924), *Money, Credit, and Commerce*. New York: Macmillan.
MARSHALL, C. F. (1983), *The World Coffee Trade: A Guide to the Production, Trading and Consumption of Coffee*. Cambridge: Woodhead-Faulkner.
MASSELL, B. F. (1970), 'Export Instability and Economic Structure', *American Economic Review* 60: 618–30.
——, PEARSON, S. R., and FINCH, J. (1972), 'Foreign Exchange and Economic Development: An Empirical Study of Selected Latin American Countries', *Review of Economics and Statistics* 54(2): 208–12.
MATHIESON, D. J., and MCKINNON, R. I. (1974), 'Instability in Underdeveloped Countries: The Impact of the International Economy', in P. A. David and W. Reder (eds.), *Nations and Households in Economic Growth: Essays in Honor of Moses Abramovitz*. New York: Academic Press.
MCCARTHY, M. D. (1971), 'Notes on the Selection of Instruments for the Two Stage Least Squares and K Class Type Estimators of Large Models', *Southern Economic Journal* 37(3): 251–9.
MCCOY, J. H. (1979), *Livestock and Meat Marketing*. Westport, Conn.: AVI Publishing Co.
MEADE, J. E. (1974), 'The Optimal Balance Between Economies of Scale and Variety of Products: An Illustrative Model', *Economica* 41: 359–67.
MEIER, G. (1989), *Leading Issues in Economic Development*. Oxford: University Press.
MIYAO, T., and SHAPIRO, P. (1981), 'Discrete Choice and Variable Returns to Scale', *International Economic Review* 22: 257–73.
MIZON, G. E. (1977), 'Model Selection Procedures', in M. J. Artis and A. R. Nobay (eds.), *Studies in Modern Economic Analysis*. Oxford: Basil Blackwell.
—— (1983), 'Review of *The Econometric Analysis of Time Series* by A. C. Harvey', *Economic Journal* 93: 254–7.
—— (1984), 'The Encompassing Approach in Econometrics', in D. F. Hendry and K. F. Wallis (eds.), *Econometrics and Quantitative Economics*. Oxford: Basil Blackwell.
MONKE, E. A., and GUISINGER, S. E. (1985), 'International Trade Constraints and Commodity Market Models: An Application to the Cotton Market', *Review of Economics and Statistics* 67(1): 98–107.
MORSINK-VILLALOBOS, M., and SIMPSON, J. R. (1980), 'Export Subsidies: The Case of Costa Rica's Banana Industry', *Inter-American Economic Affairs* 34(3): 69–86.
MURRAY, T., and GINMAN, P. J. (1976), 'An Empirical Examination of the Traditional Aggregate Import Demand Model', *Review of Economics and Statistics* 58(1): 75–80.
MUTH, J. F. (1961), 'Rational Expectations and the Theory of Price Movements', *Econometrica* 29: 315–35.

NARVEKAR, P. R. (1960), 'The Role of Competitiveness in Japan's Export Performance', *IMF Staff Papers* 8: 85–100.

NAYA, S. (1968), 'Variations in Export Growth Among Developing Asian Countries', *Economic Journal* 78: 334–43.

—— (1973), 'Fluctuations in Export Earnings and Economic Patterns of Asian Countries', *Economic Development and Cultural Change* 21(4): 629–41.

NEWBERY, D. G., and STIGLITZ, J. E. (1981), *The Theory of Commodity Price Stabilization: A Study in the Economics of Risk*. Oxford: Clarendon Press.

—— and —— (1982), 'Optimal Commodity Stock-Piling Rules', *Oxford Economic Papers* 34(3): 403–27.

NOGUES, J. J., OLECHOWSKI, A., and WINTERS, L. A. (1986), 'The Extent of Non-Tariff Barriers to Industrial Countries' Imports', Discussion Paper, Report DRD 115, International Bank for Reconstruction and Development, Washington, DC.

OBIDEGWU, C. F., and NZIRAMASANGA, M. (1981), *Copper and Zambia*. Lexington, Mass.: Lexington Books.

Organisation for Economic Co-operation and Development (1987a), *National Policies and Agricultural Trade*, Paris.

—— (1987b), *National Policies and Agricultural Trade: Country Study—United States*, Paris.

—— (1987c), *National Policies and Agricultural Trade: Country Study—European Economic Community*, Paris.

PAGAN, A. (1987), 'Three Econometric Methodologies: A Critical Appraisal', *Journal of Economic Surveys* 1(1): 3–24.

—— and WICKENS, M. R. (1989), 'A Survey of Some Recent Econometric Methods', *Economic Journal* 99: 962–1025.

PAGE, S. A. (1975), 'The Effect of Exchange Rates on Export Market Shares', *National Institute Economic Review* 74: 71–82.

PAGOULATOS, E., and LOPEZ, E. (1983), 'A Model of Agricultural Trade for Differentiated Products', Staff Paper no. 240, Institute of Food and Agricultural Sciences, University of Florida.

PARNICZKY, G. (1974), 'Some Problems of Price Measurement in External Trade Statistics', *Acta Oeconomica* 12(2): 229–40.

PATTERSON, K. D., and RYDING, J. (1984), 'Dynamic Time Series Models with Growth Effects Constrained to Zero', *Economic Journal* 94: 137–43.

PEARCE, I. F. (1961), 'An Exact Method of Consumer Demand Analysis', *Econometrica* 29: 499–516.

PERLOFF, J. M., and SALOP, S. C. (1983), 'Equilibrium with Product Differentiation', Working Paper no. 179, Division of Agricultural Science, University of California, Berkeley.

PHILLIPS, A. W. (1954), 'Stabilization Policy in a Closed Economy', *Economic Journal* 64: 290–323.

PINDYCK, R. S., and RUBINFELD, D. L. (1981), *Econometric Models and Economic Forecasts*. New York: McGraw-Hill.

PLOSSER, C. I., and SCHWERT, G. W. (1977), 'Estimation of a Non-Invertible

Moving Average Process: The Case of Overdifferencing', *Journal of Econometrics* 6: 199–224.

POBUKADEE, J. (1980), 'An Econometric Analysis of the World Copper Market', unpublished manuscript, Wharton Econometric Forecasting Associates, Philadelphia.

POYHONEN, P. (1963), 'A Tentative Model for the Volume of Trade Between Countries', *Weltwirtschaftliches Archiv* 90: 93–100.

PRAIS, S. J. (1962), 'Econometric Research in International Trade: A Review', *Kyklos* 15(3): 560–79.

PREBISCH, R. (1959), 'Commercial Policy in the Underdeveloped Countries', *American Economic Review* 49: 251–73.

PRICE, J. E., and THORNBLADE, J. B. (1972), 'US Import Demand Functions Disaggregated By Country and Commodity', *Southern Economic Journal* 39(1–4): 46–57.

PRIOVOLOS, T. (1981), *Coffee and the Ivory Coast: An Econometric Study*. Lexington, Mass.: Lexington Books.

PROTOPAPADAKIS, A., and STOLL, H. R. (1983), 'The Law of One Price in International Commodity Markets: A Reformulation and Some Formal Tests', *Journal of International Money and Finance* 5: 335–60.

—— and —— (1986), 'Spot and Future Prices and the Law of One Price', *Journal of Finance* 38(5): 1431–55.

QUIZON, J., GARDNER, B., and QUINN, L. (1988), 'Consequences of Agricultural Trade Liberalization for Developing Economies Assisted by AID', unpublished manuscript, Wharton Econometrics, and Development Economics Group of Louis Berger International.

RANGARAJAN, C., and SUNDARARAJAN, V. (1976), 'Impact of Export Fluctuations on Income: A Cross-Country Analysis', *Review of Economics and Statistics* 58(3): 368–72.

RESNICK, S. A., and TRUMAN, E. M. (1973), 'An Empirical Examination of Bilateral Trade in Western Europe', *Journal of International Economics* 3: 305–35.

RICHARDSON, J. D. (1971a), 'Some Sensitivity Tests for a "Constant-Market-Shares" Analysis of Export Growth', *Review of Economics and Statistics* 53: 300–4.

—— (1971b), 'Constant-Market-Shares Analysis of Export Growth', *Journal of International Economics* 1: 227–39.

—— (1973), 'Beyond (But Back To?) the Elasticity of Substitution in International Trade', *European Economic Review* 4: 381–92.

—— (1978), 'Some Empirical Evidence on Commodity Arbitrage and the Law of One Price', *Journal of International Economics* 8: 341–51.

—— (1986), 'The New Political Economy of Trade Policy', in P. R. Krugman (ed.), *Strategic Trade Policy and the New International Economics*. Cambridge, Mass.: MIT Press.

RIEDEL, J. (1984), 'Trade as the Engine of Growth in Developing Countries, Revisited', *Economic Journal* 94: 56–73.

ROBERTS, G. (1985), *Guide to World Commodity Markets*. New York: Kogan Page.

ROEMER, J. E. (1977), 'The Effect of Sphere of Influence and Economic Distance on the Commodity Composition of Trade in Manufactures', *Review of Economics and Statistics* 59(3): 318–27.

ROSEN, S. (1974), 'Hedonic Prices and Implicit Markets: Product Differentiation in Pure Competition', *Journal of Political Economy* 82: 34–55.

ROTTON, J. (1985), 'Astrological Forecasts and the Commodity Market: Random Walks As a Source of Illusory Correlation', *The Skeptical Inquirer* 9(4): 339–47.

ROUSSLAND, D., and PARKER, S. (1981), 'The Effects of Aggregation on Estimated Import Price Elasticities: The Role of Imported Intermediate Inputs', *Review of Economics and Statistics*: 436–9.

ROY, R. (1952), 'Les elasticités de la demande relative aux biens de consommation et aux groupes de biens', *Econometrica* 20: 391–405.

SALOP, S. C. (1976), 'Information and Monopolistic Competition', *American Economic Review* 66: 240–5.

—— (1979), 'Monopolistic Competition with Outside Goods', *Bell Journal of Economics* 10(1): 141–56.

—— and STIGLITZ, J. E. (1977), 'Bargains and Ripoffs: A Model of Monopolistically Competitive Price Dispersion', *Review of Economic Studies* 44: 493–510.

SARGAN, J. D. (1964), 'Wages and Prices in the United Kingdom: A Study in Econometric Methodology', in P. E. Hart, G. Mills, and J. K. Whitaker (eds.), *Econometric Analysis for National Economic Planning*. London: Butterworths.

—— (1980), 'Some Tests of Dynamic Specification for a Single Equation', *Econometrica* 48: 879–97.

SARRIS, A. H. (1981), 'Empirical Models of International Trade in Agricultural Commodities', in A. McCalla and T. Josling (eds.), *Imperfect Markets in Agricultural Trade*. Montclair, NJ: Allanheld, Osmun.

SATO, K. (1977), 'The Demand Function for Industrial Exports: A Cross-Country Analysis', *Review of Economics and Statistics* 60: 456–64.

SATTINGER, M. (1984), 'Value of an Additional Firm in Monopolistic Competition', *Review of Economic Studies* 51(2): 321–32.

SAVAGE, I. R., and DEUTSCH, K. W. (1960), 'A Statistical Model of the Gross Analysis of Transactions Flows', *Econometrica* 28: 551–72.

SAWYER, J. A. (1979), *Modelling the International Transmission Mechanism*. Amsterdam: North-Holland.

SCHIFF, M., and VALDES, A. (1986), 'The Impact of Sector-Specific versus Economy-Wide Policies on Incentives for Agriculture', paper presented at the Annual Meeting of the Econometric Society of Latin America, Cordoba, Argentina.

SCOBIE, G. M., and JOHNSON, P. R. (1975), 'Estimation of the Elasticity of Substitution in the Presence of Errors of Measurement', *Journal of Econometrics* 3(1): 51–6.

SHAKED, A., and SUTTON, J. (1982), 'Relaxing Price Competition Through

Product Differentiation', *Review of Economic Studies* 49: 3–14.
—— and —— (1983), 'Natural Oligopolies', *Econometrica* 51(5): 1469–84.
SHEEHEY, E. (1977), 'Levels and Sources of Export Instability: Some Recent Evidence', *Kyklos* 30(2): 319–24.
SHINKAI, Y. (1968), 'Price Elasticities of the Japanese Exports: A Cross Section Study', *Review of Economics and Statistics* 50: 268–73.
SHOVEN, J. B., and WHALLEY, J. (1984), 'Applied General-Equilibrium Models of Taxation and International Trade: An Introduction and Survey', *Journal of Economic Literature* 22(3): 1007–52.
SIMPSON, J. R., and FARRIS, D. E. (1982), *The World's Beef Business*. Ames, Iowa: Iowa State University Press.
SINGER, H. W. (1950), 'The Distribution of Gains between Investing and Borrowing Countries', *American Economic Review, Papers and Proceedings* 40(2): 473–85.
SIRI, G. (1980), 'World Coffee Prices and the Economic Activity of the Central American Countries', unpublished manuscript, Wharton Econometric Forecasting Associates, Philadelphia.
SONO, M. (1945), 'The Effect of Price Changes on the Demand and Supply of Separable Goods', *Kokumni Keizai Zasski* 74: 1–51. English translation in *International Economic Review* 2 (1960): 239–71.
SPENCE, M. (1976), 'Product Selection, Fixed Costs, and Monopolistic Competition', *Review of Economic Studies* 43(2): 217–36.
SRINIVASAN, T. N., and WHALLEY, J. (eds.) (1986), *General Equilibrium Trade Policy Modelling*. Cambridge, Mass.: MIT Press.
STARBIRD, E. (1981), 'The Bonanza Bean-Coffee', *National Geographic* 159(3): 388–405.
STEIN, L. (1977), 'Export Instability and Development: A Review of Some Recent Findings', *Banca Nazionale del Lavoro Quarterly Review* 122: 279–90.
STERN, N. (1972), 'The Optimal Size of Market Areas', *Journal of Economic Theory* 4: 154–73.
STERN, R. M., BAUM, C. F., and GREENE, M. N. (1979), 'Evidence on Structural Change in the Demand for Aggregate US Imports and Exports', *Journal of Political Economy* 87(1): 179–92.
——, FRANCIS, J., and SCHUMACHER, B. (1976), *Price Elasticities in International Trade: An Annotated Bibliography*. London: Macmillan for the Trade Policy Research Centre.
—— and ZUPNICK, E. (1962), 'The Theory and Measurement of Elasticity of Substitution in International Trade', *Kyklos* 15(3): 580–93.
STIGLER, G. J. (1961), 'The Economics of Information', *Journal of Political Economy* 69: 213–25.
STONE, J. A. (1979), 'Price Elasticities of Demand for Imports and Exports: Industry Estimates for the US, the EEC, and Japan', *Review of Economics and Statistics* 61(2): 306–12.
SUTTON, J. (1986), 'Vertical Product Differentiation: Some Basic Themes', *American Economic Review, Papers and Proceedings* 76(2): 393–8.

THEIL, H. (1971), *Principles of Econometrics*. New York: John Wiley.

THOMAS, H. C. (1988), *A Study of Trade among Developing Countries, 1950–80: An Appraisal of the Emerging Pattern*. Amsterdam: North-Holland.

THOMPSON, R. L. (1981), *A Survey of Recent US Developments in International Agricultural Trade Models*, Bibliographies and Literature of Agriculture no. 21. International Economics Division, Economic Research Service. Washington, DC: US Department of Agriculture.

—— and ABBOTT, P. C. (1983), 'New Developments in Agricultural Trade Analysis and Forecasting', in G. C. Rausser (ed.), *New Directions in Econometric Modeling and Forecasting in US Agriculture*. New York: North-Holland.

THURSBY, M., JOHNSON, P., and GRENNES, T. (1986), 'The Law of One Price and the Modelling of Disaggregated Trade Flows', *Economic Modelling* 3(4): 293–302.

THURSBY, J., and THURSBY, M. (1984), 'How Reliable Are Simple, Single Equation Specifications of Import Demand?' *Review of Economics and Statistics* 66: 120–8.

TIMS, W., and WAELBROECK, J. (1982), *Global Modeling in The World Bank*, World Bank Staff Working Papers, no. 544. Washington, DC: International Bank for Reconstruction and Development.

TINBERGEN, J. (1946), 'Some Measurements of Elasticity of Substitution', *Review of Economics and Statistics* 27: 109–14.

—— (1962), *Shaping the World Economy*. New York: Twentieth Century Fund.

TRELA, I., WHALLEY, J., and WIGLE, R. (1987), 'International Trade in Grains: Domestic Policies and Trade Impacts', *Scandinavian Journal of Economics* 89(3): 271–83.

TURNOVSKY, S. J. (1968), 'International Trading Relationships for a Small Country: The Case of New Zealand', *Canadian Journal of Economics* 1(4): 772–90.

TUSTIN, A. (1953), *The Mechanism of Economic Systems*. Cambridge, Mass.: Harvard University Press.

TYERS, R., and ANDERSON, K. (1986), 'Distortions in World Food Markets: A Quantitative Assessment', background paper for *World Development Report 1986*. Washington, DC: International Bank for Reconstruction and Development.

United Nations (1970), *International Trade Statistics: Concepts and Definitions*, ST/STAT/SER.M/52, New York.

—— (1989), *Monthly Bulletin of Statistics* 43(4): Special Tables.

United Nations Conference on Trade and Development (UNCTAD) (1976), *Resolution Adopted by the Conference: Integrated Programme for Commodities*, ID/RES/93(IV), Geneva.

—— (1989), *Handbook of International Trade and Development Statistics: 1988*. New York: United Nations.

United States Department of Agriculture (1987), *Preliminary Estimates of Producer and Consumer Subsidy Equivalents (PSE's and CSE's), 1982–86*. Washington, DC: Economic Research Service, US Department of Agriculture.

URIBE, P., THEIL, H., and DE LEEUW, C. G. (1966), 'The Information Approach to the Prediction of International Trade Flows', *Review of Economic Studies* 33: 209–20.

VALDES, A. (1986), 'Impact of Trade and Macroeconomic Policies on Agricultural Growth: The South American Experience', in *Economic and Social Progress in Latin America*. Washington, DC: Inter-American Development Bank.

—— (1987), 'Agriculture in the Uruguay Rounds: Developing Country Interests', Working Paper (unpublished manuscript), International Food Policy Research Institute, Washington, DC.

—— and ZIETZ, J. (1980), *Agricultural Protection in OECD Countries: Its Cost to Less Developed Countries*, Research Report no. 21. Washington, DC: International Food Policy Research Institute.

VASTINE, J. R. (1977), 'United States International Commodity Policy', *Law and Policy in International Business* 9(2): 401–78.

VENABLES, A. J. (1984), 'Multiple Equilibria in the Theory of International Trade with Monopolistically Competitive Commodities', *Journal of International Economics* 16: 103–21.

—— (1987), 'Trade and Trade Policy with Differentiated Products: A Chamberlinian–Ricardian Model', *Economic Journal* 97: 700–17.

VOLKER, P. A. (1982), 'On the US Import Demand Function: A Comment', *Journal of Political Economy* 90: 1295–9.

WALLIS, K. F. (1977), 'Multiple Time Series Analysis and the Final Form of Econometric Models', *Econometrica* 45: 1481–97.

—— (1980), 'Econometric Implications of the Rational Expectations Hypothesis', *Econometrica* 48(1): 49–73.

WARNER, D., and KREININ, M. E. (1983), 'Determinants of International Trade Flows', *Review of Economics and Statistics* 65: 96–114.

WELLMAN, F. L. (1961), *Coffee: Botany, Cultivation, and Utilization*. London: Leonard Hill.

WELLS, G. J., and JOHNSON, P. R. (1979), 'The Impact of Entry into the European Community on the Pattern of United Kingdom Wheat Imports: Projections to 1980', *North Central Journal of Economics* 1: 123–32.

WEYMAR, H. F. (1968), *The Dynamics of the World Cocoa Market*. Cambridge, Mass.: MIT Press.

WHALLEY, J. (1985), *Trade Liberalization among Major World Trading Areas*. Cambridge, Mass.: MIT Press.

WOLINSKY, A. S. (1986), 'True Monopolistic Competition as a Result of Imperfect Information', *Quarterly Journal of Economics* 101(3): 493–511.

YNTEMA, T. O. (1932), *A Mathematical Reformulation of the General Theory of International Trade*. Chicago: University Press.

ZELDER, R. E. (1958), 'Estimates of Elasticities of Demand for Exports of the United Kingdom and the United States, 1921–38', *Manchester School* 26: 33–47.

ZIETZ, J., and VALDES, A. (1986a), 'The Potential Benefits to LDCs of Trade Liberalization in Beef and Sugar by Industrialized Countries', *Weltwirtschaftliches Archiv* 122(1): 93–112.
—— and —— (1986b), 'The Costs of Protectionism to Developing Countries: An Analysis for Selected Agricultural Products', Staff Working Paper. Washington, DC: International Bank for Reconstruction and Development.

Author Index

Abaelu, J. N. 13
Abbott, P. C. 7, 15
Adams, F. G. 18, 189, 193, 194, 197, 201, 212 nn. 4, 5, 213 n. 8
Akiyama, T. 193, 194, 197
Allen, R. G. D. 66
Anderson, K. 277, 280–2
Anderson, T. W. 299
Armington, P. S. 7, 8, 30–1, 92, 171 n. 3
Artus, J. R. 102 n. 1
Askari, H. 212 n. 6
Atkin, M. J. 13
Augusto, S. 194, 197

Balassa, B. 16, 255 n. 1
Baldwin, R. E. 127, 243
Bale, M. D. 13, 277
Banerji, R. 127
Barker, T. S. 141, 147
Becker, G. S. 32
Behrman, J. R. 18, 121, 189, 193, 194, 197, 201, 212 nn. 4, 5, 213 n. 8
Bhagwati, J. N. 255 n. 1
Bismut, C. 164
Bond, M. E. 168
Box, G. E. P. 204
Boye, G. R. 129, 264
Boylan, T. A. 149
Brander, J. A. 247–9, 250, 256 n. 2, 257
Brorsen, B. W. 154
Brown, M. 40

Cairncross, A. K. 125
CBT, see Chicago Board of Trade
Chamberlin, E. H. 10
Chang, H. 155 n. 3
Chiang, A. C. 40, 47 n. 5
Chicago Board of Trade (CBT) 12, 13
Chu, K. 51, 56, 189
Cline, W. R. 272
Cohen, B. I. 125, 126, 135

Collins, K. J. 92
Committee of Twenty 120
Cooper, R. D. 128 n. 1
Coppock, J. D. 121
Crouhy-Veyrac, L. 21
Cummings, J. T. 212 n. 6
Currie, D. 79, 85 n. 5
Curtis, B. N. 11, 13

Das, S. P. 35, 52
Davidson, J. E. H. 73, 76, 84 n. 2
Deardorff, A. V. 7, 9, 12, 246, 257
Deaton, A. 32, 33, 34, 86
DeGraff, H. 213 n. 9
Demler, F. R. 15
Dent, W. T. 16
DeRosa, D. 128 n. 1
Deutsch, K. W. 16
Diaz-Alejandro, C. F. 120
Di Marchi, N. 73
Dixit, A. K. 7, 11, 17, 27, 34, 35, 47 n. 5, 243, 246, 247, 249, 254, 257
Dixit, P. M. 171 n. 3
Domowitz, I. 84 n. 2
Donnenfeld, S. 38
Duncan, R. C. 193, 194, 197

Eaton, J. 246, 249
Edwards, R. 202
Elbadawi, I. 84 n. 2
Engle, R. F. 76
Erb, G. F. 128 n. 2

Fair, R. 298
Farris, D. E. 197, 213 nn. 9, 10
Feenstra, R. C. 7
Finger, M. 128 n. 1
Fisher, B. S. 128 n. 2
Fisher, F. M. 193–4, 197, 212 n. 5, 213 n. 8
Flam, H. 47 n. 5, 257, 261
Fleming, J. M. 102 n. 1
Francis, J. 180

Franz, J. 13
Fry, J. 13

Gana, J. L. 17
Gannon, C. A. 32
Gardiner, W. H. 171 n. 3
Gardner, B. 258
Geraci, V. 93
Ghosh, S. 122
Gilbert, C. L. 73, 189
Ginman, P. J. 57 n. 4, 139
Ginsburg, A. L. 102 n. 1, 127
Glezakos, C. 121
Goldberger, A. S. 221
Goldstein, M. 7, 87, 139, 145, 147, 148, 149, 156 n. 4, 168–9, 180, 181
Goodwin, T. H. 102 n. 3
Gorman, W. M. 32, 46 n. 3
Granger, C. W. J. 75, 76, 140, 189
Gray, H. P. 22
Greenaway, D. 22
Greenhut, M. L. 16
Greenwald, B. C. 125
Grennes, T. 8, 13, 15, 92, 171 n. 3
Griffith, G. 203
Griliches, Z. 104, 105, 107–8, 186 n. 1, 209, 238
Grilli, E. 281
Grissa, A. 201
Grossman, G. M. 145, 246, 249, 257
Grubel, H. G. 9
Grunwald, J. 129
Guisinger, S. E. 193, 194
Gustafson, R. L. 121

Hanson, J. R. 125, 126
Harberger, A. C. 94
Hart, O. D. 34, 35, 58, 62, 63
Harvey, A. C. 73, 74, 177, 186 n.3, 187 n. 4, 206, 299, 300
Haynes, S. E. 180
Hazilla, M. 11
Heien, D. 40
Helliwell, J. F. 17, 18
Helpman, E. 9, 11, 27, 30, 33, 35, 47 n. 5, 48, 52, 243, 257, 261
Hendler, R. 33
Hendry, D. F. 73, 75, 84 n. 2, 182

Henry, S. G. B. 84 n. 2
Hickman, G. 92, 102 n. 1, 161
Hicks, J. R. 92
Hitiris, T. 147, 156 n. 6
Horn, H. 35, 47 n. 5
Hotelling, H. 10, 32
Houck, J. P. 194, 197
Houthakker, H. S. 57 n. 4, 87, 145–6, 147, 148, 168–9
Hufbauer, G. C. 22
Humphrey, D. H. 146, 147, 148
Hunkeler, J. 212 n. 6
Hunter, L. C. 7
Husted, S. 155 n. 3
Hwa, E. C. 189, 193, 194, 197
Hylleberg, S. 76

International Bank for Reconstruction and Development (IBRD) 3, 13, 213 n. 10
International Monetary Fund (IMF) 3
Ireland, N. J. 164
Isard, P. 20, 21

Jarvis, L. S. 108, 200
Jenkins, G. M. 204
Jepma, C. J. 127
Johnson, P. R. 92, 296
Jones, L. E. 27, 63
Jung, W. S. 244
Junz, H. B. 102 n. 1, 161, 168

Kaldor, N. 128 n. 1
Katrak, H. 33
Katzner, D. W. 46 n. 3
Kawata, G. 161
Keil, M. 84 n. 2
Khan, M. S. 7, 57 n. 4, 87, 139, 145, 147, 148, 149, 156 n. 4, 167–9, 171 n. 4, 180, 181
Kirmani, N. 272–3
Klein, L. R. 40, 221
Knudsen, O. 121
Koester, U. 283 n. 1
Kollintzas, T. 155 n. 3
Kopp, R. J. 11
Kravis, I. B. 94, 102 n. 1, 117, 125, 132, 161, 251, 295, 296

Kreinin, M. E. 102 n. 1, 145, 147, 148, 155 n. 2, 161
Krishnamurty, K. 56
Krueger, A. 255 n. 1
Krugman, P. R. 9, 11, 27, 30, 35, 47 n. 5, 48, 243, 248

Labys, W. C. 15, 62, 121, 134, 189, 212 n. 6
Ladd, G. W. 14, 33
Lancaster, K. J. 10, 33, 34, 35–9, 41
Larsen, F. 18
Lasaga, M. 18
Lau, L. J. 92, 102 n. 1, 161
Lawrence, C. 35, 47 n. 5
Lawrence, R. Z. 128 n. 1
Leamer, E. E. 15, 93, 94
Lee, G. 244
Leontief, W. 46 n. 3
Leuthold, R. M. 189
Lewis, W. A. 122–3, 124
Linnemann, H. 15
Lipsey, R. E. 94, 102 n. 1, 161, 295, 296
Little, I. M. D. 19, 117–18, 255 n. 1
Llewellyn, G. E. J. 153
Lloyd, P. J. 9
Lopez, E. 93
Lord, M. J. 129, 134, 164, 210, 256 n. 4, 264
Lucas, R. E. 33
Lutz, E. 277

MacBean, A. 121
MacDougall, G. D. A. 102 n. 1
McCarthy, M. D. 298
McCoy, J. H. 213 n. 9
McKinnon, R. I. 121
McNeilly, C. 146
Maddala, G. S. 300
Magee, S. P. 57 n. 4, 87, 145–6, 147, 148, 155 nn. 1, 3, 168–9
Maizels, A. 102 n. 1, 135
Manderscheid, L. V. 13
Mann, J. S. 194, 197
Markusen, J. R. 7
Marquez, J. 146

Marshall, A. 164
Marshall, C. F. 11, 13
Martin, J. J. 22
Massell, B. F. 121
Mathieson, D. J. 121
Meade, J. E. 34
Meier, G. 127
Meilke, K. 203
Miyao, T. 62
Mizon, G. E. 75, 76, 84 n. 2, 299, 300
Monke, E. A. 193, 194
Morrison, T. K. 51
Morsink-Villalobos, M. 245
Muellbauer, J. 32, 33, 34, 86
Murray, T. 57 n. 4, 139
Musgrove, P. 129
Muth, J. F. 99–100

Narvekar, P. R. 102 n. 1, 127
Naya, S. 121, 127
Newbery, D. G. 121
Newbold, P. 75
Nogues, J. J. 274
Norman, V. 7, 17, 27, 35, 47 n. 5
Nziramasanga, M. 18

Obidegwu, C. F. 18
Oliveira-Martins, J. 164
Organisation for Economic Co-operation and Development (OECD) 264, 266, 281–2
Ormerod, P. 84 n. 2

Padmore, T. 17, 18
Pagan, A. 73
Page, S. A. 102 n. 1
Pagoulatos, E. 93
Palaskas, T. B. 189
Parikh, A. 202
Parker, S. 57 n. 4
Parnes, A. 121
Parniczky, G. 56
Patterson, K. D. 79, 85 n. 6
Pearce, I. F. 46 n. 3
Perloff, J. M. 62
Pesaran, M. H. 153
Petoussis, E. 147, 156 n. 6
Phillips, A. W. 76

Pindyck, R. S. 187 n. 4
Plosser, C. I. 104
Pobukadee, J. 18
Pollak, P. K. 15, 189, 194, 197
Poyhonen, P. 15
Prais, S. J. 93
Prebisch, R. 119
Prewo, W. 93
Price, J. E. 147
Priovolos, T. 18
Protopapadakis, A. 21

Quizon, J. 258

Rangarajan, C. 121
Resnick, S. A. 92
Rhomberg, R. R. 102 n. 1, 161, 168
Richard, J-F. 73
Richardson, J. D. 16, 20, 21, 93, 127, 247
Riedel, J. 124
Roberts, G. 12, 13
Roemer, J. E. 37
Rosen, S. 33
Ross, K. Z. 57 n. 4, 149
Rotton, J. 189
Roussland, D. 57 n. 4
Roy, R. 141, 170 n. 1
Rubin, H. 40
Rubinfeld, D. L. 187 n. 4
Ryan, M. E. 13
Ryding, J. 79, 85 n. 4

Salop, S. C. 38, 41, 62
Sargan, J. D. 73, 76, 84 n. 2, 104
Sarris, A. H. 15, 16, 92
Sato, K. 102 n. 1
Sattinger, M. 34, 47 n. 5, 58, 62
Savage, I. R. 16
Sawyer, J. A. 154
Schiff, M. 245
Schmitz, P. M. 283 n. 1
Schumacher, B. 180
Schwert, G. W. 104
Scobie, G. M. 296
Shaked, A. 12
Shapiro, P. 62
Sheehey, E. 121

Sheffrin, S. M. 102 n. 3
Shinkai, Y. 102 n. 1
Shoven, J. B. 17, 258
Simpson, J. R. 197, 213 nn. 9, 10, 245
Singer, H. W. 119
Siri, G. 18
Sisler, D. 125, 126, 135
Sono, M. 46 n. 3
Sosa, S. C. 102 n. 1
Spence, M. 34
Spencer, B. J. 247–9, 250, 256 n. 2, 257
Spiller, P. T. 35, 47 n. 5
Srinivasan, T. N. 17
Starbird, E. 202
Stein, L. 121
Stern, N. 34
Stern, R. M. 7, 15, 93, 94, 102 n. 1, 153, 155 n. 3, 161, 180, 246, 257, 259
Stigler, G. J. 38
Stiglitz, J. E. 11, 34, 47 n. 5, 62, 121, 125
Stoll, H. R. 21
Stone, J. A. 149, 169, 180
Sundararajan, V. 121
Sutton, J. 12
Suvannunt, V. 33

Theil, H. 221
Thomas, H. C. 102 n. 1
Thompson, R. L. 7, 10, 15
Thornblade, J. B. 147
Thursby, J. 145, 149
Thursby, M. 12, 21–2, 145, 149
Tilton, J. E. 15
Tims, W. 18
Tinbergen, J. 15, 92
Trela, I. 259–60
Truman, E. M. 92
Tsiang, S. C. 102 n. 1
Turnovsky, S. J. 57 n. 4
Tustin, A. 65–6
Tyers, R. 277, 280–2

United Nations (UN) 4, 292
United Nations Conference on Trade and Development (UNCTAD) 3, 121

Author Index

Uribe, P. 16
US Department of Agriculture (USDA) 266

Valdes, A. 245, 258, 272, 280–2, 283 n. 1
Vastine, J. R. 128 n. 2
Venables, A. J. 35, 63, 261
Volker, P. A. 155 n. 2
von Ungern-Sternberg, T. 84 n. 2

Waelbroeck, J. 18
Wallis, K. F. 100–1, 206
Warner, D. 145, 155 n. 2
Weiss, A. A. 76
Wellman, F. L. 177

Wells, G. J. 92
Weymar, H. F. 193, 194
Whalley, J. 8, 17, 258, 260
Wickens, M. R. 73
Wolinsky, A. S. 38
World Bank, *see* International Bank for Reconstruction and Development (IBRD)

Yang, C. 281
Yntema, T. O. 171 n. 3

Zelder, R. E. 102 n. 1
Zietz, J. 280–2, 283 n. 1
Zober, Z. 33
Zupnick, E. 93

Subject Index

ACP countries, *see* African,
  Caribbean, and Pacific (ACP)
  countries
adjustment process, *see* response
ad valorem tariff-equivalents, *see*
  tariffs
African, Caribbean, and Pacific (ACP)
  countries:
  preferential trade arrangements
  for 12, 22, 265
Agricultural Basic Law (1961):
  Japan 265
arbitrage, commodity 20–2
autoregressive integrated moving
  average (ARIMA) 102 n. 3, 204–5
autoregressive moving average
  (ARMA) 100–2

Bertrand competition 9, 246, 257

CAP, *see* Common Agricultural Policy
  (CAP): EEC
cattle cycle 108, 200–1, 213 n. 10, 235
CES, *see* constant elasticity of
  substitution (CES)
CGE, *see* computable general
  equilibrium (CGE) model; models
characteristics approach, *see*
  commodities, primary;
  preferences, consumer;
  preferences, importer
Chicago Board of Trade (CBT) 12
Chow test, *see* parameter constancy
  test
CMS, *see* constant market shares
  (CMS) analysis
Cobb-Douglas:
  production function 50
  utility function 47 n. 5
cobweb theorem 209, 212, 239–40
cointegration theory 76
commodities, primary:
  concept of 29, 292
  differentiation using diversity and
    characteristics approaches 30–6
  horizontal and vertical
    differentiation of 10–14, 30, 157
  income elasticity of demand
    for 192–3
  lag structures in production
    of 199–203
  modelling effects of economic
    activity changes on 220–41
  modelling effects of export policies
    on ix, 5, 8, 18, 22–3, 242–55,
    285–9
  modelling effects of trade
    liberalization on 8, 17, 258–61,
    288–9
  protectionism in 264–8
  standards of quality for 12–14
  trade performance of 3, 115–27
  see also preferences, consumer
commodity markets:
  determinants of prices in 189–205
  effects of changes in supply and
    demand on 208–10
  equilibrium price in 61–2
  issues for research: 1970s and
    1980s 121–2
  measurement of economic activity
    changes on 221–5
  stock-adjustment process in 62
  system of equations for 65–8,
    205–8, 237–40, 284–5, 302–5
  UNCTAD IV programme for 120–1
  see also response
commodity type 41
  concept of 36
  ideal 38–9
  see also imports; preferences,
    consumer
Common Agricultural Policy (CAP):
  EEC 264–5

Subject Index

competition:
  non-price 164–5
  price 92–8, 160–2
  see also elasticity of substitution
competition, imperfect:
  product differentiation as source of 27
  see also models
competition, monopolistic:
  with imperfect substitutability 34–5
  in international trade 33, 41, 93–8, 242–3, 249–50
  large-group case versus small-group case in 8–9, 246, 285–6
  market equilibrium in 61–4
  returns to scale in 51–2
  role of information search in 38
computable general equilibrium (CGE) model, *see* models
constant elasticity of substitution (CES):
  production function with 49, 52, 56, 57 n. 3
  utility function with 40, 43
constant market shares (CMS) analysis 16, 127
consumption function 191–5
cost schedule of exporter 50–4
Cournot competition 8–9, 246–7, 257
cyclical response, *see* response

dampened smooth and cyclical responses, *see* lag coefficients; response; supply
data sources 294–8
debt, external 3–4
demand, export:
  effect of foreign market income changes on 222–4
  effect of relative price changes on 8, 19, 22, 124–5, 157–60, 244–55
  influences on 162, 164–70, 215–16
  policies with price-inelastic 286–7
  price elasticity of 43, 91, 164–70, 214
  schedules for 42–5
  steady-state and transient responses of 89–92

demand, import:
  characteristics of 86
  effect of economic activity on 123–4
  effect of multilateral liberalization on 279–80
  income elasticity of 88, 140–8, 214
  income growth elasticity of 88–9, 147–8
  influence on export demand of 162, 164
  price elasticity of 42, 88, 140–1, 148–50, 181, 214
  schedule for 41–5
  steady-state and transient responses of 86
demand, world 45–6, 98–9
devaluation policies 149–50, 164
  see also exchange rate
developing countries:
  export policies of 5, 8, 18, 125–6, 242, 245–6, 250–5
  trade performance of 3–4, 19–20, 22–3, 115–27, 181–2
diversity approach, *see* commodities, primary; preferences, importer

economies of scale, *see* returns to scale
ECM, *see* error correction mechanism
EEC, *see* European Economic Community
elasticity, foreign income:
  of exports 222–3, 233–5
  see also foreign income multiplier
elasticity, impact 195
elasticity, import growth 91–2, 164–5, 215
elasticity, income:
  aggregation of 141–2
  in consumption function 192–3
  of import demand 88
  of import demand for manufactured goods 147, 215, 220
  of import demand for primary commodities 142–7, 214, 220
  of import demand in Latin America's foreign markets 142–7
  variation in 156 n. 4
elasticity, income distribution 141–2

Subject Index

elasticity, income growth 88
  in consumption function 193
  in import demand function 147–8, 215
elasticity, price:
  aggregation of 170 n. 1
  calculation of long-run 186 n. 1
  in consumption function 193–4
  of demand 193–5
  of export demand 43, 60, 91, 94, 164–70, 214
  of export supply 51–4, 60, 172, 179–82, 214
  of import demand 88, 148–50, 181, 214
  of import supply 54–6
  inter-country differences in 148
  to measure effect of liberalization 260
  of supply 195–7
  see also demand, export; supply
elasticity of substitution:
  concept of 8, 92–8
  for Latin American exports 160–2
Engel's Law 119
equilibrium:
  in monopolistic competition 58–64
  for perishable and nonperishable commodities 61–2
  properties of 62–4
  stability conditions for dynamic 208–9, 212, 235–40
  stability conditions for static 63–4
  static and dynamic long-run 78–84
  symmetric 58, 62–4
error correction mechanism (ECM) 86, 103, 111
  and cointegration theory 76
  in consumption function 194
  in export demand function 81–4, 90–2, 95–8
  in import demand function 87–9
  in import price function 109–10
  in stock demand function 101
  use of disequilibrium adjustment term in 77, 87, 140, 162
estimation procedure 298–9
Eurodollar market 205

European Economic Community (EEC):
  effect of trade liberalization in 258, 277, 279–80, 287
  trade barriers in 264–8, 276, 283
exchange rate:
  effect on relative commodity prices 21
  see also devaluation policies
export demand, *see* demand, export
export growth and economic growth 19
export market shares 5, 117–19
export policy, *see* trade policy
export promotion 244, 247, 285–7
  effects of policies on 245–6, 255 n. 1
  unilateral strategic 248–9
  see also subsidies
exports:
  imperfect substitution of 27–8
  Latin American commodity 129–32, 138 n. 1
  markets for Latin American commodity 132–8
  quotas for 121
  uses of foreign exchange from 3
  see also demand, export; supply
export supply, *see* supply

factor proportions theory 7, 20
feedback effects 20
  in commodity markets 210, 219–20, 258
  in trade 123, 219–20, 233–5, 261, 273–4, 285, 287–9
Food Security Act (1985): US 264
foreign income elasticity, *see* elasticity, foreign income
foreign income multiplier 222–3

game theory, *see* trade policy
General Agreement on Tariffs and Trade (GATT) 274, 287
goods:
  concept of 29
government intervention:
  in exporting country trade 242–56

government intervention (*cont.*)
  to improve market information 125
  see also export promotion; trade policy

Heckscher-Ohlin model 7, 260
hedging facilities 12
hedonic price schedule 33
heterogeneity, product, *see* product differentiation
homogeneity, product, *see* product differentiation
horizontal differentiation, *see* commodities, primary; product differentiation
hypothesis testing, nested and non-nested 299–300

imperfect competition, *see* competition, imperfect
import demand, *see* demand, import
import price schedule, *see* price
imports:
  as commodity type 36
  foreign exchange earnings for 3
  influence of growth of 91–2, 162, 164, 169, 215
  see also demand, import; elasticity, import growth
import substitution policy 245–6, 255 n. 1
import supply schedule, *see* supply
income changes:
  commodity market response to 224–5
  commodity trade response to 225–35
  mechanism for transmission of 17–18, 220–4
information:
  effect of imperfect 8–9, 28, 38, 154
  importance for producers of 125
Integrated Programme for Commodities (IPC): UNCTAD 120–1, 128 n. 2
interest rates 205
INTERLINK model 18

International Bank for Reconstruction and Development (IBRD) 18
intra-industry trade 9–10, 17, 119

Japan:
  effect of trade liberalization in 258, 277, 279–80, 283, 287
  trade barriers in 264–8, 276, 283

lag coefficients:
  dampened cyclical response of 107–8, 177
  dampened smooth response of 107
  measurement of 182–6
lag distribution:
  effect of 172, 177–9
  in dynamic specification 73–5
  of import prices 150–4
  mean lag calculation of 182, 186–7 n..3
  median lag calculation of 182, 187–8 n. 4, 188 n. 5
  of production 199–203
Law of One Price 20–2
LINK project 18, 154
logistic function 86, 90–2, 95–8
Lomé Convention 12, 22, 265
London Metal Exchange 12

macroeconomic policies 245–6
manufactured goods:
  income elasticity of 147
  performance of trade in 3–4, 117–22, 147, 215
MFN, *see* most-favoured nation (MFN) status
models:
  of commodity markets 18, 189–91
  of commodity trade 65–8, 302–5
  computable general equilibrium (CGE) 17, 23, 245, 258–60, 282, 288
  econometric trade 17–18, 23, 65–8, 249–50, 259–60, 286, 302–5
  gravity 15
  of imperfect competition vii–viii, 27–8

of international trade 15–19, 23,
  65–8, 302–5
 methodologies for
  constructing 15–19
 probability 16
 spatial equilibrium 7–8, 16
 theory-based 23
monopolistic competition, *see*
  competition, monopolistic
most-favoured nation (MFN)
  status 265
multilateral trade negotiations
  (MTNs) 274
 see also General Agreement on
  Tariffs and Trade (GATT)

non-tariff barriers (NTBs) 149,
  156 n. 7
 price wedge creation by 260–1, 266
 use by industrialized countries 4–5,
  260

Organisation for Economic
  Co-operation and Development
  (OECD) 18, 225 n. 1
Organization of Petroleum Exporting
  Countries (OPEC) 4, 120, 246

parameter constancy test 140, 153,
  155 n. 2, 192, 212 n. 3, 213 n. 7,
  300–1
preferences, consumer:
 compared to importer
  preferences 31–2
 most preferred product in 35–6
 see also commodities, primary;
  commodity type
preferences, importer:
 diversity and characteristics
  approaches to 10–11, 27–8, 30–6,
  41
 indifference schedules of
  importer 39–41
 structure of 7, 30–2, 36–45
 tests of theories of 7
 see also substitution, inter- and
  intrasectoral

preferential trade arrangements 12,
  22, 160, 180, 264–5
 see also Lomé Convention;
  most-favoured nation (MFN)
  status
price:
 effect of income changes on 220–1,
  224–5
 effect of trade liberalization on
  world market 274–9, 282–3
 effect on supply of 103–10
 expectations formation of 99–102
 influence on export demand of 162,
  164
 instability 121, 277
 reasons for differences in 22
 reduced form solution for world
  market 205–8
 schedule of import 55–6
 stabilization of market 121–2, 277
 steady-state response of
  import 109–10
 see also competition; demand,
  export; elasticity, price; response
price equilibrium, *see* equilibrium
price wedge, *see* non-tariff barriers
  (NTBs); tariffs
prisoner's dilemma 247
product:
 concept of 29, 301
 product differentiation:
 effect on market power of 8
 horizontal and vertical 10–14, 30,
  157
 with question of homogeneity 20–2,
  32
 research in 9
 as source of imperfect
  competition 27
 types of 10–14
 see also commodities, primary
production controls 12, 121
production function 48, 195–203
production schedule, exporter 48–54
protection, trade:
 estimates for prices in markets 208
 policies of industrialized
  countries 264–8

protection, trade (*cont.*)
  policies of USA, the EEC, and
    Japan 264–8
  see also non-tariff barriers (NTBs);
    tariffs

random walk 189
rational expectations 99–102
response:
  dampened cyclical 107–8, 177
  dampened smooth 107
  of export supply from Latin
    America 179–86
  of import prices to world market
    price 150–4
  of imports to trade
    liberalization 268–72, 279–80,
    282–3, 288–9
  of market prices to income
    changes 224–35
  of prices to trade
    liberalization 274–9, 282, 288
  steady-state and transient 75–6,
    79–81, 86, 89–92
returns to scale 28, 48, 51–4, 63

SIMLINK system 18
specification:
  dynamic 73–5
  long-run equilibrium solution of
    dynamic 78–84, 89, 91, 95–6, 99,
    101, 105, 109–10
  restricted 75–8
  see also error correction mechanism
    (ECM)
stochastic difference equation 73–6,
  111
stockpiling, commodity 121
stocks:
  demand for and supply of 61–2,
    190–1, 203–5
  expected price formation in
    determination of 99–102
  to production or consumption
    ratio 203–4
subsidies:
  effect of export 242, 244–5, 247

effect on domestic output of 257
optimal export 250–2
use by industrialized countries 4–5
substitution, *see* constant elasticity of
  substitution (CES); elasticity of
  substitution; import substitution
  policy
substitution, inter- and
  intrasectoral 37–41, 43–5
supply:
  adjustment to price change of
    export 182
  dampened cyclical response
    of 107–8, 177
  dampened smooth response of 107
  disturbances in export 51
  influences on export 172
  lag coefficients for price variable in
    export 105–8
  price elasticity of export 51–4, 60,
    172, 179–82, 214
  relation to export price of
    export 103
  schedules 48–56
  see also production schedule,
    exporter; stocks

tariffs:
  application of 149, 257
  effects of reduction in 258, 268–83
  price wedge creation by 260, 261,
    266, 297–8
  see also non-tariff barriers (NTBs);
    tax
tax:
  conditions for use of export 246,
    286
  on consumption 257
  on exports 244–5, 249
  optimal export 252–5, 286–7
technology changes 49, 51, 119, 197–8
Tokyo Round, *see* General Agreement
  on Tariffs and Trade (GATT)
trade, international:
  analysis of imperfect competition
    in 27
  barriers to 4–5, 264–8
  changes in system of 119–22

characterization of monopolistic competition in 8–9, 52
export performances in 3–4, 19–20, 115–27, 181–2
impact of trade liberalization on 279–83, 288
research in intra-industry 9–10
theories of 6–10, 30–1, 122–8
theory-based analysis of viii, 18–19, 23

trade liberalization:
adjustment process after 258, 268–72, 274–80, 287–8
modelling of 258–61
multilateral 273–83, 287–9
unilateral 268, 272–3, 287–9

trade policy:
co-operative games for strategic 247
evaluation methods for 22–3, 249–50
for export of commodities 246–5
implications for export supply incentives 182
implications of import price elasticity estimates 149–50
non-cooperative games for strategic 247, 250, 286
optimal 249–55, 286–7
strategic 243, 247–55, 286–7
see also export promotion; import substitution policy; subsidies; tariffs; tax

United Nations Conference on Trade and Development (UNCTAD I, IV, VII) 120–1
United States:
effect of trade liberalization in 258, 277, 279–80, 283, 287
trade barriers of 264–8, 276, 283
Uruguay Round, *see* General Agreement on Tariffs and Trade (GATT)

variety approach, *see* preferences, importer
vertical differentiation, *see* commodities, primary; product differentiation

Walrasian adjustment process 276
Walrasian condition for stability 63–4